THE MAKING OF A PROFESSION:
A CENTURY OF ELECTRICAL ENGINEERING IN AMERICA

CENTENNIAL TASK FORCE

John D. Ryder, *Chairman*
Donald S. Brereton, *Vice Chairman*

Nathan Cohn
Robert F. Cotellessa
Lawrence P. Grayson
Donald T. Michael
William W. Middleton
Mac E. Van Valkenburg

HONORARY CENTENNIAL COMMITTEE

Richard J. Gowen, *1984 IEEE President*
James B. Owens, *Vice Chairman*
Charles A. Eldon, *Vice Chairman*

John Bardeen
William R. Hewlett
William C. Norris
Robert N. Noyce
Simon Ramo
Ian M. Ross
Roland W. Schmitt
Mark Shepherd, Jr.
John W. Simpson
Charles H. Townes

THE MAKING OF A PROFESSION: A CENTURY OF ELECTRICAL ENGINEERING IN AMERICA

A. Michal McMahon

The Institute of Electrical and Electronics Engineers, Inc., New York

This book was set in Goudy Oldstyle.

THE INSTITUTE OF ELECTRICAL
AND ELECTRONICS ENGINEERS, INC.
345 East 47th Street
New York, NY 10017

PRINTED IN THE UNITED STATES OF AMERICA

IEEE Order Number PC01677

Library of Congress Cataloging in Publication Data
McMahon, A. Michal (Adrian Michal), 1937–
The making of a profession.

Includes index.
1. electric engineering—United States—History.
2. Institute of Electrical and Electronics Engineers—History. I. Title.
TK23.M39 1984 621.3'0973 83-22325
ISBN 0-87942-173-8

TABLE OF CONTENTS

FOREWORD

The history of an engineering society can be dull; this one is not. Dr. McMahon's history of the IEEE and its predecessors marks a bold departure in writing the history of engineering societies. The dilemma of such an act lies in the fact that the history of the society is part of a much larger history. The committees, officers, and constitutions constitute only the outer husk. They find their full meaning when seen in the context of the history of a field. It has become quite common to tell the history of an engineering discipline without reference to its professional organizations. But the history of neither is complete without the other. The engineering society not only fosters the technical development of a field, but it is the means by which a profession can express itself and articulate its values.

It is very difficult to tell the story of the IEEE and its predecessors in the context of the development of electrical engineering (for brevity I will understand "electrical" to include "electronic"). Despite some very good progress in recent years, much of that history remains to be told. It is an enormously complex undertaking in itself. But even if this information were all available the task would remain very difficult because there is just so much of it. If it were all told, the story of the IEEE might well be buried under the mass of other material. Dr. McMahon has found a solution to the dilemma by judicious selectivity. He does not, indeed could not, present anything like a complete history of electrical engineering. Rather he has selected representative figures and critical events in order to capture the essence of that history. He has succeeded so well that his work may be misinterpreted. It is not a complete history of electrical engineering. It is a series of deftly drawn vignettes which capture enough of the engineering history to illuminate the history of the IEEE and its predecessors.

But the sketches do capture much of that which makes the history of electrical engineering important. The early parts of the story contain biographical studies of engineers as diverse as telegraph electrician and technical writer and consultant, Franklin Pope; eminent engineering scientist, socialist, and General Electric "wizard," Charles Proteus Steinmetz; and the aggressive genius of the electric power industry, Samuel Insull. Dr. McMahon gives equal attention to pivotal events in the emergence of professional electrical engineering. For example, in Chapter 1, he examines the "takeoff" of power engineering and its relation to the founding of the American Institute of Electrical Engineers (AIEE). In Chapter 2, he traces the intellectual evolution of professional electrical engineering to the point of its codification in the 1902 Constitution of the AIEE.

Although Dr. McMahon discusses only two engineering educators at length, Dugald C. Jackson and Frederick E. Terman, both were enormously important pioneers. Jackson was one of the founders of electrical engineering education prior to the First World War; he strongly emphasized the role of electrical engineers in management, and to some extent therefore, he de-emphasized the purely technical side of engineering. Terman was one of the pivotal figures in the emergence of a more scientific engineering during and after the Second World War. Similarly, Dr. McMahon has taken David Sarnoff to exemplify the rise of commercial broadcasting as a big business. Now it cannot be argued that Sarnoff was "typical." Though he no doubt shared many traits with other industry leaders, he was a charismatic, highly idiosyncratic leader. He was selected because of his critical importance. In this case as in others, Dr. McMahon has chosen, wisely I believe, to discuss a smaller number of figures and events in some detail, rather than try to cram all of the significant names and dates into one book. The result is fine history. For example, the section on Sarnoff in Chapter 5 is an original and insightful interpretation of one of the leaders who helped shape both electrical engineering and modern mass media.

Nowhere do the benefits of Dr. McMahon's judicious selection of people show to greater advantage than in the extensive sections of Chapters 6 and 7 that deal with the career of Frederick E. Terman. Terman embodied to a remarkable degree the major trends of modern electrical engineering. A long-time professor at Stanford and early student of radio and electronics, Terman was among the handful of engineers who laid the foundations for the thriving electronics industry in California. Therefore, to know something of his early life gives us insight into the rise of electronics and the sorts of factors that placed this country in the vanguard of this important technology.

Terman was a statesman of his profession. He was an active member of both the AIEE and of the Institute of Radio Engineers. His professional activities shed light on the history of both societies and on their ultimate merger to form the IEEE. By examining Terman's pre-war pioneering in electronics, his war-time work, his role in raising the scientific level of his profession, and his leadership in professional society affairs allows the reader to grasp, through the life of a particular individual, important and complex changes that were taking place in the profession, the technology, and the nation.

Terman played a significant and revealing role as an organizer of research during the Second World War. The story of electrical engineering in the Second World War alone would merit a volume as long as this one. The author does, of course, sketch in a broad background: Vannevar Bush's mobilization of American science through the Office of Scientific Research and Development (OSRD), the Radiation Laboratory at MIT, as well as other developments. But by concentrating on one fairly small part of this complex whole he enables the reader to get a feel for what was happening and why it was important, both in the long run as well as the short. He examines Terman's organization of the radar counter-measures group at Harvard. Terman found that there were not enough electrical engineers with the advanced training and scientific orientation to do the sort of work that had to be done.

He filled in the ranks with physicists whose education in scientific fundamentals he deeply admired.

But while Terman admired and used physicists, he did not lose his primary loyalty to electrical engineering. The close interaction of physics and electrical engineering after the Second World War was a part of the development by electrical engineering of a more sophisticated, scientific methodology. But this very closeness has contributed to the conflation of engineers with scientists. The similarities are great; but so are the differences. Engineers are doers concerned with practical matters, a fact that Terman would not be likely to forget. Terman admired physics, but he rejected the idea that engineers should rely upon others for knowledge fundamental to the work of their profession. He led a highly successful protest against an authoritative statement on engineering education which emphasized the basic sciences, plus the more traditional engineering fields, but which placed electronics low in the ranks of the engineering sciences. Terman dissented; so too did his colleagues. In later years the number of Ph.D. degrees granted in electrical engineering rose rapidly. Terman could reflect with satisfaction that never again would his profession have to depend upon other disciplines to accomplish work fundamental to electrical engineering. To Terman electrical engineering was a proud, autonomous profession which could learn from others, but which would not accept a permanent state of dependency. The flourishing of a science-based electronics industry in "Silicon Valley" and elsewhere since the Second World War has vindicated Terman's judgement concerning the future of his profession.

The author makes it clear that one must understand the cognitive changes in electrical engineering in order to understand the forces which led to the eventual merger of the IRE with the AIEE. Indeed, one of Dr. McMahon's most important contributions is his analysis of the dynamics of engineering society development.

The IRE emphasized the growth field of electronics. But that alone did not account for its dynamism. The IRE fostered the more fundamental, scientific approach to electrical technology favored by Terman among others. And it was this scientific bent, combined with an openness to new ideas and new disciplines within electrical engineering that led to the IRE's spectacular growth and its merger with the AIEE to form the IEEE. It would be too simple to say that the formation of the IEEE represented the combination of the IRE's scientific values with the commercial ones of the AIEE. Both tendencies are present in all engineering societies: they are inherent in engineering.

The wars of the twentieth century have contributed to carrying the IEEE away from a purely technical conception of its role. The continued close association of electrical engineers with military might after the Second World War and particularly during the Vietnam War have produced reaction from some of the members that led to a heightened sense of social responsibility in the IEEE. These concerns for the social effects of engineering became linked to the welfare of the employee-engineer, especially after large cutbacks in aerospace and defense were made in the late sixties. This growing concern for social issues and social responsibility marked a major turning point in professional evolution. Out of these events has come wide acceptance within the profession of the tenet that social forces impinge on electrical engineering as well as technical constraints.

Some of the issues raised by the new professionalism are controversial and potentially divisive. It is to the credit of the IEEE that it has been able to accommodate differences of opinion. It is to Dr. McMahon's credit that he has faced these often delicate issues of professional development squarely and in an objective, fair-minded way. Because of this essential honesty, I believe that his book gives an accurate guide to the professional problems as well as to the technical triumphs of electrical engineering. Dr. McMahon has captured both the intellectual excitement and the moral dilemmas of a developing profession. And because of the honesty with which he deals with these issues, this book carries valuable lessons for the present and for the future.

EDWIN T. LAYTON
University of Minnesota
Minneapolis

PREFACE

Two major goals have guided my work on this centennial volume of electrical engineering in America. First, I have sought to identify the cluster of engineering values that has gathered around the organizations of professional electrical engineering, namely the Institute of Electrical and Electronics Engineers and its predecessor bodies, the American Institute of Electrical Engineers and the Institute of Radio Engineers. To this end, therefore, this book concentrates on the object that has historically concerned the engineering societies themselves: the state of the profession. Second, besides ferreting out the central issues of professionalism, I have sought to relate the main currents in the history of the profession, from the rough beginnings in telegraphy before the Civil War and the emergence of an embryonic electric lighting and power industry in the late nineteenth century to the rise of an ubiquitous electronics and the organizational merger of the discipline in the late twentieth century.

The two aims are mutually supporting. The approach by way of the profession's leading values makes manageable the history of the country's largest professional engineering group. (Nearing 260,000 today, in 1910, the AIEE's 7000 members made it already the country's largest technical society.) Just as electrical engineers have formed the largest engineering body in the nation for most of this century, so has the American social landscape been dominated by their technical concerns: telegraphy and telephony, electric power generation and distribution, radio, electronics, and microelectronics.

Though I begin with the telegraph, the profession took off when electric power emerged as a technical field and an industrial pursuit during the decade prior to the founding of the AIEE in 1884. By the 1910's, when the power industry was well established, radio had entered its second decade of fundamental development, leading to the founding of the Institute of Radio Engineers in 1912. Maturing between the world wars, radio broadcasting gave way in the 1930's to the rise of electronics, as hundreds of new uses were found for the vacuum tube. The final technical events that frame this history rested on the commitment to military scientific and technological research and development made during World War II and after. Among the results of that new departure were the commercialization of nuclear energy in the 1950's and, more momentously, the microelectronics revolution that has channeled the profession's main interests since midcentury.

To tell a story so potentially vast, I have followed the careers of representative engineering figures and examined pivotal events in the history of the engineering societies and the collective profession. Examples of the latter are the stories of the participation of electrical engineers in the research efforts of the two world wars. Since the contributions of the engineers to these efforts have been virtually ignored in the leading histories of twentieth-century science and technology, the prominent roles played by electrical engineers during the First and Second World Wars made these discussions necessary for comprehending the electrical engineering experience.

In each case selected, whether an engineering life or an event in the history of the societies or the profession, I have chosen so as to reveal the major contexts in which the profession has taken shape. Thus, while focusing on the professional lives of the electrical and electronics engineer, I have also examined the role of corporations and, after 1940, the part played by military and governmental agencies in setting the terms of those lives. Additionally, I have looked at education, a central professional concern, especially in relation to the character of engineering knowledge and to the reigning image of the engineer within the profession.

Awareness of these elements of the political economy and of the nature of engineering knowledge is essential to understanding the constraints on the individual in the making of the profession. However, because individual actions also shape history, my perspective is often from the position of a particular engineer, such as Dugald C. Jackson, a leading consulting engineer and head of the electrical engineering department at MIT in the first quarter of the century, or Frederick Emmons Terman, an engineering educator at Stanford University for nearly half a century after 1925 and a director of a World War II research laboratory.

It is, thus, a dual perspective that the title seeks to convey. On the one hand, the profession has been made by forces beyond the awareness and control of engineers themselves. And yet, on the other hand, to paraphrase the English historian, E. P. Thompson, the electrical engineers of the IEEE have been present at their own making. Charles Proteus Steinmetz, the engineering scientist who fled Germany in the late 1880's and within a few years dominated the General Electric Company's engineering staff, also helped shape the consciousness of the emerging American profession. And Donald Quarles, an engineer who followed a career at Bell with a series of defense posts under President Dwight D. Eisenhower, guided the AIEE's efforts in the early 1950's in lobbying for the commercialization of nuclear fission technology. In such ways have engineers acted decisively and effectively in defining the character of engineering and shaping the context in which electrical engineers play out their careers.

An additional explanation of the title needs to be made. It is the use of the word "America." Unlike the earliest society, the AIEE, the Institute of Radio Engineers consciously omitted the word "American," wishing to express not only the stateless quality of technology but also the intended international character of the IRE. In again avoiding the use of American in its title, the IEEE, when formed in 1963, continued that tradition, and, truly, like the IRE before it, the IEEE has an international membership. Nonetheless, as I followed the long history of organized electrical engineering represented by these bodies, I found the leadership, the vast majority of the members, and the issues that have motivated their actions to be distinctly American.

I have incurred many debts in the writing of this history, not the least of which have been to archival and library staffs at Stanford University in Palo Alto, California; the Dwight David Eisenhower Presidential Library in Abilene, Kansas; the Massachusetts Institute of Technology in Cambridge, Massachusetts; Case Western Reserve University in Cleveland, Ohio; the University of Texas in Austin; the David Sarnoff Library in Princeton, New Jersey; and the American Telephone and Telegraph archives in New York. Finally, my work was considerably aided by the staffs at the Van Pelt Library and the library of the Moore School of Electrical Engineering at the University of Pennsylvania in Philadelphia.

Research for the early part of this study was supported by a grant from the National Science Foundation. The IEEE provided the major portion of support and gave, as well, a grant of scholarly independence. Specifically, I wish to thank John D. Ryder, former dean of engineering at Michigan State University and chairman of the Centennial Task Force, which commissioned me to write this history, and Reed Crone, managing editor of the IEEE PRESS.

To John Ryder and James Brittain, of Georgia State University, I am indebted for reading the entire manuscript. Bruce Sinclair, of the University of Toronto, and Robert Freidel, director of the IEEE's Center for the History of Electrical Engineering, and Edwin T. Layton, Jr., of the University of Minnesota, read specific chapters and offered perceptive comments. Additionally, I must thank Charles Layton, a friend and journalist, for spending many lunches patiently listening and sharing his wisdom so that I could return to my work with a renewed sense of purpose. To the experienced scholars who heard me deliver portions of this work at conferences and who encouraged me—especially Melvin Kranzberg, Eugene S. Ferguson, and Edwin Layton—I express my thanks and gratitude. Finally, to my wife, Lynne, and my sons, Sean and Jeffrey, who bore with me and supported me throughout this effort, I am deeply thankful.

A. MICHAL MCMAHON
Philadelphia
February 1984

1/AT THE DAWN OF ELECTRICAL ENGINEERING

In 1884, the time had come when the advantages of congregation as opposed to segregation were to be demonstrated; when the lonely investigator was to be brought into contact with his brother toiler and taught the advantages of organized work and a free exchange of ideas.

Edwin J. Houston, Inaugural Address, 1894 [1]

A gathering of "practical electricians"

The year 1884 was ripe for recognizing the arrival of the electrical era. Electrical science and technology was not only maturing rapidly, it was also becoming increasingly clear that the field was veering off sharply from America's traditional engineering culture. Even so, its celebrators gathered around an invention of the nineteenth-century: a manufacturing exhibition. Besides exhibiting their products at the nation's first specialized electrical fair held in Philadelphia, technical groups used the opportunity to hold formal meetings. A telephone association and the Railway Telegraph Superintendents held meetings during the fair. And just six months before, the American Institute of Electrical Engineers (AIEE) had been organized so that America would have a formal electrical engineering society to greet the "foreign electrical savants" expected to meet at the International Electrical Exhibition [2]. However, rather than simply hosting members and visitors in rooms assigned by the exhibition's sponsoring organization, the Franklin Institute, the AIEE was about to hold its first annual conference.

In important respects, it was not unlike later conferences. The members of the AIEE delivered papers on a variety of electrical engineering topics and later discussed the papers, all of which would be printed in the society's first *Transactions* early the next year. The presence of both theoretical and practical interests stood out in the first two papers delivered at that inaugural conference. Edwin J. Houston read a paper on the Edison effect, entitled, simply, "Some Notes on Incandescent Lamps." Houston discussed "the peculiar high-vacuum phenomenon observed by Mr. Edison in some of his incandescent lamps," because, Houston explained, he wanted Institute members to "puzzle over it." The second paper by W. M. Chandler on "Underground Wires" was directly aimed at the practical interests of the engineers [3]. Chandler's discussion of the technical problems associated with electrical transmission also signified the emerging field of power engineering.

The electrical exhibition held in Philadelphia in 1884 provided a site for the AIEE's first annual conference, and a point of convergence for many of the interests and ideals of professional electrical engineering.

However, a truer representation of the electrical engineering community—one more crowded with what would be the fundamental issues of the profession—could be found at the United States Electrical Conference hosted by the exposition. Organized by a coalition of members of the United States Congress and the presidential administration of Chester A. Arthur, it was called to consider what the burgeoning electrical industry would mean to public policy [4]. This "scientific commission" was to organize and "conduct a national conference of electricians." The country's most distinguished physicist, Henry A. Rowland, a professor of physics at Johns Hopkins University in Baltimore, headed the commission, which included a number of leading scientists and engineers: Simon Newcomb, a professor of mathematics and astronomy at Johns Hopkins University and an astronomer in the U.S. Navy; J. Willard Gibbs, a distinguished physicist at Harvard University; and Edwin Houston, a prominent electrical engineer from Philadelphia and a cofounder of the Thomson-Houston Electric Company. Houston was the only technologist on the commission. A professor at Central High, he had taught Elihu

Thomson and, later, collaborated with him in designing the arc-lighting system around which the Thomson-Houston Electric Company was formed. Thomson-Houston rivaled Thomas A. Edison's companies during the 1880's and after 1890 merged with a number of electric companies, including Edison's, to form the General Electric Company (GE). Yet, even Houston failed to represent adequately the emerging electrical engineering community.

Certainly the presence of Houston, who would serve as president of the AIEE a decade later, was important in terms of the conference's goal. That goal had been stated in Newcomb's opening statement to the conference when he urged the conferees to take "advantage . . . of this opportunity to [join] such experts in the science of electricity as could enlighten us on the subject of its practical application." The conferees therefore gathered, he instructed them, as "practical electricians." That there was such a strong representation from the engineering community helped support that idea. Among the eighty-nine American representatives were at least a score of recognizable leaders in the electrical world, nine of whom were future presidents of the AIEE. Of these, several were physicists and professors. Over half worked for the telegraph and telephone companies or for the electrical manufacturing firms. A few were traditional nineteenth-century philosopher mechanics like Houston or telegraph electricians like Franklin Pope who had been influential in founding the Institute. Yet most of the representatives belonged to the society's future. Massachusetts Institute of Technology (MIT) physicist Charles R. Cross and Lewis Duncan, a former doctoral student of Rowland's, would serve the AIEE in the 1890's, when engineering scientists largely ran the society. Elihu Thomson, who had parted from Houston by 1884, was a leading inventor-entrepreneur who became wealthy from his patents. He remained at GE for several decades, later serving as a technical elder statesman. Edwin W. Rice, then an engineer at Thomson-Houston, was the last of the group to serve as president of the AIEE, doing so at the time of the First World War. He became an influential corporate engineer, serving as president of GE also, from 1913 to 1922.

With the presence of these future Institute presidents, the national meeting closely approximated the mixture of engineering types who would constitute the profession. And yet, the conference also captured the mixture of ideas and issues that would absorb the energies and, sometimes, arouse the passions of future AIEE members. It was, thus, telling that Professor Rowland discussed in his keynote address not only the relations of science and technology but also the importance of practical applications and the kind of "public measures" that might appropriately be presented "to the Government." Rowland called for "the perfection of our means of measuring both electrical and magnetic quantities" while simultaneously asking the conference to consider the need for a Bureau of Physical Standards. In addition to the topic on standards, Rowland raised other central issues by urging the importance of education and research, ending his address with a rousing call for national support in these areas. He proffered an image of electrical engineering as a constellation of professional interests: "Let physical laboratories arise. . . . Let the professors be given a liberal salary, so that men of talent may be contented. Let technical schools also be founded, and let them train men to carry forward the great work of applied science." His elaboration on education was to find echoes throughout the history of the profession. Rowland asked that the schools

> not be machines to grind out graduates by the thousand, irrespective of quality: But let each one be trained in theoretical science, leaving most of his practical science to be learned afterward. . . . Life is too short for one man to know everything, but it is not too short to know more than is taught in most of our technical schools. It is not telegraph operators but electrical engineers that the future demands.

Actually, Rowland's regard for the education of the electrical engineer was of secondary concern in his address. Instead, the bulk of the theme of his long paper was spent on establishing the necessity of pure science to advance engineering. Pure science, he asserted, brought greater spiritual rewards while leading to the material gains of applied science. Rowland claimed ancient traditions in his support, beginning with Archimedes' denigration of his own engineering accomplishments. Quoting Plutarch, Rowland explained that Archimedes, whose mechanical devices delayed Rome's conquest of Syracuse for three years, "would not deign to leave behind him any commentary or writing on such subjects; but, repudiating as sordid and ignoble, the whole trade of engineering, and every sort of art that lends itself to mere use and profit." Thus, Rowland noted, "at the dawn of science," the competing values of pure and applied science had been argued. Rowland's purpose, it soon became clear, was to argue it again and to insist that his audience of engineers and scientists were "forced to admit that Archimedes was right." Rowland went almost as far as Archimedes in denigrating the value of the technical artifact. He attacked the popular tendency to value technical objects more than ideas—thus, to make heroes of inventors rather than scientists. As an illustration, he lamented the excitement that surrounded the invention of the Leyden jar in the eighteenth century: "It is only to the vulgar and uneducated taste that the tinsel and gewgaws of an electric spark appeal more strongly than the subtle spirit of the amber."

More remarkable than his critique of the function of engineering work in culture was the reception of his comments. No engineering speaker took issue with Rowland's address, and his assertion about the relative value of telegraph electricians even won grudging approval from the editor of the telegraph magazine, *The Operator.* Though he thought Rowland had put it too harshly, the editor admitted that Rowland's judgment about the future belonging to the engineer was sound [5]. Thus, despite the presence of a score of eminent engineers at the conference, it was left to the British physicist, Sir William Thomson (later, Lord Kelvin), to temper the harshness of Rowland's remarks. Admitting that Archimedes had despised his own accomplishments in military engineering, Thomson thought that the Greek scientist must have, nonetheless, "looked forward to a better time when science should be applied, and great engines should be devised for other purposes than that of infliction of wounds and death." Thomson even reminded Rowland that Isaac Newton "passed the last half of his life as an engineer"—though the physicist thought

Henry Rowland (left), the American physicist, disdained the dominance of the inventive spirit in America; Sir William Thomson (later Lord Kelvin), the British physicist, celebrated both the spirit and the country.

it a great loss that Newton's time at the mint had prevented him from finishing important fundamental work. Thomson even felt it necessary to balance the American's effusive praise of Michael Faraday and other English scientists with his own commendation of Joseph Henry's comparable achievements.

However, Thomson quickly moved beyond correcting Rowland's arrogance to the practical matters for which the conference had been organized. As was true for Rowland, Thomson's central concern was the need for an international system of electrical measurements and nomenclature. Thomson's interest in standards, however, went beyond the more narrow concerns of Rowland and the scientific community. To his mind, standards had to do no less with industrial needs than with the requirements of scientific investigation. As a practical physicist, he looked with an engineer's eyes to the future and saw, in addition to the laboratories, universities, and schools, great technical systems engineered for expansion. Thomson did not speak, of course, of national and international radio broadcasting systems or networks of powerful electronic computers exchanging quantities of information on a scale and at a speed unimaginable to even the best-trained physicist or engineer of the 1880's. Yet he envisioned the technical system that would dominate the world the new engineer was about to enter: the "networks of power" just beginning to put their stamp on the electrical engineer's world [6]. Thomson looked to a time

> when we have a central station, such as that which is so admirably worked out by Edison in New York, and which will be worked out more and more in all cities of the world, by which one large station shall give light to every place within a quarter of a mile of it. . . .

To describe Thomson, but not Rowland, as a practical physicist does not put the Johns Hopkins professor on a pedestal. Rowland was not so disdainful of the technical object as he suggested in his address. He could not be, since his great reputation largely rested on laboratory apparatus. His European reputation had been considerably enhanced in 1876 by work carried out with an electrically charged rotating disk in Hemholtz's laboratory in Germany. And his fashioning of a curved grating to analyze the spectrum of starlight had revolutionized the study of light spectra [7]. Before the century was out, Rowland added to his fame through corporation-sponsored court appearances during the hard-fought patent battles of the late nineteenth century. What separated the two men and gave to the Englishman an engineering vision was Thomson's longtime intimacy with the telegraph industry. Thomson had begun to mix engineering work with his science when he accepted Cyrus Field's invitation to join the board of the Atlantic Telegraph Company and help solve the technical problems confronting the builders of the first Atlantic cable.

That Thomson remained to advise on the problems confronting an extensive submarine telegraph system only further honed his engineering perceptions. It also earned him the title of "father of electrical engineering." That sobriquet justly went to the British physicist, not only for his contributions but for his country's as well. England was a major source of telegraphy, and the cable was primarily British. Yet beyond Thomson's personal role in telegraphic development, the telegraph was a primary source of the early electrical community.

The early electrical engineer emerged from other technical areas as well. The first work with commercially applicable dynamos was derived from an electroplating industry, which came to America in the 1840's. But, for the making of a profession of electrical engineering in America, the telegraph served as an initial seedbed. Its roots were in science as well as in technology; its organization was corporate and its extent national. No more nor less could be said of electrical engineering. The first roster of officers and managers of the AIEE amply established the early society's connection to the telegraph. Nevertheless, to seek the source of the AIEE in its prehistory, more than telegraphy must be explored. Indeed, the personal paths of the founders, as they created and shaped a new discipline and a national profession, issued, even more profusely, from the explosion of the lighting and power field during the 1870's. This was the critical event in instigating the AIEE. Even so, the story of electrical engineering began at least forty years earlier with the first attempts to invent an electromagnetic telegraphic system.

Telegraphy and the early electrical community

The technology of telegraphy effectively began during the 1830's when the American, Samuel F. B. Morse, initiated work on a practical electric telegraph system. His achievement rested on more than technical innovation. From the beginning, Morse was joined by both inventive partners and financial associates. In addition, representatives of federal agencies and the mechanics' institutes, having recently organized in America to promote industrial growth, helped in creating the telegraph industry. With Morse's first successes, both the Franklin Institute and the American Institute in New York joined in encouraging the federal government to aid this revolutionary new communications technology. It was an ambitious quest, for these promoters of an advancing technology sought to forge a technical system at once novel and vast. They envisioned a communications network whose scope rivaled the nation's growing railroad complex.

The first task was to devise a practical system of electromagnetic telegraphy, and this was Morse's work. Morse, however, did not invent the technical elements. Having already grasped the fundamentals of electric telegraphy, scientists and inventors in Europe and England were striving to achieve an electric telegraph. Around 1830, Joseph Henry, a Princeton physicist who later became the first secretary of the Smithsonian Institution, put together the essentials for a practical electric telegraph by setting up an intensity electromagnet that could be actuated by a distant intensity battery. On such inventive achievements, Morse based his work, begun while sailing home from France in 1832. He then envisioned the system introduced commercially a dozen years later. The essential breakthrough came on shipboard when he conceived his binary code for transmitting messages. Morse's code made the breakthrough in telegraphy possible because it reduced to a single strand the four to five wires still being experimented with in Europe. Borrowing from Henry, he utilized the direct pull of an electromagnet, creating simple and durable instruments far superior to the complex arrangements of his European competitors. Thus, Morse did more than conceive a code: he built a complete system by adding the apparatus that utilized it. Within a year of his appointment to the fine arts faculty at Columbia, the forty-five-year-old artist devised both a transmitter and receiver. Working with instruments built by Alfred Vail, a New Jersey machinist who joined him as a partner late in 1837, Morse and his associates built a system capable of transmitting over a ten-mile wire—the length Morse had learned was required between relays [8].

The intense technical activity upon which the telegraph industry rested led to more than new opportunities for the entrepreneurial talents of the nation. Out of it came a source of further technical work and a training ground for men who would help form the early electrical engineering community. In stringing lines across the country, the telegraph companies drew young men from towns and cities and thrust them into the world of the telegraph office, where the work of operating the keys offered, to the technically curious, the additional challenge of maintaining and improving the telegraphic system itself.

While working in one such office, Thomas Alva Edison began his first fruitful researches, laboring in the evenings after working the keys by day. Before success came in Boston in 1868, he had already been an operator for five years, first in Port Huron, Michigan, then in cities like Cincinnati and Louisville. In Boston, however, his work moved decisively beyond the boundaries of a key operator when the twenty-one-year-old inventor discovered and absorbed the two volumes of Michael Faraday's *Experimental Researches in*

Electricity. Edison then devised what was perhaps his first invention in the form of a duplex telegraphic system. In January 1869, according to a telegraph trade magazine, Edison left the keys to "devote his full time to bringing out his inventions." That year, he began his first business in New Jersey when he formed a partnership with telegraphic electrician Franklin Pope. Until 1876, when he established his "invention factory" in Menlo Park, New Jersey, and turned to the problems of electric lighting and power, Edison worked almost exclusively on making improvements on the telegraph [9].

As illustrated in his work on the telegraph, an original designer of electrical apparatus like Edison represented the inventive segment of the emerging profession. Yet in addition to inventors, the telegraph industry gave rise to a class of telegraphic workers called electricians. Electricians maintained the lines and installed and kept the transmitting and receiving apparatus in working order. These were the engineers of the telegraph industry when there were no formally trained electrical engineers. Almost invariably beginning as operators, through a mixture of self-directed study and random instances of formal learning, these men often left the keys to become technical advisors to management, supervisors, and independent consultants.

The careers of telegraph electricians, like Edison's early partner, Franklin Pope, and that of George B. Prescott, suggest the character of a large percentage of the technical corps of the industry. Like Edison, Pope started young. After one term at Amherst and an enthusiastic study in natural philosophy and geography in 1855, the fifteen-year-old Pope became a telegraph operator in his small Massachusetts hometown. Two years later, he moved to Springfield as circuit manager of the Boston and Albany Railroad Telegraph Lines. It was a way station, a low level of telegraphic activity, yet important skills could be learned there. Pope spent the Civil War mapping line routes along the entire east coast for one of the major telegraph companies; during the same time, he evaluated and set standards for the company's equipment. Toward the end of the war, he assisted the engineer in charge of the Russo-American Telegraph Company. Until the laying of the Atlantic cable, Russo-American planned to build a line across the Bering Strait to introduce transcontinental telegraph communication [10]. Out of his long experience with telegraphy, Pope began, in the late 1860's, to work as a publicist for the industry. He served for about six months as editor of *The Telegrapher,* then, in 1869, published his first book, *The Modern Practice of the Telegraph.* After Western Union (WU) absorbed Pope and Edison and Company in 1870, Pope superintended all WU patents for several years before resigning from the company to enter private practice as an electrical consultant and as an electrical writer.

A fellow electrician, George B. Prescott, began his career as an operator in 1847, the year of Edison's birth. In much the same way as Pope, Prescott also moved rapidly from the often "small, dirty, and poorly ventilated" second floor where the operators worked into the ranks of the technical supervisors. With the consolidation of Western Union Company in 1866, Prescott took a position as "electrician" with the company. In doing so, he joined a class of trained electrical workers who were, in both respects, numerous and prestigious simply because, in the first decades after the Civil War, telegraphy towered over the electrical industry. As an electrician for the nation's reigning telegraph company, Prescott evaluated new inventions, introduced innovations into the system, and supervised the construction of lines and equipment installations [11].

Again like Pope, Prescott left active telegraphic work to become an electrical writer, serving as an interpretor and historian of the industry, first of telegraphy, then later of the telephone and the dynamo. Prescott also developed a critical perspective, arguing for the importance of the electrician as a creative technologist. In his 1860 book, *The History, Theory and Practice of the Electric Telegraph,* Prescott defended the scientific character of telegraphic work and sought to establish the stature of the telegraph electrician. On the basis of his examination of the social and technical aspects of the French telegraph system, Prescott called for certain changes within the American industry. He argued that since the "Electric Telegraph" constituted a "true science," the electricians of the telegraph companies were scientists. In France, this was so "even for the subordinate employés" responsible for operating the equipment. Similarly, Prescott extended the title of electrician to all technical personnel, placing the operator in the category of apprentice. He believed that "besides being able to receive and transmit dispatches," the electrician "ought to possess a knowledge of the technical part of his service." With "a fund of general knowledge," the operator — the potential electrician — could "meet all emergencies." Out of this necessity for "general knowledge" came Prescott's prescription for the education of an electrician:

> It is, then, indispensable that he be initiated into the laws and properties of electricity, that he may render himself entirely competent to comprehend all the laws respecting the transmission of electric currents, and that he know perfectly all the details of construction of the batteries, instruments, etc.

Impressed that the chief electrician of the French lines published telegraphic observations from the operators, Prescott "hoped that the American companies would yet see the propriety" of appointing "thorough electricians" [12].

The presence of several classes of electrical workers — inventors, electricians, and operators — rested on the necessity for both original and routine technical activity to sustain the industry. Combined, the electrical workers devised and maintained an electrical communications system that made possible a national network of merchants, jobbers, and purveyors of stock transactions. Within Prescott's scheme, moreover, the work of all was made possible by the constant advance of the knowledge and techniques of telegraphy. This progressive view placed the task of innovation at the center of the enterprise. So, although the technical achievements for a national system had been achieved early, original work on telegraphy continued to issue from the international scientific and technical communities. Inventors not only persisted in devising novel systems, but their work also found acceptance in the marketplace. In 1846, Edinburgh inventor Alexander Bain patented an electrochemical system in England and, three years later, in the

United States. In time, nearly 1200 miles of wires were strung in the States because transmission with Bain's system was so much faster than Morse's. Two letter-printing systems devised by Americans competed even more effectively with Morse's telegraph. Though slower, the devices of Royal E. House, in 1846, and David Hughes, a decade later, were effective. Hughes' apparatus was used on the Continent until the twentieth century, just as the automatic printing system of British physicist and pioneer telegraphic inventor, Charles Wheatstone, found wide acceptance in England [13].

Wheatstone's participation in developing the technology of telegraphy, like Joseph Henry's in America, once again illustrated the involvement of physical scientists in the evolution of electrical technology. Yet, the most spectacular feat of the telegraphic era — the laying of the Atlantic cable — rested heavily on the work of physicist Sir William Thomson. The problem of insulation had been solved by placing armor over the gutta-percha used to cover the cable. Nevertheless, when the American Cyrus Field and others began, in 1856, to plan a telegraphic connection between North America and Europe, there still remained the more basic need for an adequate receiver. A paper by Thomson on the problems of underwater signaling, published the previous year, led to his appointment as a director of Field's Atlantic Telegraph Company. Subsequently, Thomson's improved galvanometers of 1858 and 1871, plus the work of others who had improved the cable's insulation and sharpened the cable's signal by adding a condenser at each end, made practical Field's efforts to lay a cable across the ocean floor [14].

Technical advances in electrical communications continued through the last decades of the nineteenth century, yet in the 1870's, the line of development shifted to the telephone. So, although the telephone originated with electric power that decade, its technical nature and early bureaucratic organization made it a child of telegraphy. Though its primary inventor lacked long-term associations to the telegraph, his work was, nonetheless, intimately related to the original technology of the modern communications industry. This was apparent at the beginning when Alexander Graham Bell and his backers competed with the agents of Western Union and their company's leading in-house inventor, Elisha Gray. In 1873, both Bell, the English-born teacher of the deaf and researcher in acoustics, and Gray, a midwesterner whose technical achievements in developing the telephone compared with Bell's, began their experiments in multiple telegraphy that would lead each to the notion of a workable telephone.

The twenty-three-year-old Bell arrived in Quebec in 1870 and soon moved to Boston to establish a career as a professor of vocal physiology and a teacher of the deaf. From research in harmonic telegraphy and out of the heady technical atmosphere of post-Civil War Boston, Bell was drawn into the search for a telegraphic system capable of transmitting multiple messages. Bell had earlier been made aware of the connections between vowel sounds and electricity by the work of the German physicist Herman von Helmholtz. That experience had touched the eighteen-year-old's curiosity no less than Edison's had been as he operated the keys in those second-story rooms of way stations in the American Midwest. For Bell, theory and practice merged eight years later in his work on a system of harmonic telegraphy. By February 1876, with the help of Professor Charles Cross and others at MIT, Bell was able to apply for his first telephone patent [15].

Gray began his telegraphic investigations in the 1860's and, by 1874, had established a business on the basis of his patents. After forming a partnership with the owner of an electrical firm, the company moved to Chicago and prospered. In 1872, Western Union purchased the firm and renamed it Western Electric Manufacturing Company. Gray's work also led to the problems of speech transmission. However, unlike Bell, Gray concentrated on creating a multiple telegraph system and did not pursue his work on the telephone. Differences between Gray and Bell did not derive from their inventive abilities. Gray was a professional inventor who continued to devise innovations in telegraphic and telephonic technology long after the younger man had left the field; but, in 1876, when Gray's entire career seemed to him and to his backers to hinge on the successful development of a multiple telegraph system, Gray accepted the judgment of the telegraphic experts that the speaking telephone was no more than a toy [16].

The dramatic finale to Gray and Bell's struggle for primacy occurred at the Centennial celebration late in the summer of 1876 [17]. Arrangements had been made for the electrical judges—which included Sir William Thomson—to view Bell's apparatus after attending to Gray's harmonic telegraph at the Western Electric exhibit in the main hall. Gray amazed the judges by simultaneously transmitting eight messages over one wire. Bell was helped, however, by an unexpected circumstance. Accompanying the judges was Dom Pedro, Emperor of Brazil, who had joined President U. S. Grant earlier to open the exposition by releasing a valve to start George Corliss's great steam engine. Dom Pedro had met Bell during a demonstration at a school for the deaf several weeks earlier on a visit to Boston. As Gray finished his presentation, Dom Pedro spotted Bell among those present, and, as Bell later recalled, "the Emperor then took my arm and we walked off together followed by the judges and the crowd." During Bell's presentation, after he had transmitted two signals simultaneously over his harmonic telegraph, Bell turned to his telephone, explained the theory behind it and pointed out that it was, nonetheless, a mere "invention in embryo." However, after Sir William Thomson, then Dom Pedro, and then others listened to Bell's words through the 300 feet of wire, they enthusiastically applauded. Dom Pedro leaped up and cried, "I hear, I hear." Even Gray had to admit the response. As he later related, while listening carefully to Bell reading from Hamlet, "I finally thought I caught the words, 'Aye there's the rub.' I turned to the audience, repeating these words, and they cheered."

Thomson's final report to the organizers of the exposition capped the story of Bell's success at the Centennial. After conducting additional tests the following day, he judged Bell's achievement of the electric transmission of speech as "perhaps the greatest marvel hitherto achieved by the electric telegraph." This clear comparison of Bell's telephone to the harmonic telegraphic systems, which he had also observed at the exposition, indicated that Thomson, unlike the advisers of both inventors, foresaw with Bell the practical potential of the "toy." "With somewhat more advanced plans, and more powerful apparatus," Thomson concluded, "we may confidently expect that Mr. Bell will give us the means of making voice and spoken words audible through the electric wire to an ear hundreds of miles distant" [18].

Western Union and the making of an electrical context

The telegraph and telephone were distinguished by their revolutionary character as instruments of instant communication. Yet the corporate organizational form assumed by the communications industry was equally notable. In the United States, by 1866, the intense organizing efforts of the antebellum years had produced in the telegraph industry the earliest sign of the corporate revolution (what historian Alfred Dupont Chandler describes as "the first nationwide multiunit modern business enterprise in the United States") [19]. This meant that well before 1880, the largest segment of the electrical industry was controlled by a single business firm: the Western Union Company.

It was not apparent to the builders of the first telegraphic lines that the industry was to be privately organized. The difficulties of raising capital led many to think that a working telegraphic system would be realized only through governmental assistance. Aid to build an experimental line came in 1843 when Congress appropriated $30,000 for Morse and his associates to link Washington with Baltimore. Morse had began his quest for public aid in the late thirties, receiving enthusiastic support from the nation's technical and industrial communities. In 1838, the chairman of the Franklin Institute's committee on invention arranged for Morse to demonstrate his system in the hall of the Institute. The inventor had already shown his electric telegraph to the members of a similar organization in New York and was headed for Washington to seek governmental aid. Like his New York supporters, the Philadelphians had more in mind than an evening's entertainment when the committee chairman asked Morse "to stop for a short time at Philadelphia on your way, and let it be seen by the Committee of Science and the Arts." Responding to a general request from the Secretary of the Treasury, its members had submitted a report to the federal government on the subject of telegraphy and, thus, had "taken an interest in the subject of telegraphs." Having considered only "visible systems" of telegraphic communication, however, the committee's members were anxious to see Morse's electrical system [20].

Because the demonstration was to take place while Morse and his associates were en route to the national capital, Alfred Vail wanted also to ensure that more was gained from the delay than satisfying the curiosity of the Philadelphians. He asked, therefore, that the Institute's committee meet at the time of the demonstration and adopt a report that same evening. The result was a statement by the Science and Arts Committee urging that "the Government . . . provide the funds for an experiment on an adequate scale." The committee explained that Morse's use of a combination of dots and lines of different length was simpler than the leading European systems designed by the German Carl F. Gauss and the Englishman Wheatstone, making "it worthy of the patronage of the government" [21].

Attaining "patronage," as these earnest technologists understood it, meant raising developmental capital to meet the costs of early experimental work and initial construction. The hesitance of private interests to invest made public monies necessary. As an early Morse associate complained, "it will be a memorable fact in telegraphic history that, vast as are the advantages and promises, scarcely a merchant or capitalist in the great cities of New York, Philadelphia, or Baltimore, could be induced to take a dollar of stock to encourage the noble enterprise!" Other individuals believed ownership by the government simply the wiser course. In Congress, in 1845, when the Postmaster General of the United States moved for ownership of the telegraph, the editor of the *New York Herald* urged that the telegraph system become a part of the postal service. In any case, he argued, "the government must be compelled to take hold of it." Still, in December of the following year, Congress declined to act on the Postmaster General's request [22].

Tremendous growth over the next dozen years made the question of ownership moot. Within two years of May 1844, when Morse and Vail had successfully communicated on the pioneer line between Washington, D.C., and Baltimore, Maryland, Henry O'Rielly extended the line to Philadelphia from Baltimore. With the New York to Philadelphia line that Morse and a partner had already finished, Washington was thus linked to New York. By 1851, just six years after the Washington to Baltimore experiment, eleven lines spread out from New York. The names of the western lines alone map out the expansiveness of the system: among them were the New York, Albany, and Buffalo Telegraph Company and the Lake Erie Telegraph Company. With lines such as the Erie and Michigan, the telegraphic entrepreneurs, capitalist financiers, and electricians extended their lines ever farther west. New Orleans was already the first great distant target. As early as 1846, three lines competed for the market: the Atlantic and Ohio Telegraph Company; the Pittsburgh, Cincinnati and Louisville Telegraph Company; and the Washington and New Orleans Telegraph Company. Their interest testified not only to the economic attractions of the Crescent City but also to the technical feasibility of reaching any point on land [23]. The telegraph had both eclipsed the postal service as a means of speedy communication and raced past the railroad in its drive across the continent. On the eve of the Civil War, then, nearly a decade before railroad officials drove in the golden spike, telegraph lines reached the Pacific.

With the expansion of the system came the concentration of management. By 1857, just six companies controlled the industry. They further lessened the rigors of territorial competition by organizing the Six Nations' Alliance, making each company sovereign in its geographical area. Seeking to link the technical, if not the commercial, parts of the expanding telegraphic system, each company agreed to respect the other's territory and to abide by jointly devised rules aimed at preventing the formation of new companies. However, the strictures of the Alliance did not prevent the Western Union Company

The crowding of telegraph wires in pre-Civil War Cincinnati reflected the density of competing telegraph companies during the early decades of industry.

from absorbing its largest competitors, the American Telegraph and the United States Telegraph Companies [24]. Centralized control had come in stages. After expansive beginnings, in which competing lines proliferated, a period of loose oligarchical organization was followed by consolidation under Western Union in 1866.

The lesson of Western Union—and the economic climate in the 1870's compared to the 1840's—dictated a shorter period for the telephone to achieve consolidation. Bell's patents began under concentrated ownership, with his father-in-law heading the group of owners. After offering the invention to Western Union for $100,000 and being turned down, the owners' only resort was to develop it. Though WU's backers challenged the new telephone company's commanding position, even hiring Edison to improve their system, they were more concerned with capturing an interest in the Bell enterprise [25]. WU's financial maneuvering contrasted sharply to the Bell strategy. From the beginning, a different spirit existed at Bell regarding technical matters. After a brief spell as company electrician, Bell left the position to his associate Thomas Watson, who took the responsibility for the technical development and expansion of the system. Yet, in an industry based on technical innovation, the technical sensibility of the chief officer was directly relevant to making a context for engineering work. The two men who headed Western Union and Bell, in 1884, suggest the varied responses that could be made to technical issues. The different styles of

Norvin Green as president of WU and Theodore Vail as general manager of Bell, in short, dictated different approaches to technical research and, thus, different levels of support for engineering staffs.

Norvin Green joined Western Union as a vice president, as part of the 1866 reorganization. He was not a technologist, a fact made clear in the path he took through the managerial ranks of pre-1866 telegraphy. Green had entered the telegraph business following a twenty-year career in medicine and politics. Then, after saving a failing line in the Southwest, he gained the presidency of the Louisville-New Orleans Line. Yet Green was next in the succession for a larger presidency, one that rested, in part, on his own labors during the years of consolidation. He had played a critical role in merging the major lines operating between the Midwest, the upper South, and New Orleans, and had been a signer of the agreement creating the Six Nations' Alliance. In 1878, he assumed Western Union's presidency. Holding half of his considerable wealth in the company, he worked to ensure the strength of WU stock, a chore that demanded he act as chief spokesman for the industry at a time when competition and concern about the social role of telegraphy was an issue in the Congress [26].

Under Green's direction, Western Union continued to absorb smaller telegraph companies. Green's first major challenge came when Jay Gould, Russell Sage, and other owners of the rival American Union Company attempted to capture a part of Western Union. Green first felt the competition from American Union in the middle of 1880; by the next year, the mounting costs of the commercial struggle ended when WU purchased American Union at twice its actual value, giving to Gould and his associates twenty-one percent of the larger company. When these acts aroused the interest of leading newspapers and the United States Senate, Green spent an increasing amount of time in defending both the industry and private ownership and control. First, in a testimony given in 1881 when a suit was brought against the company, then, before Senate hearings held during the summer of 1883, and finally, in an article on "The Government and the Telegraph" that appeared in the *North American Review,* Green urged Andrew Jackson's principle "that country is governed best which is governed least." Green relied on more than Jackson's famous assertion. He first cited the articles and amendments of the Constitution that he thought pertinent, then bolstered his argument with a recounting of the industry's rise in the United States, and concluded by favorably comparing its position to the European industry. Finally, he argued that it was contradictory for critics to call for government ownership to unify the nation's telegraph system "under one general direction" when "one of the very grounds of complaint is that one company has done so much toward unifying the telegraph service." Green's logic rested on the undisputed fact, as a veteran Chicago telegrapher put it, that Western Union had already "absorbed every one of the original lines" [27].

It was WU's campaign to consolidate the industry that absorbed Green's energies. Theodore Vail's struggles at Bell, however, drew on a different mixture of personal energies and company goals. The picture the Bell

Company president painted of Vail, when he decided to hire him to manage the company, suggested an apt person to build a national technology-based company: He is "a thousand horsepower steam engine wasting his abilities in the United States Railway Mail Service." A cousin of Morse's early partner, Alfred Vail, Theodore Vail had begun his career as a key operator for WU in the year of the final merger. Yet he distinguished himself with the mail service, rising within ten years to the position of general superintendent. At thirty-three, having achieved the senior ranks of one of the country's oldest bureaucracies, Vail accepted Gardiner G. Hubbard's offer to manage the year-old Bell Company. He became a consummate manager of corporate growth, shaping the company's strategy and structure to ensure continued expansion. As a biographer summarized his achievements,

> Vail organized the expanding telephone system; he merged the rapidly multiplying local exchanges into more efficient companies; he put into effect a practical system of financing the telephone indirectly; he provided for anticipatory technical development & for improved and more economical manufacture of telephone apparatus, with the Western Electric Company as the manufacturing unit [28].

Yet Vail's expansionist policy was not automatically accepted by Bell's directors. The acquisition of Western Electric in the 1880 settlement with Western Union exemplified the strategies for growth that led Vail to clash with investors during the early years. Vail wanted to concentrate the company's efforts to achieve both the technical and organizational breakthroughs needed to accomplish a national telephone system. His opponents were strengthened, however, by the absence of a technical solution to the problems of long distance transmission—problems which kept the telephone industry a local concern into the nineties. At the heart of Vail's dilemma was the problem Morse faced in 1838: the desire of financial backers to reap maximal profits immediately. Continued opposition to Vail's plans for building a national telephone system led him to resign in 1885. But his commitment to achieving a national system suggested a way to retain the talented manager, and, thus, that same year, Bell's directors organized the American Telephone and Telegraph Company to build and operate long lines and gave Vail the presidency [29].

In comprehending the technical needs of a national system, Vail also grasped the connections between research and future profits. He set out early to establish research capabilities in the company, naming Thomas Watson General Inspector of the National Bell Company in 1879 and making him responsible for the instrument and electrical departments. Under Watson were nine workers, two with the title of electrical engineer. Though basic research was not involved at this time, Watson's engineering department carried out experimental work in pursuit of technical improvements in the areas of laying cables, improving their operation, and collecting experimental data on transmitters and magneto telephones. In spite of this, much of their work was still aimed at the examination of patents, a particularly pertinent task in which the central technical strategem had become, even before Bell settled with WU, the acquisition of the "thousand and one little patents and inventions" relating to the operation, improvement, and expansion of the system. A Bell employee advised the company president that it was "exceedingly important that we apply for patents at once to cover all that is patentable about the apparatus that we are using in the different branches of our business" [30].

Vail's sensitivity to technical matters brought him close to the emerging world of electrical engineering. Yet, like Green, Vail was not a technologist. He could not have conducted the tests his engineers did, nor was he equipped to make invention and innovation his full-time work. Vail's role in the electrical community was, in the words of the first AIEE call for members, that of an "officer" of a company "based upon electrical inventions." That designation retained a measure of authority within the society. As a description of membership eligibility, however, it was to be a source of conflict within the AIEE, even at times rendering ambiguous the status of the Institute as a professional engineering society.

Takeoff: The rise of electric lighting and power

Managers like Green and Vail were important in shaping a strategy and structure for the electrical industry. But in the making of the AIEE, equally influential individuals came from other industrial arenas. For the technical surge behind the takeoff of the electrical age came not from communications technology but from the sudden emergence of power engineering, in the decade before 1884. From this field would come also the majority of the members of the new profession. Power engineering was a busy arena. Between 1874 and the founding of the AIEE, the rush of innovation took the dynamo from commercial introduction in the American electroplating industry to the vision of an ever-expanding power production and distribution system. During these years, many individuals were setting out to make a place for themselves. But early on in the area of power and lighting, a small circle of inventor-entrepreneurs stood out in the crowded field. The electrical age was initially defined by these individuals and their technical achievements: Edward Weston's early dynamos and electrical instruments; Charles Brush's arc-lighting system; Elihu Thomson's arc-lighting innovations; and Edison's direct-current generators, incandescent lighting system, and central-station design. By any reckoning, the activities of these men dominated the electrical world during the decade in which the electrical engineering profession began to take shape.

The rapidity of electrical development appeared starkly in the distance traveled from the industrial exhibition organized by the Franklin Institute in 1874 to the electrical exposition it organized in 1884. Though only a decade separated the two expositions, the artifacts displayed in 1874 reflected a markedly different picture of the country's technical landscape. The event celebrated the Franklin Institute's fiftieth anniversary, and, in the minds of its instigators, served as a model for the coming Centennial celebration. As it had for its first exposition of 1824, the Franklin Institute described the affair as an "Exhibition of Arts and Manufactures." Yet the affair specifically honored the age of the mechanic arts. When lead-pipe manufacturer William P. Tatham announced the exhibition, his generalized description masked the practical absence of electricity in the technical world of the early 1870's:

> The Exhibition will embrace all MATERIALS used in the Arts, in every stage of manufactures, from their natural condition to the finished products, and all TOOLS, IMPLEMENTS AND MACHINES, by which the gifts of nature are changed and adapted to the use, the comfort, or the enjoyment of mankind [31].

Throughout the exhibition hall, the mechanical expressions of the nation's technical heritage prevailed within a vast miscellaney of the products of American manufacture. Heavy industrial items, such as agricultural implements and military arms, stood among thirty-nine classes of manufacture, which included boots and shoes, carpets and china, coach work, combs and brushes, surgical instruments, drugs and chemicals, musical instruments, and examples of the fine arts and photography. Hats, caps, furs, and ladies' fancy goods were there as well. Above all, however, stood the technology of a mature industrial society. Of thirty-nine formal categories, only "Class XXIX: Models and Machinery, Lubricating Oils, and Engineers' Supplies" contained subdivisions. But the sixteen subgroups of "Models and Machinery" pervaded the hall with generators of power, machinery of transmission and transportation, hydraulic machinery, machine tools, textile machinery, sewing machines, paper and rope machinery, sugar machinery, and gas machinery. A final subclass was reserved for engineers' supplies and lubricating oil, described by the judges as "the humblest but most important agent in all the varied mechanic arts" [32].

In his formal report to the Franklin Institute, the organizing committee chairman exulted over the exhibition's mechanical riches. The products were of "peculiar excellence." Outstanding among them were the photographs, chemicals, "the wonder working of the sewing machine, . . . the rapidity of the printing press, [and] the precision of movement of the machine tools." Most impressive were the generators of power and the efficiency with which they served the organizers of the exposition: "The number of steam boilers in operation was 9, of 316 horse-power in the aggregate, consuming 267 tons of coal. There were 3 steam engines driving shafting, 22 driving pumps, and 11 driving particular machines. The whole number of steam engines at work, or in motion, was 46. The whole number of machines in motion was 281." Though the 1874 exposition lacked a dramatic symbol comparable with George Corliss's great steam engine that stood in the middle of Machinery Hall in 1876, steam power and the artifacts of what Lewis Mumford has called paleotechnics, or, the "coal-and-iron complex," permeated the exhibition. The special qualities of the exhibition, however, were best expressed in the closing address of Franklin Institute president William Sellers. Born in the year of the founding of the Franklin Institute, Sellers' remarks issued from that fifty-year perspective and from what he saw before him in the exhibition hall. "What surprising advance is shown!" he exclaimed. "How amazing has been the progress!" Sellers reacted to a display of the "iron servants of man's will" amid a "forest of belts that give motion to these machines" [33]. To him, these were the highest achievements of American technical genius. Yet what he failed to recognize — or perhaps to see in the vast hall — was the role of electrical technology in the life of America.

Even had Sellers looked closely at the small display of "electrical apparatus" exhibited among the "Philosophical, Mathematical and Optical Instruments," it is unlikely he would have found signs of a revolution. Of twenty objects, all but a "Farradic coil" and three "improved" lightning rods relied on the low-energy electrical current then available from the state-of-the-art source: the electrochemical cell. Although technically crude and physically clumsy by the standards even of a decade later, the battery adequately served the existing electrical industry. In addition, inventors and small manufacturers had entered a half-dozen alarm systems designed for warding off and apprehending burglars or signaling fires. Especially interesting to the judges was an electric clock that turned off a fire alarm system just before a shopkeeper's arrival in the morning. A more exotic entry was an "Electro-magnetic Mallet for Plugging Teeth" [34]. Clearly, the work that made the dynamo a practical source of power to drive machinery and fuel lighting systems had scarcely begun.

Upon reflection, though, the small exhibit contained a few hints of changes to come, represented chiefly by the battery. For more than alarms and dental mallets relied upon battery power; the device made possible the country's "first commercially successful applications of the electric current": the electroplating industry and the electromagnetic telegraph. As the most visible industrial activity utilizing electrical energy, the technology of the telegraph was amply displayed; telegraphic instruments alone made up a third of the electrical items. Counting the batteries, whose relatively low voltage amply fit the needs of the industry, telegraphic items constituted one-half of the exhibit. There was a printer, described by the judges as "a simple and rapid worker, combining bell, dial and printed signals"; an "improved Morse register" with "an ingenious self-starting and stopping arrangement"; a relay, switch, and sounder with "several new features"; and a sample of "Patent Covered Telegraph Wire." The judges referred the wire to the Institute's Committee on Science and the Arts to be evaluated for originality as well as quality of manufacture [35].

Although no item pointed directly to electroplating, the battery had evolved to such an extent that, by the 1830's in England and the 1840's in America, it gave rise to that industry. Shortly after 1800, when the Italian, Allesandro Volta, showed that his electrochemical pile could provide a source of constant electrical current, investigators learned to use the chemical effects of an electrical charge for assessing the elements of a chemical compound. By 1838, an English firm had begun electroplating with zinc and, soon afterwards, purchased a process for gold- and silver-plating. Subsequently, electroplating came to America through the Scoville Company of Waterbury, Connecticut, and by the early 1840's, Scoville and other firms began to use the process to manufacture plates for use in daguerreotypy [36].

Twenty years later, the English immigrant, Edward Weston, found his first job in such an enterprise and soon after joined an electroplating firm. Within a few years, he had begun to explore the possibility of replacing batteries used in electroplating with a specially designed dynamo. In 1875, after building several electroplating dynamos, he entered fully into electrical work in Newark, New Jersey. Two years later, Weston successfully demonstrated an arc-lighting system and was able to raise capital to found the Weston Dynamo Electric Machine Company. His skill was such that a writer in a contemporary English journal later judged that Weston had achieved "one of the most successful dynamo-electric machines . . . ever . . . introduced for electro-plating" [37].

With the exception of the battery, however, no display at the 1874 Exhibition actually pointed to electroplating or to imminent developments in the dynamo. Only the "Farradic Coil" among the electrical apparatus referred to the main line of innovations about to issue forth. To find a clearer sign of the new age, one must look to an individual rather than to the hardware, specifically to the figure of Elihu Thomson, who was among the judges assigned to the electrical artifacts [38]. Although not yet an active innovator, the twenty-one-year-old Philadelphia schoolteacher was already preparing for that role. Thomson had immigrated from England to Philadelphia with his family in 1858 and had begun to experiment with electricity before reaching his teens. His first device was an electrostatic generator made from a wooden stand, pieces of scrap leather, and an old wine bottle. After graduating from Central High School in 1870—an advanced preparatory public school in Philadelphia—Thomson accepted a teaching position there. In 1876, he was appointed professor of chemistry.

During the next quarter-century, Thomson ranged a far distance from the halls of Central High, achieving fame and wealth as an inventor-entrepreneur. The Thomson-Houston Company rested on his inventions, with Houston mainly contributing his stature in the local and national technical communities. Unlike Houston, Thomson left teaching for a life of engineering research, becoming General Electric's resident sage in the next century. It was from that vantage point that Thomson judged the 1876 exhibition to have been most responsible for spurring the advent of the electrical era. Certainly, he believed it had influenced him. In a long essay on "Electricity During the Nineteenth Century," published in the *New York Sun* in 1901 and then reprinted in the annual report of the Smithsonian Institution for that year, Thomson credited the Centennial Exhibition with introducing to him the potential of the dynamo. Though several were displayed, the machines designed after the style of the Belgian, Zenobe T. Gramme, most impressed him. Gramme had introduced the first commercially practical dynamo into the electroplating industry in 1870, inspiring Weston's work in 1874. He had devised a ring armature that, Thomson believed, made his machine "relatively perfect." Thomson recalled the Gramme display as "a remarkable exhibit for its time." One dynamo ran an arc lamp; a larger machine had been designed for electrolytic work, "such as electroplating and electrotyping"; yet, "most novel and interesting of all, one Gramme machine driven by power was connected to another by a pair of wires and the second ran as a motor." Neither did Thomson fail to recognize the significance of Bell's exhibit. He believed the Centennial had marked the "very birth" of Bell's "speaking telephone," which was destined by century's end to be "a most potent factor in human affairs." Nevertheless, he thought the Gramme machines more significant. Their presence at the Centennial, he wrote, "was a foreshadowing of the great electric power transmission plants of to-day; the suggestion of the electric station furnishing power as well as light; and to a less degree the promise of future railways using electric power" [39].

The influence of the technical exhibitions on Thomson was a common occurrence. Beyond inspiring young technologists, the exhibitions, and the competitive tests that often accompanied them, served as important competitive arenas for the growing number of entrants to the lighting and power field. From comparative tests made of dynamos in 1877, in fact, Charles Brush gained the lead in the design and sale of arc-lighting systems. Unlike the other pioneers in electric power systems, Brush had attended college, receiving in 1869 a degree in chemistry from the University of Michigan. Though he worked for several years in Cleveland as a chemical consultant, he soon began to study and experiment with electricity. By 1875, Brush had built a dynamo as a first component in an arc-lighting system; he had attained a patron and had access to the foundry, shops, and skilled workers of a telegraphic equipment manufactory. A year later, he had a complete system; by 1877, he was almost ready to introduce his system on the market. But before that could occur, the firm that manufactured his system decided to submit a Brush dynamo to the Franklin Institute for inclusion in the dynamo efficiency tests to be conducted that year [40].

With the stated aim of purchasing a dynamo for its own use, the Institute had invited "all builders of such machines to send one of their make to be submitted to a comparative trial." When the manufacturers of the Weston machine, Siemens of London, and the Gramme companies failed to submit a dynamo, the Institute borrowed one of the Centennial Gramme machines. A Wallace-Farmer machine had been secured earlier for demonstrations and lectures at the society's building. Thus the submission of the Brush machine enabled the examiners to compare two American designs with the leading European machine. Though the machine was an early version of that which would eventually power his system, the Brush entry reflected the careful design that would typify his arc-lighting apparatus and make it supreme in the American market. By the end of the decade, Brush had achieved most of the essentials for a superior system. He developed a nearly constant current dynamo, which, with the high voltage of an 1879 model, could run sixteen lamps. Brush handled the problem of lamp failure and, thus, current increases by fixing his lamps to short circuit upon failure. He later fashioned an automatic regulator to maintain a constant current. Besides devising an automatic carbon feeder, Brush improved the carbon itself, settling on petroleum coke as the basic material, which he copper-plated. Also, after observing that the carbons burned smoothly only after becoming tapered with use, he sharpened the end instead of leaving it blunt as was the practice. What finally distinguished Brush's system was its simplicity, a necessary feature when few purchasers understood the devices. Thus Brush had built a superior system by concentrating on the essentials: a constant current dynamo, an automatic carbon feed, and a simple design.

The Franklin Institute examiners found this latter feature especially impressive, attributing to it the wide usefulness of the early Brush machine. "The great ease of repair," the committee reported, was derived from the "mechanical details of its construction" [41]. This evaluation reflected a long-held opinion of American mechanics — one consistently reflected in the Franklin Institute's technical judgments: a successful technical device must be understandable and capable of repair by the average mechanic. That consideration was accepted by all the leaders of the first generation of electric-power system builders. Indeed, two who were among those who rendered the judgment were Elihu Thomson and his Central High colleague and entrepreneurial partner, Edwin Houston.

Though Thomson had begun to build his own dynamos before the Institute's tests, those tests, and the contact they brought with Brush, first aroused his interest in arc-lighting systems. Then a visit to Paris in 1878, where he saw even more arc lights in use at a technical exhibition and at a railroad station, reinforced his desire to design and market his own system. Within a year after a Philadelphia manufacturer provided support for Thomson and Houston to develop a system capable of competing with Brush's system, they installed a nine-lamp system in a local bakery. Two years later, a group of New England investors formed a company around Thomson's patents. When the machinery it manufactured proved clearly inferior, Thomson took his talents to a new group of backers, who organized a firm in Connecticut in 1880. Thomson's concern for the fate of his creative work was characteristic of the inventor-entrepreneur who presided "over the life of an invention from birth to social acceptance" [42]. Needing to develop a competitive system, he readily agreed to serve as an electrician to the new company and moved to Connecticut to begin improving his system. He had already devised an automatic current regulator to produce a steadier current and to operate more efficiently than Brush's system. To provide a system that could handle more lamps and thus operate at less cost, Thomson developed a larger dynamo. The challenge, he found, was not increasing voltage so much as protecting the system against the threat of higher voltages posed to the commutator. A critical link in the system, the commutator converted the alternating current issuing from the armature into the direct current used in the lighting circuit. Thomson devised a method of insulating the commutator from high voltage; by the end of 1881, he had developed a competitive product [43]. Out of six years of solid technical work, Thomson's enterprises had come to rival both Brush and Edison's company.

Brush's and Thomson's achievements in arc lighting established their technical reputations. Their arc-lighting systems, moreover, continued through the century to be important economically. From 6,000 arc lights in service in 1880, the industry continued to grow, so that nearly 400,000 lights operated in 1902. As entrepreneurs, however, it was necessary for them and

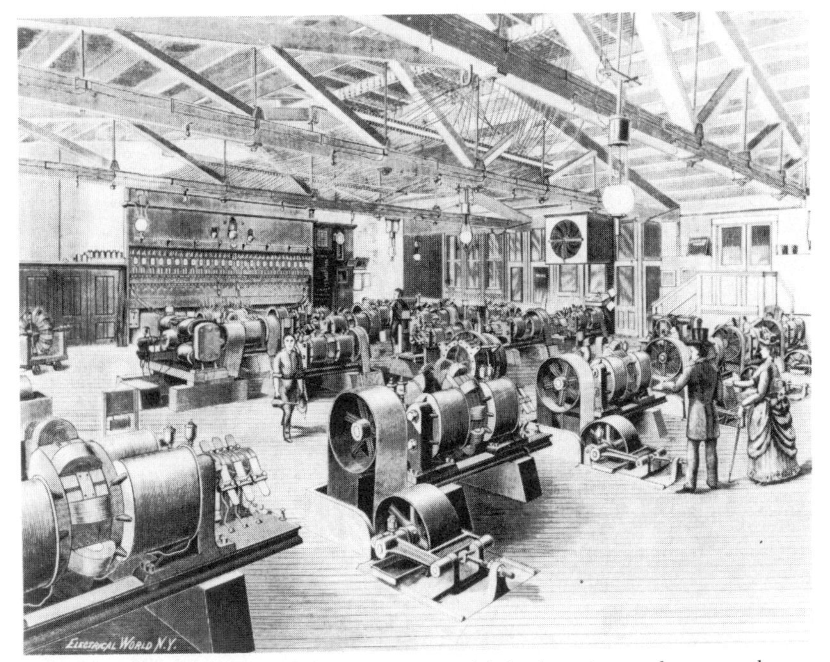

Just as the industrial exhibitions drew visitors, so did this late nineteenth-century dynamo room of the Brush Electric Lighting Station at Philadelphia.

their companies to respond to the new developments in lighting. This led the Brush Electric Company, in 1883, to purchase the American rights to an English incandescent lamp; about the same time, Thomson-Houston began to sell incandescent lighting to operate on the company's arc-lighting system. Such an approach to incandescent lighting was precisely what Edison rejected when he set out to develop "a comprehensive system." As Edison explained some years later, he followed the principal that "the failure of one part to cooperate properly with the other part disorganizes the whole and renders it inoperative for the purpose intended" [44]. Therefore, in 1877, when the researchers in power and lighting struggled to perfect competitive arc-lighting systems, Edison began his experiments with incandescent lights. He continued to work on incandescent lighting until he achieved a successful lamp at the end of 1879. Having set up his invention factory in 1876 at Menlo Park, New Jersey, Edison, two years later, hired a mathematical physicist who had been trained at Princeton and in Germany under Hermann Helmholtz. With these innovations in technical research and development, Edison was thus able to move steadily toward his goal of a comprehensive system of incandescent lighting.

Edison's idea of an adequate system, as with any successful inventor-entrepreneur, extended beyond technical invention to considerations of the marketplace. He, therefore, sought a system costing no more or no less to operate than the existing urban gas-lighting systems. He was confident that he could develop a lighting system whose brightness, unlike the strong glare of arc lights, could be controlled and, thereby, adapted to residential and office purposes. Because Elihu Thomson doubted that an incandescent system would be profitable, he continued to work on his arc lights, even after witnessing a successful demonstration at Menlo Park. But because arc-lighting systems used high voltages to light city streets and large interior spaces, Edison built a new dynamo that would operate with the low voltages necessary for electric lighting in homes. He also replaced the serial circuits of arc lights with a parallel system and devised a light with high resistence to reduce the current. This latter step avoided excessive energy loss in transmission and reduced the amount of expensive copper that was used.

At this time, Edison's system promised no more than the isolated electric power and lighting plants that characterized arc-lighting installations. From the beginning, however, Edison's work contained the germ of a revolutionary idea: the central power station. The early arrangements at Menlo Park were mounted for demonstrations, not regular service and, therefore, contained only the possibility of a power station. The 1882 opening of Edison's Pearl Street station in New York, however, made manifest a future in which electrical energy would be produced at one point and distributed to an indefinite number of other points. Yet, however promising the plan to build a central station near Wall Street in lower Manhattan, the bankers who controlled the Edison Electric Light Company continued to favor the isolated plants already in production. The dispute between Edison and the owners of his patents clarified the fundamental differences of these choices. The issue was essentially that which aggravated Vail's relationship with the early Bell Company backers: the choice between more profits now or the promise of greater profits later from a more efficient system [45].

Pearl Street represented far more than the inventor's continuing conflict with businessmen concerned with revenue; it also pointed to the persistent creativity of Edison's technical group. In important ways, Pearl Street represented the final act of wizardry for the leader of the group. But on the occasion of this struggle over company policy, that inventive act had not been completed. Though Menlo Park was a model for the permanent installation to be made in New York, the arrangements in New Jersey had been makeshift. For the station begun in Manhattan in the fall of 1881, more durable and efficient apparatus was required. Pearl Street required not only more of everything — ten generators, for example — but also different arrangements: The wires were laid underground, bamboo-filament lamps were fabricated, and screw sockets were devised to replace the cruder arrangements at Menlo Park. For the power source itself, Edison developed the largest dynamos constructed at that time (forecasting those that Edison's personal assistant, Samuel Insull, would have GE engineers construct a decade later in Chicago). Edison coupled these "Jumbos" directly to steam engines mounted on the same base. Technical innovations continued to be necessary. Two months before its opening, the Pearl Station was almost wrecked by tests when the steam engine's governors proved ineffective. New steam engines equipped with governors designed to eliminate the torsional vibration solved that problem, as similar corrections solved other problems.

Building Pearl Street had tried the wizard of Menlo Park on several levels. The tension induced within Edison during these months of trial, coupled with the importance of the experimental station to his larger vision, all came to bear on the inventor on opening day: "Success meant world-wide adoption of our central-station plan," Edison later recalled. "Failure meant loss of money and prestige and setting back of our enterprise. All I can remember of the events of that day is that I had been up most of the night rehearsing my men and going over every part of the system. . . . If I ever did any thinking in my life it was on that day" [46].

Edison's emotional state at the opening of the Pearl Street station derived not so much from a fear of technical failure as from the need to achieve both technical excellence and quick commercial success. Such tension was natural for the technologist turned businessman on the eve of a new undertaking. He had much at stake, yet as an inventor-entrepreneur, this was a necessary feature of his work [47]. The successful inventor-entrepreneur, in short, developed a special sensitivity to the challenges of both effective marketing and technical design. Though Edison's talents shone in both regards, men like

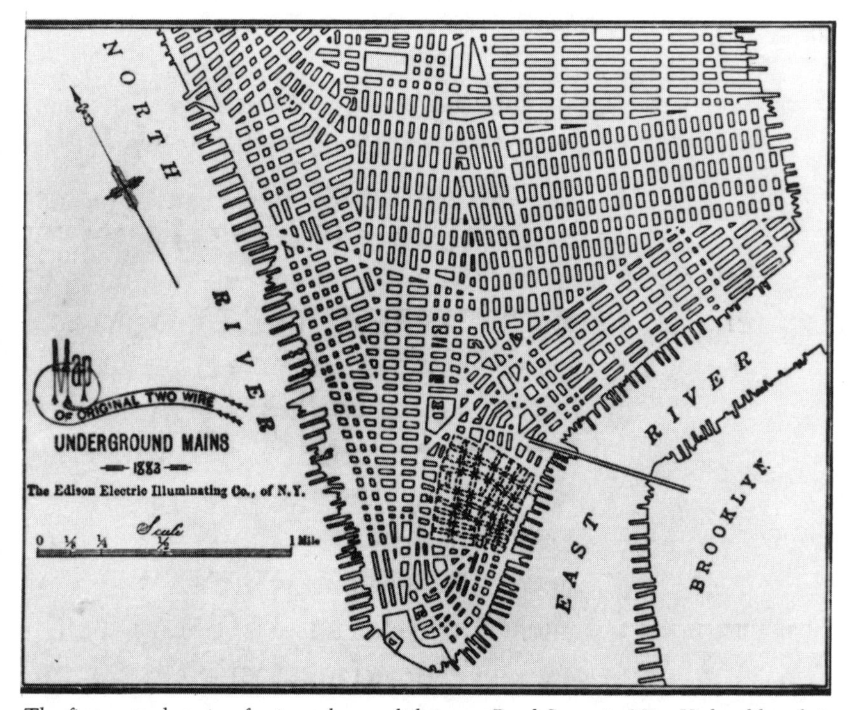

The first central station for incandescent lighting at Pearl Street in New York: although it reached less than fifty square blocks, Pearl Street served Edison well as a working model for his sales force.

Brush and Thomson were equally attuned to the demands of the marketplace. Acceptance of market requirements led these inventors to place great importance on exhibiting and demonstrating their inventions. The industrial exhibitions were stages on which to establish the credibility of their systems. Manufacturers' displays, thus, shared characteristics with the long familiar technical and scientific demonstrations, offering a method for gaining a stamp of approval from an impartial jury of peers. Thomson learned, firsthand, the importance of such approval when the Franklin Institute's superior rating of Brush's dynamo contributed greatly to its success in the marketplace. The same tests led the makers of the Wallace-Farmer machine to remove their dynamo from the market. Brush followed his Philadelphia success with an exhibit at the Mechanic's Fair in Boston the next year, where he again won out over Wallace-Farmer. At the end of the fair, Brush sold his machinery to a Boston merchant, whereas Wallace and Sons took its apparatus home to Connecticut [48].

Edison sought similar approval for his incandescent lamp at the 1881 International Electrical Exhibition in Paris. In the efficiency tests there, Edison's entry won out over two English submissions and one other American lamp. Thomson equally desired an objective appraisal of his arc-lighting system. So in 1883, when the Cincinnati mechanics' institute announced that its annual exhibition of western manufactures would feature electric lighting, Thomson submitted his arc-lighting apparatus. His arc-lighting system won the top award, and Edison, competing against the Weston lamp—the leading American rival to his device—won the award for incandescent lights [49].

The importance of the exhibitions—and the different understandings of their purposes—became sharply apparent as Elihu Thomson and Edwin Houston planned for the 1884 International Electrical Exhibition. Though six years younger than Houston, Thomson possessed a sensitivity to commercial matters that Houston did not. Houston, after graduating from Central High in Philadelphia, studied in Berlin and Heidelberg before returning to Central's faculty as professor of physical geography and natural philosophy. The need to invent that dominated much of Thomson's career never possessed the professorial Houston. During the next several decades, Houston wrote textbooks and prepared a dictionary of electrical terms in addition to serving as president of the AIEE. Such interests led him to view the coming electrical fair differently from his partner. Since Thomson was responsible for planning the Thomson-Houston Company's exhibit for the 1884 exhibition, Houston's request to a company officer to include some favored artifacts was passed to Thomson for a response. He wrote Houston that the request to introduce "extraneous matter . . . puts me as electrician of the Co." in the position of approving exhibit items not seen. To do so would undercut the "due proportioning and . . . good mechanical design" of the company's exhibit. In explaining to Houston the reason for exhibiting at all, Thomson articulated the basic principles of advertising: "What we show must be regarded as an advertisement by the Co. of the goods it does sell, or would have sold at some period of its existence, enhanced in display... calculated to excite interest and fix attention" [50].

Inventor-entrepreneurs played critical parts in the successful takeoff of the power-engineering field. Important personal qualities were necessary: ambition, creative ability, the capacity to work long and hard hours, and, as demonstrated in the conflict between Houston and Thomson, a sure grasp of marketing needs. Such qualities had been exhibited, first in telegraphy, then in telephony and the electrical manufacturing and power industries. The success won in these fields had led directly to the founding of new industries and, in turn, to the consolidation of those industries. At the same time, however, the knowledge and skills of the technical makers of these new industrial arenas began to be consolidated. It was thus that the success of entrepreneurs led also to the rise of a profession and to the founding of an engineering society.

Founding time: The American Institute of Electrical Engineers (AIEE)

A half century of electrical development created, therefore, not only an industry but the seeds of a profession as well. That the profession rested no less on the industry than on technical systems and artifacts was concisely symbolized by the setting of that first conference of the American Institute of Electrical Engineers. And though manufacturers' exhibitions were not unusual, the density of electrical apparatus exhibited in 1884 documented a technological revolution. In 1874, the Franklin Institute had required nearly forty categories to organize the miscellany brought together at the exposition; yet the coherence of the items gathered in 1884 reduced that number to seven. Four areas were given to the basic divisions of electrical technology: production, conductors, measurements, and applications. The latter area contained subsections for apparatus requiring electric currents of "low power" and items needing "great power." Three sections contained displays relating to terrestrial physics, historical apparatus, and miscellaneous topics—chiefly electromechanical apparatus, educational exhibits, and bibliographical items. From the nearly 200 exhibitors, the largest displays predictably came from the largest companies and bore the names of the founders of the modern electrical industry: Bell, Brush, Weston, Edison, and Thomson [51]. These names, which carried the central message of the electrical exhibition, denoted not only the leading inventors in the fields of communications and electric power but also the chief characteristics of modern engineering. The names pointed to the fruitfulness of directed research that was attuned to the marketplace, to the development of precise electrical measurements, to the importance of engineering design aided by mathematical tools, and, most indicative of the immediate future of the electrical engineering profession, to the need to understand more precisely the challenges of designing, constructing, and maintaining large technical systems.

Yet the names of the inventor-entrepreneurs that gave impetus to the age of electricity did not solely define the age. Their names certainly did not reflect the presence of executive officers like Green and Vail or engineering administrators like Edwin Rice. Nor did the inventors' names reflect the role of physicists and professors like William Anthony of Cornell and Charles Cross of MIT, who already had begun to establish the first university programs in electrical engineering that would train future members of the AIEE. The men whom the country early made heroic inventors, in short, and who dominated the manufacturers' exhibition in Philadelphia, were necessary, but not sufficient to depict the emerging profession of electrical engineers.

Still, it was the Franklin Institute's announcement of an International Electrical Exhibition that spurred members of New York's electrical community that spring to organize a national society. The first of three "calls" published in the *Electrical World* urged the formation of a professional body to avoid the "lasting disgrace to American electricians if no . . . national electrical society" could receive the great number of foreign visitors expected. The self-described "electricians and capitalists," joined by "others prominently connected with electrical enterprises," announced a meeting for early May at the rooms of the nation's pioneer engineering body, the American Society of Civil Engineers. An organizational meeting had been held already, and a committee had been appointed to report back on matters that concerned the society's name, objectives, membership criteria, dues, management, and appropriate standing committees, to consider, in short, "the whole scheme of organization" [52].

Basic issues of membership standards and organizational aims had been broached already in the circular that proposed the new society. New York electrometallurgist Nathaniel S. Keith explained that the advantages of membership were available not only to electrical engineers, electricians, and teachers but also to "inventors and manufacturers" and "officers . . . of all companies based upon electrical inventions." Keith added that "all who are inclined to support the organization for the common interest" were eligible. Moreover, the slate showing the officers and managers who were elected at the meeting held on May 13 amply demonstrated the commitment to such a broad-based society. Thus, the first president of the American Institute of Electrical Engineers was Norvin Green. Among the vice presidents were inventors of the stature of Alexander Graham Bell and Thomas Edison. They were joined by physics professor Charles Cross of the Massachusetts Institute of Technology, an engineer from the Western Electric Company in Milwaukee, and two veteran telegraphic electricians, George Hamilton and Franklin Pope.

The list of twelve managers extended the diversity represented by the president and vice presidents. Charles Brush was elected along with Western Union's chief inventor, Elisha Gray. Besides Hamilton and Pope, there were two additional telegraphic electricians, one of whom was the publicist George Prescott. Other managers were Edwin Houston, Edward Weston, Theodore Vail (Bell's general manager), two electrical men from Florida and Indiana, an engineering professor from New York, and the general superintendent of

New York's telephone and telegraph company. The initial roster of the Institute's leaders, in short, included inventor-entrepreneurs, college-trained engineers, physicists, teachers, managers from the communications and electrical manufacturing industries, and a sizable contingent of telegraphic electricians. Although none of these electrical men came from as far away as the West Coast, the geographical spread represented by the elected leaders in this second month of the new society's existence suggested that only a short time would pass before the AIEE was a national body in fact as well as in vision.

Selecting the leadership from a broad spectrum of electricians and capitalists was matched, moreover, by the broadly conceived objectives proposed for the new society in that first call. The American Institute of Electrical Engineers would serve members by publishing papers, discussions, and transactions from the meetings and would establish contacts with other technical societies. A museum and a library were also contemplated. Additional aims suggested a broad conception of the social and political role of a professional engineering society. Not only would the American Institute of Electrical Engineers settle "disputed electrical questions" within the industry — a sign of the importance given to uniform industrial standards from the beginning — but it would also protect its members "from unfavorable legislation" [53].

That the electrical community consciously entered into this act of professionalization was evident in these broad programmatic ideals. That a broad-based, national society was contemplated appeared also in the diversity of the electrical scientists and technologists who gathered in Philadelphia. Yet to call themselves electrical engineers was to beg the question. This became apparent during the next two decades as the AIEE's diversity became the fundamental challenge in organizing and defining the nature and purposes of the fledgling society. And yet, in handling the conflicts induced by the differences within the Institute — differences that, in some instances, altered the status of some of the founding officers and managers — the new organization would become at one and the same time a society of both professionals and engineers. And in the process, members would begin to understand the special meaning of engineering professionalism.

2/THE NEW ENGINEERING AGE

We are in an engineering age, an electrical age, with its physical, commercial, industrial and social changes, with its new conditions, new opportunities and new responsibilities.

Charles F. Scott, 1903
"President's Address" [1]

Who shall be members?

Three years after the founding of the AIEE, retiring president Franklin Pope reflected on the increasing "prosperity of our organization." He linked the technical advances in the electrical industry with the Institute's prosperity. Thus, the quick success of the society rested on "the rapid development" to be found "in every department of the science and art" of electricity. A few years earlier, Edison's Pearl Street station had represented the state of the art in the field of lighting and power, Pope explained. Yet, just since taking office, he had observed a number of "revolutionizing" achievements. Electric motors were being applied to "minor branches of manufacturing in large industrial centres, and to the propulsion of street railroad cars." Long-distance incandescent electric lighting had been introduced to make "practicable and profitable . . . the employment of alternating currents and induction apparatus." Important advances had also taken place in storage batteries and electric welding. But it was the promise of electric power that most impressed Pope. As an adviser to George Westinghouse on his purchases of European electrical patents, Pope knew this field first hand. He was one of several electrical experts Westinghouse had hired to help him assemble an alternating-current electric power system capable of high-voltage, long-distance transmission. Thus, Pope derived the mission of the AIEE from his sense of the great significance of this work. The AIEE's special work was to act as a clearing house for technical advances in the field by keeping "fairly abreast with . . . its members and co-workers in their individual and professional capacities" [2]. In turn, the AIEE would transmit information to members about advances in the discipline.

Such a mission required appropriate organizational machinery, for which the acts of 1884 had provided only a foundation. Responsibility for constructing the machinery passed, then, to the Council. This working group of officers and managers ran the Institute, meeting monthly to pass on special membership questions, to organize committees as the need arose, and to handle the myriad problems that challenged the young society. In planning for the increased geographical spread of the society, the first Rules of the AIEE had given the Council authority to appoint an Executive Committee should it become difficult to gather a quorum. Yet until 1897, the close proximity of most directors made such a step unnecessary. Thus it was the full Council that began in 1884 to construct the organization. At its second formal meeting, the Council had set up a dozen working committees to oversee the interests of the organization. Eight of these committees were concerned with technical matters; four dealt with the administrative affairs of the Institute. Within a year, only the administrative committees remained.

In its attempts to establish the American Institute of Electrical Engineers as a technical arbiter in the country, the technical committees failed. It was simply too early to create a full-blown program of technical activities aimed at giving technological leadership in larger spheres. A prior task was to establish a network of programs to unite the membership. This was done in a number of ways, including the annual conference, publications, and monthly meetings. The 1884 constitution authorized the Council to print in a transactions or to have read before the Institute any papers submitted "when they think it desirable." Thus, within months after the Philadephia conference, the first *Transactions* appeared. The Institute's publishing venture proceeded smoothly, stumbling only upon inflated expectations regarding the society's growth: the 1000 copies of the *Transactions* printed in 1885 became 500 the following year and by 1889 were reduced to 350 bound in cloth and 150 in "paper untrimmed." In addition, the Rules established an annual conference in the style of the Philadelphia gathering to be held each spring [3].

Yet, even before the Philadelphia conference, some members wanted more than annual meetings. Bunched in the New York area as they were, the idea of regular meetings for members to hear and discuss papers came up early. Papers had been read at times during the early monthly Council meetings, the first coming in June 1884 on the subject of the patent office. By 1886, however, several members had prodded the Council into establishing separate monthly meetings in New York after the practice of the New York Electrical Society. The first of these special monthly meetings started shakily when "a complication of business" prevented an engineer employed by the newly formed Westinghouse Electric Company from reading his paper on "Incandescent Lighting from Central Stations." President Pope saved the occasion by arranging for a new speaker to speak on the same subject. At the end of his presidential term, Pope looked with special pride on the "series of special meetings" held that year. They had strengthened the Institute, he believed, by "increasing the interest" of members who lived beyond "the immediate vicinity of New York" [4].

Once the New York meetings were begun, the logical step was to export the model to other cities. This move was given impetus at the meeting held during the Columbian Exposition in Chicago, in 1893, when the vitality of the local members led Institute secretary, Ralph Pope, to urge expansion of the monthly meetings. It would increase "the value of membership" and help the "Institute . . . attain still higher standing." The branch meetings would not only bring new members into the society, but they would also keep members abreast of technical changes in the field. Subject matter was to be carefully chosen and distributed to all members. Though local papers were encouraged, members meeting in other parts of the country were to read and discuss papers previously delivered at the New York meetings. The stated intent was to raise the professional level of the entire membership. As a device for increasing membership, the monthly meetings were eminently successful. Indeed, the growth of both the profession and the industry can be traced in the spread of the geographical branches. For a time after 1900, twenty branches a year were organized in towns ranging from Allegheny, Pennsylvania, to Ames, Iowa; to Madison, Wisconsin; and to major western cities like Cincinnati, Denver, and St. Louis [5].

Developing programs during the early years was among the least difficult tasks for the experienced AIEE leadership. Western Union president Norvin Green actively presided over the Council's monthly meetings during his presidency, even agreeing to a second term of office. Furthermore, on the board were heads of manufacturing firms and telephone exchanges, editors, and founders of industrial and technical magazines. Yet, while meetings and publications could rely on experience, drawing the conceptual lines of the discipline and the profession was a far more difficult task. Part of the difficulty would come from complications arising from demographic changes in the membership brought about by the rise and incorporation of the electrical manufacturing and power industries. For although the corporate revolution produced an explosion of opportunities for many in the electrical field, it also brought stress and hard times to some. The "new conditions" that Charles F. Scott spoke of in 1903 affected even those engineering types who formed the core of the profession—professors, consultants, corporate engineers, and engineering managers. Executive officers were pushed to the sidelines, making way for a new class of managers. Telegraph electricians faded out, and inventor-entrepreneurs moved from center to periphery. The traditional electrical experts gave way to formally trained power engineering consultants who designed and oversaw the building of isolated and central power plants. Professors who were at first drawn from the physics departments evolved into teachers who were trained to teach in the new electrical engineering departments. And the physicists gave way to the engineering scientists—often with European training in mathematical physics, but sometimes products of leading American schools—who were persuasive voices in the Institute during the 1890's.

In all these cases, individuals gave concreteness to the general events: Franklin Pope, who made the transition from telegraph electrician to electrical expert until he collapsed before the accelerating changes; inventor-entrepreneur Elihu Thomson, who struggled with his financial backers over questions of quality control in production; Professor William Anthony, whose status as a physicist delayed his election to the Institute's presidency; the immigrant mathematical physicist Charles Proteus Steinmetz, who carried the scientific revolution in engineering to General Electric and into Institute meetings; and corporate engineers like Benjamin Garver Lamme and Hammond V. Hayes, both of whom helped shape the engineering role at Westinghouse and the Bell Telephone Company. Some of these men were in Philadelphia in 1884; others joined the AIEE during its first decade. By the end of the century, however, all had played a part in bringing the Institute and the profession to maturity.

It was, then, Scott's "physical, commercial, industrial, and social conditions" and the ambitions of individual engineers from which would come the profession and the AIEE. Given that the members of that profession were changing — metamorphosing from traditional to modern, from telegraphy to power — the Institute, too, had to adjust, shaped by the makeup of its membership as it was. In such a situation, the problem of definition was paramount. It required both a disciplinary definition and a precise statement of membership criteria, as well as a grasp of the qualities of the professional engineer. From the start, the members who struggled with this problem understood, as Pope's remarks suggested, that new technical areas were being propagated in freshly cultivated fields. Nonetheless, it would be some years before they realized that pure and applied electrical knowledge was also entering an intellectual revolution. Because it would take time to perceive the nature of the new knowledge base of electrical engineering, the process of definition itself would take time.

The issue of membership criteria was at the heart of the problem of definition and, consequently, arose in the first year, its controversial colors appearing almost immediately. As soon as membership grades were established, individuals relegated to the lower grade protested. Their protests centered around the "Associate" grade, a class of membership below the rank of full "Member." The first definition of the Member grade held that "Members and Honorary Members shall be professional electrical engineers and electricians. Associates shall include persons practically and officially engaged in electrical enterprises, and all suitable persons desirous of being connected with the Institute. . . ." In October, 1884, when the Council passed a motion raising all members of the Council "from Associates to Members," the protests increased. At the December meeting, President Norvin Green told the Council that they would soon have to consider making changes in the rules governing the election of Associates. At the annual meeting the following May, Green's prediction proved true when members charged the constitutional definition as being "misty." No clear distinction existed, they argued, between individuals engaged in electrical enterprises and electrical engineers presumably also engaged in such enterprises. The Council responded with a resolution that, rather than distinguishing more sharply, expanded the "Member" category. It would include, in addition to the "electrical engineer or electrician," any person "so intimately associated with the science of electricity, or the art," that their membership, as determined by the Council, "would conduce to the interests of the Institute." Such a category easily fit Green and other purely managerial types among the officers and directors. The resolution, in effect, gave the Council a blank check to allow disgruntled Associate members into the Institute's top membership grade. This failed to satisfy members still classed as Associates, who rightly claimed that the resolution contradicted the Institute's formal rules [6]. As a result, the Council's resolution never entered the constitution.

A solution was found when a Board of Examiners was named, in 1885, to assist the Council in passing on the requests for transfer to full membership. Examiners were to rule on the merits of all applications for transfer from the Associate grade and were "to provide rules to test the qualifications of applicants." This removed the Council from being the sole authority over admissions, since the Council appointed members to the Board but could not serve. The extent to which slights could be felt over issues of status was demonstrated at the naming of the first members to the Board of Examiners. After three members were appointed, Frank W. Jones, a veteran telegrapher, complained that there was no "professional electrician" on the committee. The first board thus came to include five persons. Not only was a telegraph electrician added, but a university professor, a group that was also omitted from the initial list of Examiners, was appointed as well [7].

There was clearly no simple answer to the question of who was to be a member, as demonstrated by the anomalies that appeared during the next several years. Elihu Thomson joined the Council in 1887, yet did not become a full Member for four years. When consulting engineer Cyprien O. Mailloux, a charter member, learned that he had been classed as an Associate, he resigned his membership on the Board of Examiners until the Council ordered the secretary "to correct the records." The same measure was resorted to when another member found his status different from what he had thought. By the end of the decade, the confusion over grade began to dissolve before changing realities in the profession. In 1889, Theodore Vail asked the Council to rescind his full-Member status and assign him to the Associate grade. Norvin Green's anomalous position was resolved when a group of Members and Associates petitioned that he be made an Honorary Member. It was not a perfect solution, for although the grade lacked voting privileges or opportunity for office, it was, nonetheless, constitutionally reserved for electricians or professional electrical engineers. However the Council accepted both moves as reasonable solutions to a problem that appeared to be receding. These actions highlighted the practice of passing most problem-cases to the Council. For example, one member's request for transfer to the Member grade was rejected when the Council doubted "his having had sufficient experience." Yet in the case of Moses G. Farmer, a petition signed by a number of Members and Associates led to his elevation to full-Member status in 1890 [8].

Although the power of the Council to act on its own authority when dealing with membership transfers was unquestioned, the path of constitutional revision was open for codifying, more precisely, the standards for membership. Yet of the four constitutional revisions of the criteria for full-Member status, made between 1885 and 1896, the designation of the professional engineer was only slightly altered. The Institute's statement of its objectives also remained unchanged until 1894, continuing to promote electrical science and technology that was "connected with the production and utilization of electricity" and the "welfare" of the electrical workers "employed in these industries." There was, in short, little conceptual room for the professional engineer in the constitution. In defining "Member," the place of the engineer had even receded: in 1885, Members were "professional electrical engineers and electricians"; ten years later, this grade included "Electrical Experts, Electricians, or Electrical Engineers possessing such knowledge of the principles of electrical science and such familiarity with the practical applications of electricity in its several branches, as those branches imply" [9]. The inclusion of experts and electricians tended to dilute the place of the engineer, as it did also the definition of electrical science. By the mid-1890's, it had become clear that formal assertions would gain precision only as the profession absorbed the internal debates over the place of the electrician and the nature of the professional engineer. Only later, when its leaders saw the shape of the new "electrical age" as clearly as had Charles Scott, from his vantage point of the new century, would the words finally come together to mark the maturity of professional engineering. To get there, however, would require a new generation of engineers and a host of new institutions. In all, it would require the incorporation of a new world [10].

Ambiguous engineers: Telegraph electricians and inventor-entrepreneurs

The intensifying debate over the nature of the engineer began as early outside as within the AIEE, and as with Henry A. Rowland's disparagement of the telegraph electrician at the National Electrical Conference, often with a negative cast. The initial attempt was not so much to say what the new engineer was as with declaring who was not an engineer. Though soon after the founding of the AIEE, it became apparent that the title "electrician" existed rather shakily in the Rules of the AIEE alongside that of "electrical engineer"; the ambiguity in the meaning of the word remained to the next century. In 1884, "electrician" frequently described the chief technical adviser of a company; usually, however, it denoted an employee of a telegraph company who oversaw and evaluated new methods and apparatus in the field and supervised the maintainance of the system. Despite this widespread specific use, the title of "electrician" applied also to the most heroic of the inventors of the era. It denoted the role of inventors as diverse as Alexander Graham Bell and Elihu Thomson in companies organized to exploit their inventions. As electrician for the National Bell Company, for example, Bell's task was to perfect the telephone while managers and executive officers developed its commercial potential. In his first contract with the American Electric Company, Thomson's title was "resident electrician." In addition to the stock that was given to Thomson for his patents, he received as electrician an annual salary of $2,500. The term was still being used in the 1890's as titles for highly trained engineering scientists such as Charles P. Steinmetz ("Electrician" for GE in Schenectady, New York) and Arthur E. Kennelly ("Electrician" for the Philadelphia electrical consulting firm he operated with Edwin Houston) [11].

In the early years especially, the label mainly pointed to the electrical workers in the telegraph industry. Since the year of the AIEE's birth, signs appeared of the telegraph electrician's receding function in the electrical engineering community. The telegraph magazine's grudging acceptance of Henry A. Rowland's harsh assertion at the electrical conference in 1884 that "telegraph operators" were not in demand was sign enough. Yet five years later, a writer in *Electrical World* provided a brief historical review of their demise. An article on telegraph electrician George A. Hamilton, who was once a telegraph operator, referred also to other "'old time' operators, among whom may be mentioned T. A. Edison, P. B. Delany, F. L. Pope, T. D. Lockwood & Gerrit Smith . . . ." Like them, Hamilton

> fought his own way up from the key to well assured technical position and reputation. Men are now arising into eminence in electrical science and industry in this country who have never had the remotest connection with telegraphy, but it is the fact that but a few years ago there was not a single well-known electrician in the front rank who was without telegraphic affiliations, or had not himself, in pursuit of a livelihood, mastered the mysteries of the Morse alphabet and of the operating room . . . . [12]

Even the veteran telegraphers had to recognize the change. Thomas Lockwood, who left the keys to become a technical adviser to the patent department at Bell, thought the label no longer described the character of a professional electrical worker. Most men who "advertised themselves as electricians," Lockwood told an AIEE gathering in 1892, "were not electricians in any sense of the word." They were "mechanical bell hangers" who had taken up electrical work as one more "means of obtaining a livelihood." Because they were so numerous, Lockwood explained, "the word fell into some disrepute, and . . . it was necessary to coin another and more euphonious one. And thus it came about that before we had any institutions for learning in that line, we had electrical engineers" [13].

More bluntly, however, the industrial changes in the country brought obsolescence for the early telegraph electricians. Their decline as an engineering type was dramatically illustrated by the last years of Franklin Pope's electrical career. Having spent the greater part of his adult life as a telegraph electrician, Pope, nonetheless, successfully made the transition to the era of lighting and power. Around 1880, he left his position as patent counsel for Western Union to work as a private patent attorney and to resume editorial work that he had first undertaken during the late 1860's when he edited *The Telegrapher.* This time, however, he widened his sphere of interest, working with two of the leading engineering publications of the era: the *Electrician and Electrical Engineer* and the more broadly cast *Engineering Magazine* [14].

Pope's transition into the new technical era went rapidly and smoothly. When Westinghouse entered the hotly competitive electric power field and hired Pope as a primary adviser, it appeared that the veteran telegrapher was successfully making the transition to power engineering. At the end of the decade, Pope published his first book since his history of early telegraphy, this time, however, on the *Evolution of the Electrical Incandescent Lamp.* However, Pope's career had already begun to fail. His immediate decline could have

Franklin Pope's career rose with the telegraph industry and faded with the growth of the giant electrical companies in the 1890's.

stemmed from his advice to Westinghouse not to develop an alternating current system because of the inherent danger from the high voltages used. But Pope later changed that view, and besides, all of Westinghouse's associates and employees apparently advised against investment in alternating current. Nearly a year later, moreover, Pope accompanied Westinghouse and his brother on a trip from Pittsburgh to Great Barrington, Massachusetts, to inspect the first alternating current power plant. The company's chief engineer, William Stanley, had designed the system and overseen construction of the plant [15]. With Stanley and others, Pope had weathered the negative advice given to Westinghouse and, as demonstrated in his 1887 presidential address, was a staunch supporter of Westinghouse's plans for alternating current. Pope's decline came for other reasons, reasons disclosed in 1895 as he corresponded with John E. Hudson, president of the American Bell Telephone Company, during a search for employment.

A decade since his AIEE presidency, the years had moved Pope far from Pittsburgh and his role as one of Westinghouse's leading electrical experts. He had returned to Great Barrington, his birthplace as well as that of the first alternating current lighting system, and was out of work. A year earlier, Pope had written the Bell president, informing him that he planned to move to Boston where he wished "to make some arrangement with the American Bell Telephone Company." Now in June of 1895, Pope was prepared to decide on his "plans for the future." Pope explained to Hudson that the decision to seek work at Bell came because of conditions in the electrical industry. "You know something of the . . . commercial warfare which has been waged for some years between the leading electrical companies." It had made the "peculiar type of expert service . . . required in their litigation . . . distasteful." Dissatisfied with his work for the U. S. Patent Office, Pope's consulting career had also come to a dead end. "Outside the service of the large companies," Pope lamented, "the amount of profitable work is rapidly decreasing" [16].

It was not the end of an era of private consultants that Pope was experiencing. It was the end of the old-style electrical expert, an example of whose work Pope described to Hudson. His work as a patent counselor had included "making investigations and collecting evidence for counsel in litigated cases, . . . preparing and prosecuting applications for patents; [and] testifying as an expert in court cases." Though he did not remind Hudson, Pope had served as "expert witness" in at least two previous cases for the Bell Company. Because he knew that Bell was now "well served" in such matters by Lockwood, he informed Hudson that he "should expect to work in entire cooperation with him, or should you prefer, . . . under his direction" [17].

Hudson responded in late July, apologized for not writing "more promptly," and promised a decision after his vacation "some time in September." In mid-September, Pope sent a brief note asking if there was a "probability that an opening may be found . . . during the coming fall or winter. Of course if you are not likely to need me you will not hesitate to say so. But if you could make good use of me, I should be very glad to serve you." Within a few days, Hudson responded, reminding Pope that he had said he would answer when he returned from his vacation, from which he was "only just back." Hudson had still not been able "to look into the matter" but "it may not turn out to be possible to arrange the matter on the lines of your letters." Pope did not go to work for Bell; less than a month later, he died in the basement of his Great Barrington home, killed by 2000 volts of electricity when he touched a line leading to a transformer [18].

The electrocution of a man with Pope's experience raised serious questions within the electrical community. He had entered the electrical field at fifteen as an operator with the American Telegraph Company in Great Barrington. In 1895, the fifty-year-old electrical expert was at home again in western Massachusetts, with reason to believe that his thirty-five-year career had come to an end. At a time when the safety of alternating-current systems was being questioned, Pope's death prompted the AIEE to appoint a committee to investigate. On the committee were early associates of the electrician, including men of the stature of Edward Weston. After an investigation that included a visit to Great Barrington, the committee concluded that the system was at fault. But because it was well known that poorly insulated wires carrying high voltages were still common, the publicity given the electrocution of an ex-president of the AIEE helped achieve a safety code for wiring and grounding, something many had wanted for years [19].

The pertinent truth of Pope's death was the private experience preceding his death: that of Pope looking for work and finding none. Instead, he found a radically changing industry and a profession being inundated by a new generation of engineers. Most of these new engineers had no knowledge of the life of the electrical expert or the independent engineer. Nor would they, since employment in the new corporate workplace was increasingly the rule in the electrical engineer's world. For long before Pope's death, the telegraph electrician was in eclipse. That fact was perhaps never so cogently expressed as when Elihu Thomson spoke in 1889 at a banquet in England given by the Lord Mayor of London. It was a general meeting of English engineering societies, and since Thomson was president of the AIEE, his hosts offered a toast to the American engineering society. In response, Thomson took the opportunity to define the electrical engineer. He first ruled out the electrician as a serious contender for the title of engineer. In searching for the origins of the "electrical engineering profession," which was "a very recent . . . division of engineering," he looked for professional precedents. But examples older than a decade were difficult to find. Although it was "true that the telegraphic engineer was in a somewhat restricted sense an electrical engineer," it was "more true" that electrical engineering rested on a level of "scientific and mathematical" work beyond the ability of telegraph electricians. The search for early examples of the field must look to a man like Sir William Thomson, whose "genius made . . . ocean telegraphing an engineering success" and gave the "electrical engineer instruments which are as his rule and square and compasses." To the AIEE president, Sir William Thomson represented a "true scientist" and, as such, was "the father of electrical engineering" [20].

The point of Elihu Thomson's compliment to his hosts was that if the telegraphic field held engineering precedents, they were not to be found in the work of the telegraphic expert. The profession, in short, owed more to the highly trained scientist than to the self-educated electrician. Still, the electrical engineer was a complicated professional creature, made the more so by the "constant expansion" of the field. This made the modern electrical engineer an amalgam of talents, uniting "the qualities of the mechanician, the chemist, the physicist and general technologist." In addition, the field was just opening up. The electrical engineer had already won "victories" in the now familiar areas of "electric metallurgy, telegraphy, telephony, arc lighting, incandescent lighting, motive power transmission and electric railways." And "who is to say where the growth and development will stop?" Thomson envisioned the supplanting of the traditional combination of steam engines and dynamos by hydropower directly producing electricity "to propel our railway trains, to do our mental work, to light our streets and buildings, to run our factories, and to effect our chemical operations such as bleaching, tanning and others." It was an expansive vision, yet Thomson looked even further afield when he asked: "Shall we even dare to hope that electrical communication on the Atlantic may be maintained with our friends and dear ones

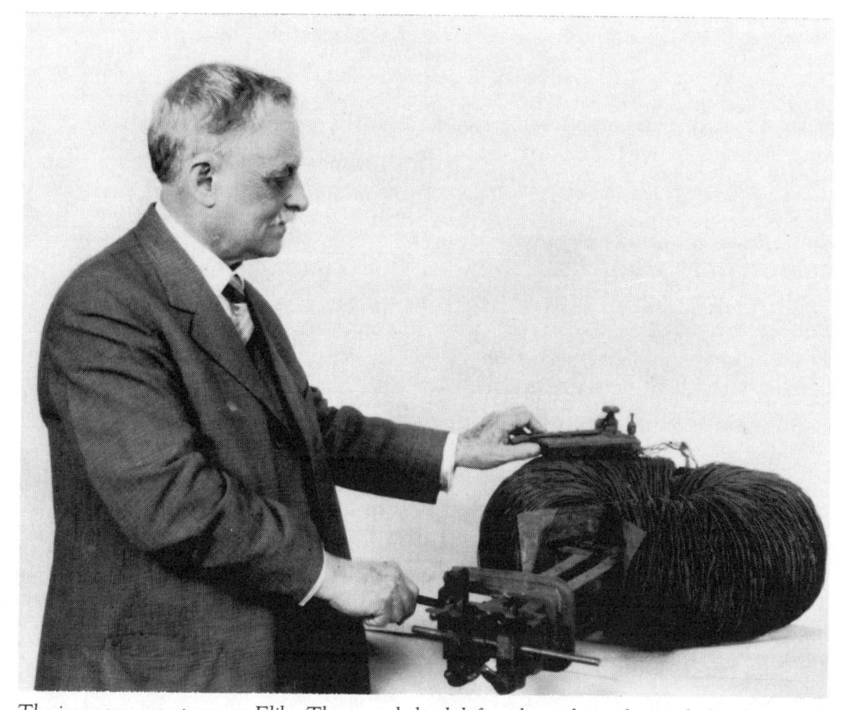

The inventor-entrepreneur Elihu Thomson helped define the early profession before he moved to the periphery.

ashore?" With that question alone, Thomson demonstrated how far and how rapidly the profession was moving beyond the abilities of the old-time telegraph electricians.

Elihu Thomson described a dynamic electrical world, which was leaving behind older segments of the profession. Yet, the fact of national consolidation promised little more for the inventor-entrepreneur. Given the pattern of industrial consolidation, which was concentrating both the manufacturing and communications sectors of the electrical industry, inventor-entrepreneurs were time-bound, tied to the early phases of the development of a new industry. That fact would be repeated as new technical areas were developed, and except for a rare Elmer Sperry, who moved into new arenas as older ones were consolidated, most inventor-entrepreneurs were affected similarly by concentration. Their problems, in fact, often began early in their careers, as they sought to have their creative achievements introduced into the marketplace. An example was Thomson's 1882 contract with the American Electric Company, in which an attempt was made to protect both parties. Not only would Thomson continue to serve the firm as chief engineer, but the company also promised to avoid unnecessary delay in manufacturing the inventor's arc-lighting system. However, within two years of the agreement, Thomson was blaming American Electric's president for his "lack of energy" and the "feeble and puerile undertakings" that hindered attempts "to carry out the original contract" [21].

The quality of the company's work was equally unsatisfactory. Thomson was especially concerned with the quality of carbons being used. He had understood that the company would manufacture its own carbons; instead, it was using carbons made for the Wallace-Farmer system, as "miserably impure as they were." Several times, Thomson advised the company's managers that the firm would go out of business if they persisted in using inferior carbons. To Thomson, the problem was the ineptness of management, since "I had made excellent carbons with my own hands and no machinery." The case of the carbons, moreover, was "a sample of hundreds of like facts." Thomson's special feelings for his inventive offspring poured out of his frustration with this first large business venture:

> All my efforts have been given to the working of the system in the best sense and I can not afford to allow a system which I believe is without parallel to be crushed out of existence by a continuance of the business policy of the past two years; a policy . . . to make the electric business a bolster for a weakly hinge factory [22].

Like Elihu Thomson's, Elmer Sperry's early struggles typified those of the inventor-entrepreneurs. After leaving his parent's Ohio farm, around the time of the Centennial, to begin a life of electrical and mechanical invention, Sperry required a dozen years to achieve a position that supported his inventive impulses. Rather than concern himself with the quality of his company's product, Sperry wanted to continue inventing without having to give undue time to problems of production, construction, and maintenance. Having to attend to such tasks frustrated his ambitions during most of the 1880's. Though he had achieved a competitive arc-lighting system in 1883, the Chicago company set up to produce it was plagued by undercapitalization, making it necessary for him to assume the position of company electrician. The company's troubled state left him too anxious, as he put it, to "turn out the inventions." During these years, Sperry produced fewer patents than at any other time in his active career. By 1888, the twenty-seven-year-old inventor had learned that "no man can work and worry too." He had achieved, by then, the success requisite to concentrated work. Sperry's dilemma, as his biographer, Thomas Parke Hughes, has explained, was the company's desire to use "its talented young inventor in routine engineering." To avoid this, Sperry developed a pattern of spending about five years in a field, then leaving it as "an in-rush of inventors, engineers, managers, and corporations" took place [23].

Elihu Thomson's response, however, compared more closely with that of other successful inventors. For even the most successful of the power engineer-entrepreneurs—Thomas Edison, Elihu Thomson, Charles Brush, and Frank Sprague especially—merging their individual companies into the General Electric Company in the early 1890's not only struck their names from company mastheads but also altered or dissolved their entrepreneurial roles. In time, each entered a sort of elegant retirement accompanied by ambitious technical or scientific experiments. Edison sank millions into an attempt to make a new fortune with an ore-separation process he was experimenting with in New Jersey. On his great estate in Swampscott, Massachusetts, Thomson built an observatory to study the stars. He remained as technical sage at GE and, after 1900, frequently lent his stature to AIEE committees. Brush retired to Cleveland to serve as a philanthropist to the city and eventually to give active support to the eugenics movement. Of all the leading groups of electrical men during these years, the inventor-entrepreneurs experienced the fewest obstacles to acceptance within the Institute. Not only were they the heroic inventors of the era, they were also among the founders of their profession. Elihu Thomson's 1889 presidency was one sign of their assured place within the Institute; others were Alexander Bell's election to that office two years later, and Frank Sprague's the following year. Whereas industrial change meant obsolescence for the telegraph electrician, it meant the opportunity for new ventures for the inventor-entrepreneurs. In either case, their importance to the profession was radically altered. They had become nonexistent or, as with the inventors, peripheral.

Practical physicists, engineering education, and the Institute

As the careers of the inventor-entrepreneurs suggested, the education of the engineer during the formative years of the profession often took place in private studies, home laboratories, and small shops. There was usually a private or public library where an aspiring technologist could learn the intricacies of the field. Elihu Thomson credits his first working knowledge of electricity to a gift his mother gave him in 1857, *The Magician's Own Book.* From this, he learned to make his electrostatic generator from wine bottle, leather, and wood. As late as the 1870's, he experimented with apparatus on his kitchen table, as he documented in a self-portrait made with a new camera. Elmer Sperry, on the other hand, read the U.S. Patent Gazette at the YMCA library in his hometown of Cortland, Ohio. And however unique Edison the inventor, reading Faraday in his rooms early in his career cast him with many more young men in the country.

From the 1870's, however, the training of the electrical engineer increasingly took place in the colleges and technical schools. Electrical engineering instruction generally began in physics departments, whose members had created much of the new engineering knowledge. In this manner, electrical engineering departments appeared within a few years in many scattered places. So although the first programs were established just prior to the AIEE's founding, physics departments had already begun to train electrical engineers. In 1880, the physics department at the new Johns Hopkins University graduated William W. Jacques to become an engineer in the even newer Experimental Department at the Bell Company. Even in 1883, when Professor William Anthony set up one of the nation's first electrical engineering programs at Cornell, Anthony's physics department already had begun to train electrical engineers within its experimental physics program. Physics departments were not the sole source of electrical engineering programs. Like the field itself, its educational roots grew as much out of the world of industry and entrepreneurship as from the heady world of nineteenth-century electrical science. When Michael Pupin arrived at Columbia University in 1889, he was greeted by Francis Bacon Crocker, a "practical engineer," as Pupin, the newly graduated doctor of mathematical physics described him. Crocker received his doctorate in 1885 at Columbia's School of Mines. At the college's request, he then initiated an electrical engineering department. Until then, separate electrical engineering departments and programs generally came out of the physics departments, or, as in Columbia's case, the school hired electrical scientists like Pupin with advanced training in physics. A spate of electrical engineering programs had thus appeared at MIT, Harvard, and Yale in New England; at the Stevens Institute of Technology in New Jersey; and at other schools along the Atlantic seaboard [24].

Yet, as pathbreaking as the courses in applied electricity at these older schools were, the rapid expansion of electrical engineering into schools west of the Appalachians demonstrated, even more sharply, the dynamic state of the field. Two new departments in new western universities suggests the logic of physics as a primary source for electrical engineering education. Soon after the founding of the Case School of Applied Science in Cleveland in 1881 and the University of Texas in Austin in 1883, courses, and later, full programs in electrical engineering emerged from the physics departments.

The initial electrical courses at Case appeared in the second year among the classes announced by the physics instructor, Albert A. Michelson. Michelson had made his scientific reputation at the end of the 1870's while an instructor at the U.S. Naval Academy in Annapolis, Maryland. He continued his experiments on the speed of light at Case with chemistry professor Edward W. Morley. For his physics students, however, Michelson developed a program of studies including, besides basic courses in classical physics, a number of classes in the engineering fields. Drawing came during the first two years, with mechanical and civil engineering taught during the third and fourth years. These were replaced in the physics curriculum, in 1885, with specialized electrical topics. A course entitled "Electricity and Magnetism" was offered in the second year, along with related laboratory work. More electrical courses came the following year, and attention was given to "theory and practice" and such "practical problems" as could be illustrated through the study of batteries; the measurement of currents, resistances, and electromotive forces; the "location of faults in telegraph circuits; laws of electromagnets; intensity of magnetic fields; efficiency of electric lamps and dynamo-machines." Also that year, the school acquired for the department a three horsepower dynamo driven by an Otto gas engine [25].

Finally, in 1887, the year before he departed for Clark University, Michelson offered a full "Course in Electrical Engineering." Electricity and Magnetism was introduced in the second and third years, and in the fourth: thermodynamics, engineering construction, details of practice and design, electrotechnics, and laboratory work in electrical testing. For the second term of the senior year, the school catalogue announced its "electrical engineering" course along with a statement of its practical aims: "In view of the important advances in the application of Electricity and Magnetism to electric lighting, electro metallurgy, and electric transmission of power, a course will be given in Electrical Engineering."

Physical conditions and economic interests in Texas during the 1880's called more for civil and mining engineers at the new university. These areas were readily included in the curriculum for the Department of Applied Mathematics in the School of Mathematics in 1884. The civil engineering curriculum included drawing, roads and railroads, "field practice," and, in the senior year, mechanical, civil, and mining engineering. In the 1885 catalogue, however, a "Course Looking toward Engineering" was listed under "Academic Degrees." Students in the program would take two classes in "electrical engineering" in each term of the senior year. The prominence given this new field in the school's general studies did not come from any pressing need in the community. Rather, it accompanied the arrival, in 1885, of a University of Edinburgh-trained associate professor of physics, Alexander MacFarlane, a mathematical physicist who later joined Steinmetz and others in solving the theoretical problems of high-voltage transmission. To the "courses" offered as concentrations for academic degrees, MacFarlane added, in his first year at Texas, a "Course in Electrical Engineering," involving advanced study in "natural philosophy." A year later, the course adopted as its text Clerk Maxwell's volumes on *Electricity and Magnetism.* The university's physics program steadily advanced under MacFarlane until, in 1889, natural philosophy became "mathematical physics" in the curriculum. MacFarlane's offerings proved heady fare for the Texas diet. Consequently, when he left for Lehigh University in 1894, instruction in electrical engineering ceased at the University of Texas until 1904, returning as part of a newly established Engineering Department. The reason given for its reinstatement was the ample local opportunities offered in the field. Besides the riches of Texas mineral deposits, the department's bulletin explained that "the climate of Texas in comparison with that of northern and eastern States indicates possibilities in long-distance transmission of electric power heretofore unequaled" [26].

Despite the contributions of physicists to both electrical technology and electrical engineering education, their relationship with the founding generation began awkwardly. A physicist's presumed distance from industrial concerns made him suspect within the Institute. For this reason, when Cyprien Mailloux learned, in 1886, "there had been a movement to introduce some physicist into the chair of President," he bluntly questioned a physicist's right to full membership and thereby the right to hold office. He doubted that such an individual could sufficiently grasp the needs of the electrical industry. "We should have a practical man at the head of things," he told a meeting of the society. Mailloux's attitude was common. In 1884, a writer in the *Electrical World* declared that even if "Edison's mathematics would hardly qualify him for admission to a single college or university, . . . we would rather have his opinion on electrical questions than [that] of most physicists." Mailloux's doubts had official sanction, being appropriate to the Institute's stated goal of promoting "the Arts and Sciences connected with the production and utilization of electricity." Though professors in general were not prohibited membership into the Institute, neither were they explicitly included. Members were defined in the 1884 Rules simply as "professional electrical engineers and electricians." In the 1885 version, an applicant not only had to know "the principles of electrical science," he also had to be familar "with the practical applications of electricity." Constitutional silence on professors continued through the century [27].

Mailloux explained that he did not mind a physicist so much as one insensible to practical matters. To clarify his position, he called attention to the membership of "several men who occupy professorships in the different colleges who would be a great honor . . . as President or Vice-President." Noting further that he had learned of the rumored candidate's connection with "the School of Electrical Engineering" at his school, Mailloux nominated for the presidency William Anthony, a professor of physics at Cornell University. When Anthony did win the presidency in 1890, he had left Cornell's physics department to become an electrical engineer with a business firm in Manchester, Connecticut. As if to fix the meaning of Anthony's presidency, Mailloux was chosen to extend the Institute's thanks at the end of Anthony's term and to unveil a portrait of the retiring president. Mailloux expressed his admiration for the professor as a man, as the society's presiding officer, "and as an American electrician." He wished to stress "the word electrician, because we should know that Prof. Anthony had a reputation as an electrician and as a physicist long before the American Institute was known." Mailloux referred to Anthony's role in building the first Gramme machine in the United States, which he exhibited at the Centennial Exhibition. Anthony had brought his reputation to the Institute, standing by it when the Institute "needed the encouragement." In the early days, "the American Institute . . . was regarded by scientific men more as a trade organization than as a scientific body. . . . Prof. Anthony did the Institute great honor at that time, when certainly it was not any brilliant honor to him" [28].

Mailloux had come not only to accept the leadership of a professor, but had also come to believe in the intrinsic relationship of knowledge to the modern engineer. In doing so, he extended and clarified Elihu Thomson's idea of the engineer as an amalgam of fields and areas of skill. After speaking so highly of the need for physicists in the society during the business meeting, Mailloux later joined in discussing a paper by Professor Francis Crocker on stationary electric motors. His remarks demonstrated a growing appreciation of the changes taking place within his profession: "We are at last entering on a phase where the electrical engineer is superseding the inventor, where the necessity for paying strict attention to electrical and mechanical engineering requirements is becoming obvious to us all." Though he spoke from the perspective of the power engineer, Mailloux perceived the historical importance of the changes taking place around the Institute, and more than most of the founding generation, saw the increasing importance of formal study to the profession. As he told his fellow AIEE members, such matters as the design of an armature have "become a question of good common sense, coupled with engineering knowledge and skill" [29].

Deepening engineering content: The engineering scientists of the 1890's

If the 1880's was a decade when doubts about physicists and professors were laid aside, the next decade was a time of insistent affirmation. To make the point, the AIEE followed Anthony's presidency with a series of AIEE presidents possessing credentials as physicists. The first were representatives of the experimental tradition in physics—from Anthony's term, in 1891, to Franklin Institute professor of natural philosophy Edwin Houston, who served two years in the middle of the decade. Though honored in the evening of their days as technical leaders, these men eased the way for the new engineering scientists. Mostly trained as mathematical physicists, having combined studies in mathematics, physics, and electrical engineering, the engineering scientists dominated the last half of the decade and the early years of the twentieth century. Louis Duncan, who received his doctorate in physics from Johns Hopkins in 1885 and then headed the electrical engineering program there, served as president for two years after Houston's term. The next presidencies were those of two leading mathematical physicists: Arthur Kennelly, from 1898 to 1900, and Charles Steinmetz, in 1901. Though serving their terms at the turn of the century, Kennelly's and Steinmetz's energies and interests deeply marked the Institute throughout the nineties. They were often joined by Michael Pupin, who made an unsuccessful bid for the presidency against University of Pennsylvania professor and electrochemist Carl Hering in 1900. They consistently set the course and level at Institute discussions, which followed the papers at society meetings. As part of the printed transactions of the Institute, the discussions helped move the Institute ever closer to a commitment to rigorous standards.

All were immigrants—Pupin from Serbo-Croatia, Steinmetz from Germany, and the India-born Kennelly from England—and all began their careers in America within a two-year span at the end of the eighties. Only Kennelly lacked a formal education; in spite of this, he joined with Pupin and Steinmetz, both of whom had pursued doctoral studies in physics and math, to transform the furthest edges of engineering knowledge and the status of research in the field. The three men typified the character of the new engineering science and, also, the worldly spheres in which it took root. Pupin was a university professor and a researcher in communications, Kennelly was an independent consultant and researcher in power engineering, and Steinmetz was a corporate engineer and chief researcher at GE.

At the meetings and in the publications of the Institute, the physicist-engineers helped to induce an intellectual revolution in American electrical science and technology. Work was just beginning in the long-distance, high-voltage transmission of electric power and in the long-distance transmission of electromagnetic waves for telephonic communication. The importance of their contributions thrust them into positions of leadership in engineering affairs. Steinmetz and Pupin were accepted into the Institute at the Council's March 1890 meeting. Kennelly had joined nearly two years earlier, ten years later becoming a full Member. Pupin did not advance from the Associate grade until 1915. Steinmetz, however, attained full-Member status in 1891.

Membership grades did not always indicate superior accomplishment, as demonstrated in Pupin's delayed ascent; he simply never applied until 1915. In Steinmetz's case, his quick ascent accurately reflected the rapidity of his professional rise. Shortly after arriving in America in 1889, Steinmetz began to attack the technical obstacles that blocked the development of alternating current power systems. The year before, he had completed his doctoral work at the University of Breslau in Germany—though before he could receive his formal degree, he had to flee the country to avoid arrest for his political activities. Steinmetz's studies in theoretical physics, electrical engineering, and higher mathematics, plus contacts with German immigrants already established in the American electrical industry won him immediate employment [30]. And yet, as was clear in the printed discussions in the *Transactions* and, later, in the reminiscences of fellow engineers, it was not only Steinmetz's research but also his role in the AIEE that established the intellectual agenda for the decade. He first commented at an AIEE discussion in 1890, and before the end of the year, he had both read his initial paper to the society and published a brief "Note on the Law of Hysteresis" in an electrical magazine. The fruits of this work appeared in two papers in the *Transactions* in 1892. The next year, Steinmetz became a research engineer for General Electric after that company had absorbed the small New York firm that employed Steinmetz.

The power of Charles Steinmetz's presence stayed in the memories of many engineers who saw him during these years. The career of Edwin W. Rice, Jr., had placed him in close proximity to a number of the giants of the electrical age. Rice had been with Edison and then gone to work for Edison's financial wizard, Samuel Insull, who had moved to the presidency of a Chicago electric power company after the GE merger. Rice soon returned to GE where he worked closely with men like Elihu Thomson. Yet, besides these associations, he nonetheless remembered the distinctive force of the engineering scientist: there was Steinmetz, with "his small frail body, surmounted by a large head with long hair hanging to his shoulders, clothed in an old cardigan jacket, cigar in mouth, sitting crosslegged on a laboratory worktable." Steinmetz brought the power of this presence to the AIEE meetings as well. One engineer attended especially to hear the GE scientist. He remembered that "when Steinmetz spoke, no one else was heard." Charles Scott recalled, forty years later, his first impression of Steinmetz at the annual meeting in 1894, still awed after a career that had included a long period as chief engineer at Westinghouse, the presidency of the AIEE, and, finally, the chairmanship of the electrical engineering department at the Sheffield Scientific School at Yale University. He remembered a discussion on the magnetic field in induction motors. It "was . . . dragging a bit. Then a distinct, resonant voice amply loud from a corner of the large room, shielded from my view by a pillar, gave a comprehensive summary statement of the whole subject. . . . The treatment was as complete and the language as perfect as if the speaker were reading from an article prepared for the *Encyclopedia Britannica.* It was Steinmetz talking extempore" [31].

Steinmetz's interest in dynamo design was not new in the country, but his theoretical approach was newly received by the electrical community. Before the 1880's, Henry A. Rowland had been unable to publish his research papers in an American journal, though they represented the most advanced work in electrical theory in the country at the time. As early as the 1860's Rowland had become interested in electromagnetics and had built an electrical generator. Shortly after receiving his bachelor's degree from Rensselaer Polytechnic Institute in 1870, he completed his first major paper and submitted it to America's oldest scientific periodical, the *American Journal of Science.* Rowland had taken an idea from Faraday, subjected it to mathematical analysis, and added a discussion of methods and results. However, the Yale University-based journal rejected it and, shortly after, on the advice of the physics faculty, declined to publish a second paper. Rowland sent this paper to James Clerk Maxwell at Cambridge University where the professor of experimental physics quickly recognized its value and had it published in the *Philosophical Magazine* in 1873. Though the American physicist did not apply his work to

The young General Electric engineer Charles Proteus Steinmetz had fled Germany as a political fugitive in the late 1880's.

technical problems at this time, the paper made significant contributions to the search for an adequate design for large dynamos [32].

The absence of American engineers and physicists who understood the mathematical analyses in Rowland's papers was his problem in the 1870's. Physicists like William Anthony used an experimental and descriptive approach. Thus, Anthony's gifted student, Harris J. Ryan, who replaced Anthony when he left Cornell and who later headed the electrical engineering program at Stanford University, was stymied by a lack of training in mathematics. Though accounted a sophisticated experimentalist, and though he had begun research on hysteresis before Steinmetz, Ryan was unable to follow the paths along which Steinmetz's mathematical analysis led him. How far matters had come since Rowland's troubles with the American physics community was demonstrated, during the nineties, by how readily AIEE members supported Steinmetz's views.

Though often blunt, Steinmetz gained respect through his energetic participation in Institute discussions. Yet he had a mission beyond enhancing his own reputation. His larger purposes were partially illustrated in 1896 during the discussion of a paper on recent developments in "vacuum tube lighting." He disagreed with a number of points in the paper, he told the group. "The foremost criticism . . . I have to make" is that it contains "a number of vague claims and statements, without offering any proof for them." Not only was "the use of external electrodes in vacuum tubes old," but also some of the author's conclusions were "erroneous." Several other statements were "unintelligible" and other calculations "too fantastic to pass any scientific scrutiny." When several members defended the paper against his harsh criticism, Steinmetz explained that he did not mean to condemn the paper; what he and others "condemned was the entire absence of numerical data" [33]. Steinmetz spoke for new elements in the society, doing so with force and clear argument. In this instance as in others, he won verbal support from other speakers. Arthur Kennelly was a consistent supporter, and on this occasion, William Anthony echoed Steinmetz's comments in explaining that "the absence of [precise] information is what we are criticising." Steinmetz concluded that the requests for specific data sought to clarify "what such a paper, to be suited to such a body as the American Institute of Electrical Engineers, should give" [34].

The expanding corporate context: Research laboratories and engineering departments

That Steinmetz was also a compelling personal leader added force to his assertions and in turn compounded his influence on the Institute. But his work rested upon the rising interest in engineering science in the large companies. However, the interest in research spread beyond the companies. American physicists were becoming more research oriented during these years. From the engineering scientists in the universities also came attempts to establish research traditions. During his presidency in 1898, Arthur Kennelly called for a liason between the AIEE and the engineering programs in the colleges, with the Institute suggesting research topics that needed investigation. But the story was more mixed in the schools. Anthony's inability to wrench funds from Cornell's new president had led him to become a consultant in the 1880's. However, Johns Hopkins' commitment to train research physicists was yearly adding to the ranks of engineering scientists, and Michael Pupin was initiating research laboratories and traditions at his school. When Pupin joined the newly formed Department of Electrical Engineering at Columbia in 1889 as "Teacher of Mathematical Physics," the facilities paled before what he had known at the Polytechnic School in Berlin. The department's building was a small brick structure. What students called the "cowshed" contained a small collection of laboratory equipment consisting of "a dynamo, a motor, and an alternator, with some so-called practical measuring instruments." As William Anthony had once done at Cornell, Michael Pupin and Francis Bacon Crocker sought more laboratory items through giving a series of twelve lectures to New York businessmen and lawyers interested in the electrical industry. They raised just $300 from their efforts. Pupin found it equally difficult to interest affluent New Yorkers in the laboratory needs of the new discipline. When he asked a wealthy lawyer, who sat on the board of trustees of a large educational institution, to contribute to a fund for new laboratory apparatus, the man failed to see the necessity. To his mind, "graduate schools in science needed only a lot of blackboards, chalk, and sponges, and a lecturer who could prepare his lectures by reading books" [35].

Conditions in the large electrical firms were significantly different. In 1889, when Pupin and Crocker were lecturing to fund the purchase of laboratory apparatus, Bell had been making, for a decade, tentative moves toward establishing research on the problems of long lines, and Westinghouse was developing apparatus ranging from the large hydropower installation at Niagara Falls to the electric motors on which Nikola Tesla was working. The fruits of these efforts not only gave support to Steinmetz, but also helped provide the receptive audience he found at AIEE meetings. For Steinmetz, Pupin, and Kennelly were joined at these meetings not just by men like Alexander MacFarlane, whose paper on the use of complex quantities in alternating current analysis preceded Steinmetz's on the same topic at the 1893 meeting in Chicago. The high engineering standards held up by these men were also shared by a number of young engineering researchers educated at American schools. Prominent among them was the group of Ohio State University graduates, including Ralph D. Mershon, Charles Scott, and Benjamin Lamme, who went to work for Westinghouse around 1890. Lamme took only a bachelor's degree; yet he was remarkably good at "figures"—and at that time at Westinghouse, unique, as the head of the engineering department quickly discovered and exploited. Mershon graduated with a master's degree, and Scott spent a year at Johns Hopkins. These men had entered the power field; yet other Johns Hopkins' graduate students like John Stone Stone turned to communications. Stone attained a position at Bell and later became a pioneer in the radio field. It was clear from the activities of the American engineers, as of the immigrant engineers, that support for research was rising. Not only was it reflected in Institute discussions, but it was solidly present at companies like Westinghouse, General Electric, and Bell.

Though no large, formal laboratories appeared until the end of the century, the commitment to industrial research grew slowly out of the eighties and nineties. For whereas the university researcher had to struggle simply to acquire equipment and a place for his work, the industrial setting already contained such a place. Beginnings could be made in the already-established engineering departments, and during these years, companies like Bell and Westinghouse were establishing their own testing and experimental shops. Research was to remain in a raw state, as when Lamme was considering his first job with Westinghouse and a "man, who had pretty close connection with Mr. Westinghouse, told me it would be a mistake to go with the Westinghouse Electric Company, as there was no field there for an educated engineer" [36]. Nevertheless, Steinmetz's investigations around 1890, first at Eickemeyer's in Yonkers, New York, then at GE in Schenectady, received ready support. He did this work, moreover, in the absence of a research department of the modern sense. When Steinmetz joined GE in 1893, he went to the Calculating Department. And though considered an experimental branch in contrast to the Standardization Department founded three years later, its work also came under the general engineering directorate at General Electric.

By no means was Steinmetz conducting pure research. He engaged, rather, in directed industrial research in which the benefits were calculable but not always immediately applicable. He investigated the properties of electrical apparatus to achieve an optimal design for a marketable product. More extensively as an engineering researcher, Steinmetz was concerned with devising an optimal design for alternating current machines. This was similar to the work at Westinghouse where Nikola Tesla and Charles Scott were working on an efficient induction motor. Again, their investigations sought to develop specific apparatus for production as well as to understand the design parameters for electric motors. Research units were, thus, not so much absent from the large companies as they were called by other names. Lamme remembered that, in the nineties, Westinghouse had "no real Engineering Department. . . . The Laboratory was supposed to be something in the nature of an Engineering Department, but it really corresponded more nearly to a Research Department" [37].

Clearly, research existed in the engineer's new workplace. The kind of engineering research pursued in industry always contained several elements, of which basic research was only one. A more continuous and intimate form of research was of an engineering kind, immediately applicable to technical problems. Yet, in any case, as the historian of research at General Electric, Kendall Birr, has argued, whether industrial investigations "be fundamental or applied research, [they] are connected in one way or another with industry and are directed primarily toward improving technology and maximizing economic satisfactions. Industrial research in the long run is utilitarian." A formal commitment to basic research came in the electrical manufacturing industry in 1900, when GE established its Research Laboratory in Schenectady [38]. In the field of communications, Bell made an official commitment to basic research in 1907, a step directly linked to the founding of Bell Laboratories in 1925. Yet, those moves did not spring full blown from the minds of executive officers. Steinmetz had been encouraging such a step at GE for several years; at Bell, even during the period Theodore Vail was absent, long-range research initiatives appeared from time to time.

One such instance was John Stone Stone's attempts, between 1890 and 1898, to do what Bell officials later called "fundamental research." His efforts and the ambivalent response given them by both administrative and engineering managements presented a clear picture of the world of the corporate engineer in the electrical companies established in the late nineteenth century. A wealthy, young engineering scientist, Stone began his tenure at Bell by means of a letter from a friend of his father's to Bell president John Hudson. He was well educated and had several years of advanced training. After two years of study at Columbia in the areas of electricity, physics, and mathematics, he went to Johns Hopkins in 1888 to continue his scientific and engineering studies under Henry A. Rowland. In 1890, a recommendation from the former chief-executive Gardiner Hubbard led Hudson to give the twenty-one-year-old Stone a position in the Mechanical Department, then headed by the Harvard-trained physicist Hammond Hayes [39]. A half century later, Hayes remembered Stone as the first of "my associates to show interest in the theoretical principles underlying the telephone art." Not only did Stone's career at Bell indicate that Hayes certainly did not consider the new man an associate, it also showed Hayes' unwillingness to support Stone's interest in researching theoretical and practical problems in telephony without a close relationship to the immediate technical needs of the company. Only a year after joining the department, Stone was insisting on the need for such long-range research and persisting in spending his time on it as well. In his 1891 report to Hayes, Stone criticized the methods used to "measure the capacity of lines, cables, etc." as "useless and misleading." Stone believed in "the necessity of [acquiring] accurate knowledge" and had been seeking means to measure "the exact effects of capacity and self-induction on telephone currents in cables and long lines." For this kind of work, Stone reported, "much time has been required." But it was time that Hayes did not want to relinquish from his department's other responsibilities. In his own report for 1891, Hayes informed Hudson that he had abandoned the work "given to the theory of the propagation of alternating currents" and to developing "laws" regarding "period, distortion and attenuation of the telephone currents." Hayes believed that such theoretical work could be done best, and "more economically," by students at MIT and Harvard [40].

As Hayes sought to limit theoretical work in his department, Stone persisted, throughout the nineties, as he pursued research interests acquired during his studies at Johns Hopkins. Having been strongly impressed with the English physicist Oliver Heaviside's theory on the propagation of electromagnetic waves along wires, Stone was using Heaviside's work in his studies of long lines for Bell. In 1894, he invented, what was called in a patent of

1897, an "air-line equivalent cable." Though this was assigned to Bell, his researches led him to a post-Bell patent in 1900. Concerning what later was called "loose couplings," it led directly to his full involvement in radio by the end of the century [41].

Stone's persistence in following his own interests led eventually to a break with Hayes. Unable to get Stone to perform the work assigned to him, Hayes complained to Hudson, in 1896, that the engineer's absences from his normal duties sometimes lasted as long as six months. He wanted Stone "to be of some direct assistance on the problems which are before this Department." Hayes' memorandum responded to Stone's recent request that the company "relieve him from work" in Hayes' department and provide an independent office away from the "rules and regulations as to attendance and work . . . in the Mechanical Department." Hayes' final argument had to do with the question of the level of research the company ought to be supporting. The idea of pure research was not at issue. Rather, the issue was basic research, which was still directed at the solution of problems of interest to industry, but directed in areas whose true usefulness might not be realized for decades, and, even then, in a different area of communications. This was the nature of Stone's work; so Hayes recommended that Hudson deny the request on the basis that "a man so situated could not keep informed of the requirements of the business and would soon become an inventor engaged in the development of unnecessary apparatus." But Stone refused to compromise, explaining that "it was not likely" that he would "attend to his work any more regularly than he had. . . . In fact, . . . he would probably be more irregular in his attendance." Complaints of this nature went on for several years before Hayes asked, in 1899, "to be allowed to discharge Mr. Stone" and to be authorized to hire a new "assistant" [42].

Though certainly tinged with personal conflict, Stone's problems at Bell also stemmed from Hayes' hesitance at committing his staff to research not immediately applicable. Hayes, however, fully supported investigations of the technical obstacles, which at that time prevented the Bell Company from successfully completing a national system. Two years before Stone left, Hayes hired an engineer with degrees from MIT and Harvard and advanced training in mathematics and physics from Vienna and Paris. Hayes replaced Stone with a trained engineering scientist who had earned a master's degree in physics from Harvard and had then remained at Harvard for two additional years of study in physics and mathematics. However, the work of these men, as with engineering researchers throughout the electrical industry, concentrated on the immediately applicable, attending primarily to the problems involved in centralizing the company's operations. During these years, primacy always went to activities serving this goal. Placing Hayes' experimental Mechanical Department under the company's chief engineer, in 1893, only reinforced an old emphasis. The career of Joseph F. Davis, in fact, followed the main stream at the company more directly than did that of Hayes. Davis' first position was, in 1880, as an "engineer." By 1891, he had become chief engineer [43]. The rise of Davis and the Engineering Department made clear that research at Bell, whatever its character, would be done by men trained as engineering scientists in physics and mathematics and that their work would be subsumed under general engineering.

This had been true for some time. When, in 1880, Thomas Watson's engineering division had established an Experimental Department, he hired Rowland's student, William Jacques, to run it. Arriving just a few months after Davis, Jacques took charge of what earlier had been the Electrical Department and would later become Hayes' Mechanical Department. Under Jacques, the department's work largely dealt with general engineering tasks, but some research was always present. In a "plan" developed in 1883 on his "work for the immediate future," Jacques listed a potpourri of assignments. Besides examining the "technical value" of inventions submitted from outside the company, Jacques performed other duties that "naturally fall to an electrician." Among them were investigations for "use in legal cases" and experimental work "to perfect an instrument for louder speaking." At the same time, he was staying alert to "any incidental inventions that may occur." When Hayes took over the department, his first report showed similar regard for experimental work aimed at improvements to the system. During 1886, the department's twelve members largely spent their time reporting upon inventions submitted, testing materials and instruments, and doing work for the Legal Department. Other than the clerk and errand boy, all members performed technical tasks. Four inspected instruments, two worked as electricians, and two as machinists. Yet there was also experimental work by which Hayes stayed attentive to the creative potential of his staff. Of one man, he reported that his "skill and experimental ability" should rather be employed "in experimental work, than . . . as an instrument maker" [44].

The broader responsibilities of the department were not entirely to Hayes' liking, as he made clear in 1889. "General engineering questions" had taken most of his department's time, he explained, preventing attention to technical problems and threatening the development of an efficient system. Yet "the state of the art," he believed, "is to-day passing through a period of change from old and imperfect to newer and supposedly more efficient apparatus." He wanted his engineering staff to concentrate on the fundamental technical obstacles to this transition [45]. Besides capable engineering talent being needed at the local exchanges, "the processes in use" must be studied by trained engineers who do not have to give primary weight to "the question . . . of cost." The savings would come from cables that were "better electrically and cheaper than have ever as yet been produced." Hayes wanted to use his own time to work on the fundamental technical problems that faced the system, including—in addition to the matters of conduit, cable, and switchboards—the critical "question of . . . long lines." From the "electrical standpoint," this question was of "vast importance," deserving "the closest attention and experiment." The obstacles to expansion and efficiency needed to be both "practically and theoretically solved." Hayes had to report, however, that work in this area had been "unsatisfactory and desultory, owing to the mass of work before the Mechanical Department" [46].

And the work was extensive, growing more so as the building of a national system proceeded. In 1894, the forty-six-man Mechanical Department fell into three divisions: laboratory, instruments department, and shop. The laboratory itself contained under half of the department's staff, eight of them "electricians," supported by two chemists, three wire inspectors, and eight draftsmen. Over half the members, of what had earlier been the Experimental Department, spent their time testing, repairing, and building apparatus. Hayes divided the department's work "into three classes — engineering work, designing and testing." "Engineering work" involved, in addition to such specific tasks as installing specially "designed apparatus" in selected cities to compare with apparatus already in place, the continuing study of long lines, especially of "pole construction," and switchboards. The mass Hayes referred to gathered chiefly around the Instrument Department, where nearly 300,000 old and new instruments were handled in 1893. Other assignments intruded also on Hayes' research time, as in 1894, when the Shop initiated a course of instruction that took "technically trained" men or "educated men whose inclinations were technical" and trained them for positions with the local exchanges in the expanding system [47].

The work of the experimental branch of the Engineering Department precisely served the basic assignment given to chief engineer Davis, which, in the words of one of his assistant engineers, was "to completely establish the telephone business throughout the United States" [48]. This overriding objective dictated the time Hayes and his men spent improving the design of central office equipment, underground conduits, and cables. The desire to centralize also directed the work of the rest of the Engineering Department in standardizing building design, gathering statistics, and preparing special reports. In short, Hudson explained to Davis in 1899, he wanted "the engineering department to have such a knowledge of the plants of the Licensee Companies . . . as will enable it . . . to gradually but surely effect uniformity, and on advanced lines" [48]. This was the work and ultimate purpose of the engineering staff at Bell between 1880 and 1907. By subsuming both routine and creative engineering and by including both engineering research and maintenance, Bell's managerial and technical leaders were creating a new context for engineering. The only significant distinction between what was happening at Bell and at the electrical manufacturing companies was that the telephone company had been at it longer. For at these companies, similar departmental structures were being erected and familiar technical goals were taking shape. The result was no less than a new class of engineers, ranging across the engineering spectrum from engineering scientist to testing-room electrician.

The new conditions and engineering professionalism

The new class of engineers ranged across the organizational terrain of the AIEE. They were the ones who swelled the columns in Charles Scott's statistical study of the members coming into the Institute at the beginning of the new century. Scott examined the thousand new Associate Members who joined during the year before he assumed the presidency in June 1903. He reported his findings in an inaugural talk to the board. The new engineers were educated: Forty-five percent held degrees from schools of recognized standing, and seventy-seven percent had graduated within the past ten years. They were overwhelmingly young: Nearly sixty percent were between twenty-five and thirty-five years of age, and only ten percent were over forty-five. They were heavily concentrated in the employ of large companies: Of those whose training or experience qualified them as electrical engineers, fifty-five percent could be classed as corporate engineers, with thirty percent from the manufacturing companies and another twenty-five percent from operating companies. There were few signs of the electrical expert or the telegraph electrician, though sixteen percent were managers and superintendents, and ten percent were consulting engineers. Ten percent were at technical institutes and colleges, with four percent as professors or instructors. The remaining nine percent worked with small mills and mining operations or were draftsmen [49].

Scott's statistical picture depicted a professional cadre of electrical engineers far different from the hybrid group of industry representatives and proto-engineers that had gathered in Philadelphia in 1884 [50]. And just as the early Rules captured their character, the Constitution of 1901 reflected the new engineer. In what the Board described as "a practically new instrument," the society had radically, and in great detail, redefined the electrical engineer. In 1901, the term "electrical engineer" dominated the formal criteria. From two lines describing the highest grade five years earlier, twenty lines now defined a Member's qualifications. The grade contained three categories: "professional electrical engineer," "professor of electrical engineering," and persons who had done "important original work, of recognized value to electrical science." The professional engineer was required to have five years of experience, to have been in "responsible charge of work" for at least two years, and to be "qualified to design as well as direct electrical engineering works." The definition was clearly elitist; yet, it was a professional elitism based on knowledge and skill. Though still optional, graduation from a recognized school of engineering credited the applicant with a year of experience. A professors' qualifications were simpler, demanding just two years of "responsible charge of a course of Electrical Engineering." The final category of individuals who had done original work covered the successful inventor, whether self-taught or an engineering scientist. Whatever the category, the Constitution made clear that full membership in the Institute belonged only to the skilled, active engineer. To be admitted as an "Associate," on the other hand, required only an interest in or connection with the "study or application of electricity" [51].

These statements of membership criteria were not made lightly by the Institute's leadership. Such constitutional strictures, however, were intended in practice to provide general guidance to the society's Board in passing on applicants for transfers from Associate to Member. They sought, rather, to place the engineer at the head of the society. As in the early years, the Board of Directors could still veer from the rigid standards of membership. Thus in 1903, the Directors asked the Board of Examiners to recommend for "transfer . . . Associates occupying conspicuous positions in the profession," whose qualifications were not covered in the constitutional definition [52]. That the AIEE was bound as much by personal and social constraints as by constitutional ideals was demonstrated dramatically in 1903 and 1904 when first a black man and then a woman applied for membership in the Institute. When the issue of race was raised in April 1903 by black engineer Robert E. Lee's application, the Institute's Board of Directors ruled that the application "take the regular course." Further, the Board took the opportunity to pass a resolution reinforcing its nondiscriminatory position:

> It was voted that it be considered the sense of the Board that the constitution does not make any stipulation or restriction in regard to color or sex, and that it is the unanimous opinion of the Directors present that the Board of Examiners should consider all applications solely on their technical merits without reference to color or sex restrictions [53].

Yet less than a year later, in February 1904, when three members—as per the constitution—proposed Susan B. Leiter for Associate grade, the Board hesitated. Leiter undoubtedly fit the Associate grade since, as a laboratory assistant at the Lamp Testing Bureau in New York, she was certifiedly both interested in the application of electricity and connected with it. Indeed, the Directors did not question her on these grounds, deciding instead to withhold her name "from present action as a matter of policy." They wanted time to examine the policy, and so instructed Cyprien Mailloux to "ascertain from the British Institution the practice regarding the admission of women members." The Institution of Electrical Engineers (IEE) had, in fact, just five years earlier taken in its first woman member. The night she was taken in, however, turned out to be the "longest and fullest session in the history of the Institution." Mailloux thus reported in March that the IEE's president had informed him that "women were admitted to membership" in their organization. The Board, nonetheless, took special action, calling for final disposition at the April meeting with "those not attending in person . . . requested to submit written opinion." But even though Steinmetz and two other established members wrote in support of admitting women to membership, the Directors voted to create a class of women members possessing the same privileges as Honorary Members, a class without voting privileges or able to hold office. The matter was then referred to the Committee on By-Laws, and Leiter's application was not again considered [54].

The absence of women in the Institute followed from their relative absence in the profession. But there were other women involved in the electrical community during this period. Among them was Lulu Bailey, an instructor in physics at the University of Texas, who taught laboratory practice and electrical measurements to the electrical engineering students there. Benjamin Lamme's sister, Bertha, also took her degree in engineering in the 1890's from Ohio State University and went to work for Westinghouse for a few years. Though there must have been more women involved in electrical engineering during these years, as late as 1934, the Institute could list only ten women members in the anniversary issue of *Electrical Engineering,* and all had joined after 1923 [55]. Susan Leiter's attempt to join almost twenty years earlier made her a pioneer; yet it was not given to the majority of Institute leaders to pioneer in this area.

The tendency to follow cultural attitudes as well as formal rules did not lessen the importance of the constitutional changes of 1901. For that new instrument declared the electrical engineer's rising desire for professional standing. The real challenge to its ideals would come, moreover, not from "color or sex," but from the corporate business community and the ideologies of scientific management and commercial engineering. But that major challenge to engineering professionalism in the AIEE did not emerge until after 1905. The challenge in 1901, rather, was the more straightforward question of the technical goals of the engineer and, thus, the nature of engineering. The new Constitution had made some bold departures. In the preamble statement on the "objects" of the Institute, the promotion of matters relating to "the production and utilization of electricity" now came after the goal of advancing "the theory and practice of Electrical Engineering." This, then, was the question of the engineering age: whether the profession's work and body of knowledge rested on science or practice, or, if a mixture of these, in what proportions. In the two specific areas that challenged the Institute and the profession at the turn of the century—education and technical standards—the ratio was critical for the one and irrelevant for the other. Standards equated with practice, or, as the ideal was often expressed, with "best engineering practice." Yet in the educational arena, participants saw matters differently. There, the question of the mixture of science and practice in the curriculum and, concomitantly, in engineering work was crucial and controversial.

But although the long path to engineering professionalism had not ended with the century, electrical engineers had come closer to a sense of who they were and of the nature of modern engineering. Indeed, these early years of the new century formed a watershed for the profession. The place of engineering science had been established, and the engineers possessed a growing awareness of the changes promised in the incorporation of the nation. Yet the legacy left by the first two decades of the American Institute of Electrical Engineers was as mixed as would be its experience during the next two decades. For the mixture would become a split as the Institute leaders and members were torn by the tension caused by a growing imbalance between engineering science and industrial practice.

3/ELECTRICAL ENGINEERS AND THE AGE OF ORGANIZATION

The engineer does not work in isolation. . . . His work may be done through conscious co-operation, as in a corporate organization or learned society, or it may be done through that co-operation which he, like all others, exhibits as a member of organized society, and which is often most intense and effective when he is attending strictly to his own business with no thought of co-operation in his mind. [1]

John J. Carty, 1915

"Upon common ground"

Though the electrical age had come far by 1900, only in the new century did engineers begin to grasp the meaning of the national technical systems they were designing and building. During the winter of 1909, such a moment of awareness came for Bell engineer John J. Carty as he and his assistants toured the company's telephone exchanges on the Pacific Coast. The vice president and chief engineer of AT&T had been working for twenty-five years to make it possible to talk from one point to increasingly distant points. Even as he traveled, the continent was being wired for telephone service and engineering researchers were devising the technical means for efficient long-distance transmission. As a national system could be seen coming into place, the Bell entourage sought to help West Coast telephone managers prepare for the expansion that would follow the successful linking of service between the two coasts. With the long tour nearing an end, Carty was struck by the import of a completed Bell system. Standing on a Seattle street, he felt "the isolation of the Far-Western State." He would "always feel it," an assistant remembered Carty saying, "until he can talk from one side of the United States to the other" [2].

Other engineers whose positions placed them at the heart of the maturing technical fields of electric power and telephony recognized with Carty the nationalizing tendencies of the age. They, too, sensed what the new technical world implied for the electrical engineer. To General Electric engineer David B. Rushmore, the times constituted an "age of organization." In a talk on industrial needs and education in 1908, Rushmore urged professors to define their goals in terms of the new reality in the business world. He wanted colleges to help young engineering students understand and meet the requirements of large industrial enterprises. In his AIEE presidential address in 1909, Louis A. Ferguson, engineer and vice president of Commonwealth Edison of Chicago, described the period as an "Age of Centralization." Ferguson's concern was the engineer's responsibility in the industrial arena. The engineer's part, he said, was to help companies achieve the "economies" made possible "by large production and centralized direction." He challenged the engineer to recognize that "centralization leads toward standardization" and to cooperate in achieving that goal [3].

In energetically promoting centralization, the arenas of education and technical standards came easily to mind for Rushmore and Ferguson. In the one, that student engineer was initially shaped; in the other, the mature professional engineer contributed to the cooperative task of standardizing engineering work. The advice of Rushmore and Ferguson was especially appropriate for the period between 1900 and World War I. The expanding national context for engineering work during that time can be illustrated by Carty's experience. In 1900, Carty worked for a New York telephone exchange. In 1907, he moved to the American Telephone & Telegraph (AT&T) headquarters to help make a national system a reality. And in 1915, just a half-dozen years after his trip to the West Coast, Carty had reached the presidency of the American Institute of Electrical Engineers and was being pulled into participation in the national political order. This actually occurred on the eve of the preparedness movement, when President Woodrow Wilson sent letters to Carty and to the heads of the other leading engineering societies, seeking their aid in the cooperative effort of organizing the nation's technical talent for a great war. But before that expansive step took place, the country was being organized by the expanding world of the business corporation. This was the organizational context out of which Carty and his electrical engineering colleagues in the AIEE would devise a response to the age of organization. And though the tendencies of the age would bring wrenching change within the Institute itself, the society would first be absorbed in the fundamental issues of education and uniform industrial standards.

As these issues were among the central, external concerns of the Institute during the years between its founding and the world war, they drew the attention of leading figures from the electrical engineering world. Standards workers included such men as Francis Crocker, Arthur Kennelly, Charles Steinmetz, and Comfort A. Adams. For more than three decades, they and scores of other members performed the often arduous tasks involved in the Institute's standards work. In the educational arena, one man, supported by expanding ranks of professors in electrical engineering departments across the country, emerged as the leading ideological and institutional force. For a quarter of a century, from his early years at the University of Wisconsin and then during his long reign over electrical engineering at MIT, Dugald C. Jackson developed an electrical engineering program that directly matched curricula with the technical needs of the industry. Drawing, thus, from the same industrial source for both standards and educational curricula, the single ideal of "best engineering practice" came to the fore.

Some men, like Steinmetz and Jackson, remained on the Institute stage for long periods of their lives. Others played smaller roles, or stayed in the wings, contributing in important, if less intense, ways. At other times, an engineer might appear only briefly yet strike so clear a note that his influence would continue. Charles Felton Scott was such a man. Though active in setting the Institute's course only during the year of his presidency, no one did as much to chart the AIEE's future during these critical years of American engineering. The Westinghouse engineer began his presidency in 1902 by proclaiming an engineering age and declaring a double task for his term of office. In response to the great surge in membership in the late 1890's, he initiated far-reaching structural changes, adding student branches and establishing technical committees. Additionally, as his unprecedented inaugural address promised, he made his presidency a platform from which to define the character of the electrical engineer and to assess the place of electrical engineering in the new age. This latter task Scott undertook with relish. He gave three major addresses on the engineer in 1902 and 1903 and added substantive statements on the state of important facets of the engineering art when he introduced sessions at the meetings. On all occasions, Scott viewed the Institute and its future within a national context.

Scott dealt, first, with the need of the emerging national engineering fraternity for direction. Believing success would require coordinated activity from all engineers, he initially sought a headquarters building in New York to house the major engineering societies. An old dream of members of the New York contingent was to have a meeting place and a library. Yet to Scott's mind, besides its administrative value, a national center would serve as a professional home—a symbol of the character and mission of American engineers. That he thought it would, above all, unite the national engineering community became clear in his bold proposal for a joint engineering center. In a speech on "The Engineer of the Twentieth Century," delivered to the Engineer's Club in Philadelphia, Scott publicly announced his notion. At first, the club's president had hesitated to invite the new AIEE president when he learned that Scott was not yet forty. In spite of this, Scott captured the occasion with a persuasive and eloquent speech on the future of the engineer. Fearing at first that he had only an "assemblage of . . . platitudes" to offer, Scott recalled that "as I began to write the theme developed. I found that the great discovery of the Nineteenth Century was cooperation, made possible by the engineer; that he was an essential factor in our new industrial and economic life" [4].

Scott envisioned America's diverse engineering community headquartered "in a fine Capitol of American Engineering." Each group would be active within the separate engineering fields, but all would be joined in a "congress of engineers." He pictured "an eminent body, . . . powerful in advancing the common interests of engineers" and representing "the engineering profession in its relation to other professions, to pure science, to education, to legislation, to public improvements, and to the general welfare." Scott consciously drew on Institute traditions when he turned to the idea of cooperation for his theme. A decade earlier, Edwin Houston had insisted that the AIEE was "in no sense a local organization." It had become "a national body" because of the advantages of a "cooperative plan [of] directed, organized effort as opposed to unorganized, undirected effort." Several years later, when Arthur Kennelly promoted his scheme of linking college research to industrial needs, he explained it as "a system of co-operation" that would "accelerate progress in all branches of inquiry, application, and industry." When Scott spoke in 1903, the idea pervaded his address. In urging Institute members to use their collective power, he filled the air with phrases like "common unity" and "community of interest," characterizing the Institute as a "fraternity which is called [a] profession" [5].

Comity among the engineering societies was important to the engineers, Scott recognized, because of the national consolidation of industry. As a Westinghouse engineer since the early years of the company, Scott knew concretely the age of organization and had seen the consolidating tendencies he described as an intimate part of the rapid rise of the electrical industry. He had worked with Nikola Tesla on his invention of the multiphase, or alternating current motor, and had helped design the great alternators for the power plant at Niagara Falls. Within a twenty-year period, then, Scott had observed "electrical novelties" become "commercial necessities" and "simple experiments . . . great systems." A technical revolution had occurred with the ability to transmit and distribute electric power, an achievement, Scott explained, that had placed the electric motor "between [steam] engine and lathe, between the waterfall and the loom." From these achievements, the electrical industry had grown from a $1 million investment in 1884 to a $4 billion industry by the time Scott spoke. He attributed this growth to the electrical engineer, who had made electricity necessary to nearly everyone's plans for "future progress," whether business man, manufacturer, engineer, or writer of Utopian novels [6].

Ample confirmation existed for Scott's elated assessment of the industry. To arrive at $4 billion, he used a recent article from *Engineering Magazine* to recalculate data from Kennelly's presidential speech of 1898. He was also assisted by T. Commerford Martin, who was then completing the first national survey of the power industry for the United States Bureau of the Census. Martin knew the electrical industry as well as anyone in the country. Nor was he a stranger to the Institute. He had served as the last of the telegraph presidents, following Norvin Green and Franklin Pope. Born and reared in England, Martin began his electrical career as a boy on the *Great Eastern,* the ship that laid the trans-Atlantic telegraph cable in the 1860's. He came to America in 1877 and joined Edison's group at Menlo Park. Handling publicity more than invention, Martin published articles in the New York newspapers on the telephone, microphone, and phonograph. His lifelong work began in 1883, however, when he helped found the *Electrical World* and stayed to edit it for twenty-six years. His census study thus rested on twenty years of reporting on the industry as well as on fresh data gathered in the national survey. His survey established what many already knew: that the industry was marked by rapid and sustained growth. So rapid, in fact, that when Martin told Scott that the industry's size "has been doubling every 5 years," Scott was stunned [7].

Martin's report amply covered the subject of its title, *Central Electric Light and Power Stations.* It also reflected the expansion of a vast and complex industry — of which the segment that produced and distributed electricity was one component. The engineers who worked for the manufacturing companies not only designed and built machinery for the electric power companies but also furnished motors and apparatus to consumers of electric power from a myriad of other areas of the national economy. While enumerating the market for the central station's electricity, Martin depicted this expanding achievement of the electrical engineer:

> The commercial uses comprise the heating and lighting of private dwellings, hotels, business houses, and office buildings; the furnishing of power for the propulsion of electric motors attached to elevators, ventilating fans, etc., as well as those in factories and in other industrial establishments; and the supplying of current to railways for the operation of cars. The public uses relate more particularly to the lighting of streets, parks, docks, and municipal buildings [8].

Martin's primary investigative target, the central station, formed the foundation stone of the electrification movement and, thus, a measure for the total industry. Looking chiefly at the private power industry — since municipal stations were generally miniscule — he found not only bigger stations but also a widespread "tendency toward financial and physical consolidation." From Scott's "great systems" had come the technical centralization that Chicago engineer-executive Louis A. Ferguson fervently advocated to the AIEE, or what Martin later called the "systematization movement." However, Martin described this tendency in his 1902 survey after finding that, just twenty years after Pearl Street, the word "station" needed to be qualified. The word had become less and less synonymous with the word "plant," he explained, because a "station" increasingly contained two or three plants "in a single city." The rapidity of consolidation made it difficult "to obtain figures that fairly represent the real growth in number." For Martin found that in the nascent electric power industry, stations were consolidating whenever they attained "such size that economy can be effected by putting them under one management." Like Scott, Martin had discovered a central feature of the new age. However, the systems Martin described went far beyond the Niagara Falls installations that Scott had used in depicting the scale of growth in the electrical engineer's world. Martin pointed instead to a project then being undertaken by the Los Angeles power company, Edison Electric. The company planned a power system that would spread across the high inland ranges of California. It would include three "mountain water plants and four steam plants in different localities with coal, oil, and natural gas as possible fuel." Transmission lines would connect seven plants, "some of them miles apart." This was only one case in Martin's national survey; for his statistics also documented the birth of electric power as a major industry.

While Scott was inspired enough to join Martin in celebrating these engineering triumphs, he was also prompted by these triumphs in his desire for a unified profession. A "profession whose interests are so diversified and so extended" made it critical that engineering "workers should be brought together." This was where the AIEE came in. He envisioned the engineering society as a "common meeting place" where

> discoveries may be announced, inventions described, engineering schemes criticized and new undertakings presented and discussed. Here the student and professor, the investigator, the inventor, the manufacturer, the operator and the consulting engineer may meet upon common ground [9].

Scott wanted much from the Institute. After two decades of working to build a professional society, he urged Institute leaders to move more aggressively into the national arena. Scott's addresses were in effect heralds of an age of national engineering. His desire for organizational mechanisms—local meetings and technical groups, for example— in which to reach a larger and more diverse membership was thus inextricably connected to the external issues. Chief among these were the needs for a collective position on "the proper education of the engineer" and for the establishment of "standard practice" in engineering and industry. Standards work loomed important because the field of electrical engineering was "crystallizing." Recent advances in the precision of engineering work made it time for the profession to move quickly to adopt the appropriate definitions, principles, and laws governing "engineering practice." The question of electrical engineering education loomed equally large, for on its resolution rested the future of the American engineer. Scott saw the challenge as assuring the quality and character of educational offerings. He wanted electrical engineering curricula to emphasize science and the humanities as well as technical matter. Yet the crux was to determine the content of the core of the engineering course of study. Scott knew that "the purely 'practical man'" could no longer "hope to maintain himself in the front rank." As a result, he wanted a heavy dose of theory with practical work, which in turn made theoretical work "definite and certain." Scott offered no concrete recommendations for electrical engineering curricula in the colleges, but he recognized that electrical engineering education had to rest on a proper curricular mix between theoretical and practical studies.

The issues of technical standards and education, and thus the Institute, cohered because of the context in which each evolved. Not only did the ideal of industrial practice prevail in the work of standardization, as it must, but it also came to dominate educational offerings, an outcome by no means obligatory. Education and standards were common issues not for intrinsic reasons, but because of the common response to them from the profession. But there was a difference between the two issues. Questions as to how and to what end the Institute would engage in standards work were in dispute for only a brief moment. With education, however, conflict arose early and persisted. From the beginning, education presented a clear choice between preparing the engineering student in theory as well as in practice —Scott's choice—or offering a curriculum aimed more directly at the teaching of standard practice—Jackson's choice.

Educating the electrical engineer: Steinmetz's ideal course

By 1900, AIEE spokesmen gave science an esteemed place in the makeup of the field of electrical engineering. Even before engineering scientists rose to influence during the 1890's, Institute members had recognized the unity of science and technology, and it seemed to some that the relationship was growing ever closer. In his report of 1887, as secretary of the AIEE, Ralph Pope had observed that while "scientific training" was not much appreciated in 1884, it is now "generally admitted," only three years later, "that the best commercial results can be obtained in the electrical field by following out scientific theories." In his inaugural address a half-decade later, Edwin Houston argued that scientific and technical areas were inseparable. Houston even criticized the organizers of the electrical congress at the Columbian Exposition for dividing the congress into three sections: Pure Theory, Theory and Practice, and Pure Practice. He was "disposed to doubt" that pure theory could exist apart from practice, since theory rested on the facts; but he was certain that there could "be no such thing as pure practice apart from theory" [10].

The issue of the nature of engineering knowledge, however, appeared only sporadically in the early years, gathering momentum only as it was joined to the question of engineering education. Nor was it, as Houston assumed, a foregone conclusion that electrical engineering education would be joined to theoretical studies. But at the Institute's first formal session on electrical engineering education, in 1892, all the speakers agreed that a college education implied training in the basic physical sciences. University of Nebraska professor Robert B. Owens regretted the increasing "necessity of becoming specialists," by which he meant, the need for advanced studies. Nonetheless, he advocated a curriculum that included classes in mathematics, chemistry, and physics along with three years of strictly technical courses. Dugald Jackson, only recently having joined the faculty at the University of Wisconsin, justified the study of physical sciences as increasingly necessary to engineering success. While allowing that "practice has made thousands of good [engineers] without aid of the college," he believed, like Owens, that any one of these engineers would have become "more eminent" with a technical course taken with a liberal dose of mathematics and physical science. As a matter of fact, such a background was rapidly becoming necessary. During the discussion, the audience applauded a member's comment that although Western Union did not at the moment depend "upon electrically educated engineers, the time is coming and is not far distant, when it and all other companies must employ educated men or must fall behind in the race" [11].

Though, by the early 1900's, some engineers only grudgingly accommodated science in their engineering work, others enthusiastically endorsed a united science and technology background as essential to innovation and growth in the electrical fields. Attention given to research and theoretical work in technology-based corporations, like Westinghouse and General Electric, had been accepted as a necessary corollary to continued growth. It was obvious to most, in short, that a rich harvest had been gained from mixing basic science and technical knowledge. Technical advances were abundant during these years, when men like Steinmetz, Kennelly, Pupin, Stone, Lamme, Scott, Tesla, and a host of other rigorously trained engineering minds worked for industrial enterprises. It had come to seem that in spite of the obstacles, the necessity of both scientific and technical knowledge to innovation would give a firm place to each in electrical engineering education. That outcome, in fact, appeared assured when, at the end of his presidency in 1902, Charles Steinmetz devoted his presidential address to the need of recognizing the fundamental role of experiment and theory in the training of the electrical engineer. He spoke at a session held during a meeting of the AIEE on the subject of "the education of the engineer." Because no session on education had been held at an AIEE meeting since Dugald Jackson spoke ten years before, Steinmetz organized this one, and he made it a special occasion by including his presidential address [12]. Steinmetz carefully chose his speakers to cover the spectrum of opinion, mixing industrial employer and research scientist with consultant and professor. Held during the annual conference at Great Barrington, Massachusetts, the session contained six papers and was followed by a discussion, in which four participants commented on the papers. Both the topic and Steinmetz's role gave the session a heightened importance, eliciting so much comment that, when printed, the discussion ran longer than the formal papers.

Steinmetz's initiative came from a more specific source than his lifelong interest in education. Just months before the AIEE session on education, the president of Union College in Schenectady—Steinmetz's home since shortly after joining GE in 1893—asked him to establish and chair a department of electrical engineering at the college. The engineer was already accounted a teacher by the stream of young engineers and research scientists that flowed through the engineering and research departments in Schenectady. Even a style of teaching had emerged at GE: his intense, face-to-face approach to questioners and the considerable time he gave to inquiring colleagues. Steinmetz's commitment to teaching was well known. Now he offered to his colleagues his thoughts on the ideal electrical engineering curriculum, a subject then occupying him as he considered the Union College offer [13]. His goals for the electrical engineering graduate suggested the core of this curriculum. Young engineers were to leave his course of study with an understanding of the "fundamental principles of electrical engineering and allied sciences, and a good knowledge of the methods of dealing with engineering problems." He was concerned with foundations, with properly fitting "the younger generation" with knowledge to ensure an "unbroken advance" in science and engineering. Emphasis on the fundamentals followed in part from his belief that the resounding success of modern engineering rested on the rise of "empirical science" —"universally acknowledged as the source of all human progress." Empiricism had subjected the reigning metaphysical science to a "searching criticism." Now, without a highly trained cadre of engineers to subject the new science to the same "hostile" criticism, all that had been gained was at risk: "Herein lies the greatest danger to the unbroken progress of science."

The tradition of "destructive criticism" that Steinmetz wanted to continue depended upon students being well prepared, the prospect for which he found "by no means entirely encouraging." Educational institutions generally confused the task of filling their students with information with that of imparting an understanding of fundamental scientific principles and engineering methods. Steinmetz was dismayed to find that in the colleges, "memorizing takes the place of understanding." To correct this, he sketched a course of instruction that would impart understanding, what he called " an ideal electrical engineering course." It included mathematics —plain and solid geometry, arithmetic, algebra, plain trigonometry, analytical geometry, and calculus— "but no memorizing of integral formulas" that "can always be looked up in a book." Additionally, the student would gain "a thorough knowledge of general physics, especially the law of conservation of energy . . . and of chemistry, especially theoretical chemistry and the chemical laboratory."

Although thorough training in the physical sciences was essential, Steinmetz did not think the colleges should try to train the graduate to leave school as a "full-fledged engineer." Educational institutions should rather train students to assume the responsibilities of an engineering position "as efficiently as possible." Though Steinmetz wanted engineers to be broadly educated rather than narrowly trained, the student's ultimate goal, Steinmetz remarks made clear, must be "practical work" in an industrial setting. William Esty, a Lehigh University professor who spoke after Steinmetz, made the assumption explicit. It was necessary, he said, to emphasize "the intimate relations between the engineer, the manufacturer and the college." Practical work related to two items in the electrical engineering curriculum: design and laboratory work. Steinmetz thought classes in design, to be "of very secondary utility and rather objectionable." Design was better taught in the industrial setting, especially in the engineering departments of manufacturing companies where the "considerations" required of "the designing engineer" came to bear on his work. Mostly, design was unnecessary since such "a very small percentage of the college graduates enter the field of designing. . . ."

None of the professors agreed with Steinmetz's exclusion of design from the program. Comfort A. Adams, Jr., a young Harvard professor, caught the consensus when he pointed out that teaching design did not attempt "to turn out fully equipped designing engineers, but rather to teach the student the 'reason why' in dynamo design, and to cultivate in him the habit of looking for the reason, of applying his principles." Yet none disagreed when Steinmetz argued that laboratory work was important. No matter how much theory the student might learn, he said, without laboratory experience, "the average college graduate is inferior to the practical man . . . [who] frequently pushes ahead" of him. Whereas few students would enter the field of designing, most will handle apparatus, and therefore need more "to be familiar with the completed apparatus . . . than to be able to design it."

The disagreement over design was minor compared to the consensus that gathered around Steinmetz's call for rigorous training in science. Even Edward B. Raymond, a businessman who urged a "strictly technical course" and "work, hard work, constant work" while advising the student to look upon the classroom as his "office," supported courses in mathematics, chemistry, and physics. But, then, inventor-entrepreneur William Stanley criticized Raymond's other notions for restricting the engineer "to that sort of duty which a soldier follows." He argued that "the men who are to carry the burdens of the future" required rigorous training in scientific and technical fundamentals. Without such an education, "there can be no inventors, because there can be no imagination." Without the training Steinmetz advocated, Stanley insisted, "we restrict our engineering work to its very narrowest limits." Stanley's support was typical; however, Cyprien Mailloux demonstrated the consensus better than anyone when he concluded the discussion with an image of a "pyramid of knowledge." At the base was the "foundation of general education, mental training, and intellectural development or culture" that supported and held together "the mass of the pyramid." Directly above were the sciences, including mathematics. At the apex, Mailloux placed the engineer's "specialized knowledge."

Mailloux had caught the central thrust of Steinmetz's curriculum: he designed it to place engineering knowledge at the top of the pyramid. The message of the Great Barrington session, and especially of Steinmetz's keynote paper, made this point clear. The GE engineering scientist, whose training in mathematical physics at the University of Breslau rested on the solid foundation of the German gymnasium, wanted the discipline of electrical engineering to provide American students with comparable training. From his program of study, the student would emerge not only with the laboratory experience to perform the routine tasks of the operating and production engineer, but also with the "theoretical armament" to attack the most difficult problems then challenging engineering and physical scientists:

> If we are told of matter moving in a vacuum tube with velocity comparable with that of light, of small chips breaking off atoms, of free atoms of chlorine and sodium floating around in a salt solution charged with opposite electric charges, whatever that may be, of electricity being propagated not through the conductor, as we use to assume, but through the space surrounding it, . . . it is time to pause and try to understand.

Training the electrical engineer: Dugald Jackson and "best practice"

Steinmetz's ideal curriculum sought to equip the electrical engineering graduate to "understand." Not everyone, however, received Steinmetz's ideas so warmly or accepted his critique of existing educational programs. Though Charles Scott took his key ideas from Steinmetz, Dugald Jackson did not. And having missed Steinmetz's session at the Great Barrington conference, Jackson took the opportunity in a paper given the next year to strike out at the assumptions made by the German-trained engineer about American electrical engineering programs. Steinmetz had offered "proposals . . . as apparently new," Jackson charged, that had "for many years been largely included within the ideals of numerous American colleges of engineering."

Thus, he dismissed Steinmetz's call for rigor as irrelevant because misinformed. Jackson turned then to describe electrical engineering training in American colleges as he saw it and to map out the directions he believed should be taken. Jackson found the "old prejudice" against engineers with formal training he observed eleven years before had been eclipsed by "the industrial results achieved by college men." Those successful college graduates came out of programs in which adequate foundations had been laid through subjects such as physics, mathematics, and English. Like Steinmetz, Jackson rejected the idea of training students for immediate employment, since these craft skills were best learned in "the factory or field." His ideal course of study rested, rather, on "principles, principles, principles, and rational methods of reasoning" [14].

Yet Jackson's goals veered sharply from Steinmetz's. The electrical engineering education Jackson had in mind produced "industrial engineers — men with an industrial training of the highest type, competent to conceive, organize, and direct extended industrial enterprises of broadly varied character." He wanted a broadly conceived program of studies to prepare engineers to "reach the influence in the industrial world for which their caliber and training fits them." The basic education that Jackson wanted had been given a typology and a name by Robert Owens, the professor who had been with him in 1892 at the AIEE's first session on education. At that time, Owens had listed the types of engineer needed: installing engineers, who "superintended the equipment of central or isolated electric light or power plants, electric railways, mining plants, etc., and runs the same when completed"; designing engineers, who had more "to do with the manufacture of electric apparatus than its installation"; and engineers "employed by standardizing bureaus and in laboratories, to make tests of electric and magnetic constants, calibrate instruments, etc." To equip students for this work, Owens offered a number of ideas in an 1898 paper to the Society for the Promotion of Engineering Education. Part of the solution was for the professor to stop taking "refuge in the country" during the summers and to begin to spend the time in "a shop or factory." Owens also encouraged professorial consulting to ensure a closer link between "college work and outside practice." His goal was to bring "school work" in line with the "best engineering practice," by which he meant "the methods and practice of our more successful makers and best constructing engineers" [15].

Owen's ideas on fitting both the professor and the curriculum to engineering practice would prevail within the AIEE. However, it was Dugald Jackson who did most to shape this tradition. His view of the professional electrical engineer had been nurtured during his early engineering years. After graduating from Pennsylvania State College with a degree in civil engineering, he spent two years in advanced study at Cornell, teaching physics during the second year. Jackson left Cornell in 1887 with Harris Ryan. Joined by two other engineers, they formed the Nebraska-based Western Engineering Company. In 1889, when the United Edison Electric Light Company of New York bought out Western Engineering, Jackson became chief engineer of the Sprague Electric Railway and Motor Company, which earlier had been brought into the growing family of Edison companies. The same year, when the various Edison enterprises were joined under the Edison General Electric Company, Jackson became chief engineer of the Central District of Edison GE, supervising the design and construction of electric railways and power plants. The expansive young engineer did not leave the business world, in 1891, when he accepted the chairmanship of the electrical engineering department at the University of Wisconsin. He had begun a forty-year career that mixed professorial with business duties. That same year, Jackson and his brother organized a consulting firm. Once settled at Wisconsin, Jackson began to develop a philosophy of education that aimed at making the electrical engineer at once an engineer and a business manager. For if Owens replaced Steinmetz's program with instruction in engineering practice, Jackson added new components. Specifically, he told his AIEE audience, in 1903, that he wanted advanced students to study "the forms and formalities [of] the affairs of business life." This subject would comprise, with courses in general engineering, design, and industrial efficiency, the final year of study at college. For the engineer "to reach his highest influence," Jackson believed, "each man must combine in one, a man in the physical sciences, a man in sociology and a man of business" [16].

As Elihu Thomson had defined the engineer as an amalgam nearly fifteen years before, so did Jackson. But Jackson's mixture included nonengineering ingredients. Curricula were already moving in this direction. An 1899 survey of eighteen leading college programs in electrical engineering found the uniform presence of mathematics and physics, but also noted as "interesting" the growing "number of courses in which economics and law are included."

Dugald C. Jackson, a professor and promoter of commercial engineering, in his office at MIT.

In 1902, for example, a Department of Business Engineering had been established at Stevens Institute in New Jersey to provide lectures on shop accounting, depreciation, statistics, laws of contracts, and general business methods. At Purdue University in Indiana, moreover, a course in telephone engineering had been fashioned after consulting with engineers, managers, and presidents of telephone exchanges. Its organizer had come to believe that "the best training . . . for a business man is the engineering training." He planned to add a course in "commercial business engineering," believing that "all of the great work of the future is going to be very largely engineering of one kind or another" [17].

But Jackson took the idea the furthest. In 1907, shortly after moving to MIT to head the electrical engineering department, Jackson helped develop the first plan for a GE-MIT cooperative training course for electrical engineering students, a course that became the most famous of such undertakings. The idea of electrical companies training college graduates to fit company needs was an old one. During the 1890's, both Benjamin Lamme and Charles Scott taught regular classes at Westinghouse. A general engineering class instructed new engineers in the company's "practices"; Lamme saw the one on design as his "own private class." General Electric and Bell held similar courses for its new employees. So acceptable was the idea of complementing the graduate engineer's training that Scott undertook, as president, to have the AIEE participate. From that desire and the advice of a number of his "professor friends," he established the student chapters at colleges. When asked for advice, Harris Ryan, then teaching at Cornell, told him it was a "bully idea; we've already started." Scott wanted the AIEE to bolster the work of the industrial courses and to help cushion the "plunge from theory to practice on graduation" by giving students "an insight into the problems and practices of the profession" [18]. The cooperative courses Jackson sought to inaugurate were conceived to give the student industrial experience while still in college. In cooperative courses, the engineering student would alternate between the classroom and the shop.

Jackson had carried the germ of this idea from Wisconsin. Before accepting an offer from MIT to head its engineering department, Jackson sought guarantees that he would be able to shape the MIT program according to his educational ideas. After studying the electrical engineering program in MIT's 1906 catalogue, Jackson wrote MIT's president that he would accept the position only if he could make certain changes in the electrical engineering curriculum,

> the changes tending toward giving the students a broader outlook on engineering. Observation and experience have taught me that many college courses in electrical engineering tend too much toward making mere specially trained electricians instead of (what electrical engineering graduates ought to be), broadly thinking, competent young men each of whom possesses a well balanced outlook on the expanse of engineering and is trained to embrace all proper opportunities [19].

What Jackson meant by "proper opportunities" became clearer at an AIEE session, in 1908, on "the relation of the manufacturing company to the technical graduate." The session consisted of one paper on "The New Method of Training Engineers" by GE engineer Magnus W. Alexander. Alexander had begun his American career, upon arrival from Germany, as a design engineer at Westinghouse and General Electric. He soon left engineering work proper, however, to become the first director of GE's education and personnel department at Lynn, Massachusetts. From that position, he developed his ideas on the "new method," first articulated at the AIEE annual conference at Atlantic City, New Jersey [20]. Colleges and universities, Alexander explained, had in the last century trained doctors, lawyers, teachers, ministers, and philosophers in response to "the demands of the life of that time." By the early twentieth century, however, those demands had changed. Industrial development had followed the steam engine and the railroad, to which the colleges had responded first with programs in civil engineering and later by giving "particular attention to the teaching of mathematics and physics." With the increasing use of electricity in daily life, the colleges added programs in electrical engineering. Changing conditions in the industry naturally required that the schools keep a close eye on "the practical applications of electrical engineering theories in the factories," bringing increased cooperation between college and industry. "And with this grew the interdependence of the two institutions." As colleges found it more difficult to keep up with industrial developments, "practical shop and field work" became critical to the training of the electrical engineering student. It was this tendency that Alexander wished to consummate in the cooperative course —his new method of educating the electrical engineer.

Alexander's new method came more directly out of a company program that he and others had devised at Lynn several years before. The recent graduate spent two years being introduced to the company's requirements "for designing and estimating, construction and commercial engineers, and technical salesman." Supervisory committees met with the student periodically to question him on his future plans and to gain an impression of the student's abilities, including such matters as his grasp of theoretical and applied knowledge, and his alertness. The examination "does not aim to find fault . . . but rather to assist him in his work and point out to him the way to success." But the Lynn program failed to give the student "insight into the practical side of electrical engineering and into the proper relation of the economic forces of an industrial organization," knowledge that was necessary to "those who wish to take leading positions in the industrial field." Consequently, the new method rested on the belief that "the best engineering education" required "that the teaching of the theory and practice should go hand in hand." Under this scheme, the colleges would teach theory, "leaving it to real workshops to initiate the student into practical work." In moving away from the atmosphere of college, "the young man's character" was developed to meet the "stern call of practical life with its demand for cooperation of all forces." Then the engineer would become "capable of assuming responsibility" and could "therefore be placed in positions of leadership."

When he finished his paper, Alexander asked the members of the AIEE to do "a great service to the rising generation of engineers and to the industries of the country by deliberating over this problem of engineering education." His request was quickly and bluntly answered. Bernard A. Behrend, a design engineer at Allis-Chalmers in Milwaukee, characterized the proposal as "the most vicious educational innovation that has been proposed in recent years." The source of Behrend's opposition was similar to Steinmetz's: his own rigorous training in the German educational system. But the critical edge in his comment emerged from his struggles with his superiors at the Allis-Chalmers Company. As chief electrical engineer, Behrend was responsible for all engineering decisions; nevertheless, important steps were being taken by managers without technical knowledge, and without consulting with Behrend. He was a leading research engineer and had hired several engineers who themselves were capable design and research engineers. But instead of being able to pursue research to improve the electrical equipment that the company was selling in such large quantities, his engineers were often given routine tasks involving testing equipment or supervising installations. This unhappy situation had developed mainly after the president of the company was replaced by a new man, the first nonengineer Behrend had worked under [21].

Behrend's criticism went beyond personal pique. Alexander's plan seemed to him a proposal to train all engineers for routine work and to place them in charge. Yet the fundamental problem was the "starting and stopping, accelerating and retarding the boy's mind continually instead of allowing him to obtain a given velocity and momentum." Comfort Adams of Harvard admitted that Alexander's plan had some advantages, especially the social one of bringing the student into contact with members of the "laboring class" and their point of view. But he, too, thought the practical problems of moving students back and forth and arranging the curriculum to fit the moves was a disadvantage. It was compounded moreover by the "danger of overemphasizing the commercial side of engineering." This, he said, "I consider a very serious danger." Steinmetz, speaking as a "college professor" and from his other position "as user of young men," did not think it necessary for college training to impart everything an engineer might need to know. He directly rejected the direction in which Alexander and Jackson would take engineering education. Colleges should teach "the humanities" and "the scientific foundations of electrical engineering," Steinmetz concluded, and leave "subjects like factory management and business administration to be learned afterwards."

At the heart of the debate was the contrast between Behrend's conception of the engineer and Jackson's. Like Alexander, Behrend came from Germany, but he had continued to work as a design engineer. Behrend also retained a regard for the rigorous German educational system. From the tradition of "professional" or "technical middle schools" of his homeland, he distinguished between "foremen, superintendents, . . . master mechanics," and engineers. Behrend backed Steinmetz, insisting that university-trained engineers "should have full command of theoretical knowledge." Practical shop experience could come afterward. Thus, the introduction of "this alternating course . . . into our universities" concerned him greatly. He wanted the student to concentrate over a period of time on a single subject. The colleges failed to train students capable of the "painstaking accuracy and devotion to laborious detail, so essential to all really great work in engineering." Commercial work required essentially different characteristics, he said, since it consisted chiefly in taking "someone-else's thought and work and [making] it a commercial success." By discouraging college graduates from pursuing "new and important creative engineering work," the new method threatened "the stability and continued prosperity of our manufacturing industries."

To the criticisms of Behrend, Steinmetz, and the others, Jackson pointed out that "we are here looking at [engineering training] as professional education," similar to that achieved in law schools, through case studies and moot courts, and in medical schools, by association with hospitals. Engineering schools must "improve their processes" if their graduates were to "go far in the engineering industries of the future." This was the path Alexander had in mind, who responded to the critics that his plan aimed at training "high grade manufacturing and executive engineers." The criticisms must have considerably dampened the enthusiasm of the two educational reformers. And on top of that, an economic slump at GE delayed Alexander's plan to train executive electrical engineers. Not until America's entry into the world war, in 1917, when the pressure on industrial firms to increase production supported the innovations, did the GE-MIT Cooperative Course get under way.

Jackson responded, nonetheless, by continuing his efforts to establish a broadly conceived electrical engineering program at MIT. In 1909, he brought a young electrical engineer into the department whose duties soon expanded to the development of Jackson's cooperative educational projects. The assistant professor was William E. Wickenden, whom Jackson had earlier hired at Wisconsin. Wickenden graduated in 1904 from Denison University in Ohio and taught briefly at the Rochester, New York, Mechanics Institute before going to Wisconsin. MIT was his final teaching position. His developmental duties there gradually moved him into a career in educational research and administration. The work on cooperative courses at MIT led not only to the GE program of 1917 but also to Wickdenden's temporary transfer to the Western Electric Company to establish a program to orient newly employed graduates. After a year, however, Western Electric sought his services on a permanent basis. Wickenden explained to MIT President Richard C. Maclaurin in 1918 that, because "the Engineering Department is under heavy pressure from war work and feels that the intensive development of its younger men is imperative," he wished to be released to join Bell [22].

Wickenden's subsequent career assured that Jackson's legacy would be far-reaching. The GE-MIT Cooperative Program graduated its first class in 1923 and became the most famous of the cooperative schools that sought, in Wickenden's words, "the closer correlation of teaching with industry." The legacy went beyond the cooperative courses, eventually rebounding not only in Jackson's life as a teacher but also in a major educational study conducted by Wickenden. Following his first year at Western Electric, when he had been on leave from MIT, Wickenden accepted a permanent position there as personnel manager. In 1921, he shifted to the parent company, AT&T, as assistant vice president. Two years later he accepted the task of directing a major study for the Society for the Promotion of Engineering Education (SPEE). From this project came a number of studies that sharply criticized, if they did not attribute to Jackson, the impact that Jackson's lifelong goal of combining business and technical expertise in the engineer had on electrical engineering education in America [23].

Vannevar Bush later recalled the personal results of the heavy dose Jackson had taken of what the era called "commercial engineering." Bush had entered graduate training at MIT after taking a bachelor's degree at Tufts College in Boston. He was fond of Jackson for having successfully intervened when Kennelly, then a joint professor at Harvard and MIT and director of Bush's doctoral thesis, wanted to increase the requirements. Bush remembered also his experience as a young MIT professor when an unusually large enrollment led Jackson to take one of the sections of Bush's freshman course on electric circuits. The first conflict came when Bush had to defend his right to teach the course by his own lights rather than by those of Jackson's textbook on the subject. Soon after the classes had begun meeting, Bush faced a rebellion from the students in Jackson's section. They complained that "they were learning a great deal about public utility companies and their management, but they were not learning a doggoned thing about electric circuits" [24].

Nonetheless, the impact of the early emphasis on practice in electrical engineering education went far beyond the personal. Robert A. Millikan, a professor at the California Institute of Technology and one of the nation's leading physicists between the wars, put the matter succinctly in 1921. "The technical schools of the country" had blundered in recent decades by "sacrificing fundamentals in the endeavor to so train men in the details of industrial processes that they are ready to be producers the moment they leave school." That was a devastating criticism; yet the same sentiments came even more strongly from within the electrical engineering profession, and were documented. The source was Wickenden and the series of educational studies he directed from 1924 to 1929 for SPEE with a grant from the Carnegie Corporation. The aim of nearly twenty bulletins issued sought to determine "the objects of engineering education and the fitness of the present-day curriculum" [25].

The conclusions Wickenden drew in the final chapter of a comparative study published in 1929 came close to the criticisms aired in Steinmetz's 1902 presidential address. Wickenden did not believe the colleges were meeting their responsibility in areas that were critical to the health of the discipline. He found that American engineering colleges were burdened by the need to engage in "quantity production for the ordinary technical, supervisory and commercial needs of industry." Efforts to create selective schools to impart "superior scientific standards" had generally "fared badly." Wickenden concluded, then, that engineering colleges and institutes were largely unfit to train engineers for the research end of the engineering spectrum. The immediate result was that technical research in the country had "depended in large measure on men of European training or upon men trained in pure science." Though Wickenden thought the conditions in the schools were being corrected, he found engineering in America "far from being self-sufficient on its higher intellectual levels." Not all the problems rested with the colleges, since the "limitations of our secondary education" left students unprepared for serious scientific study. The colleges, however, had failed to establish programs of "extended study of the general sciences by selected superior students." The teaching of design suffered equally from too little time alloted and too few "fully qualified teachers." Yet here, "systematic training within industry" had provided a solution, especially in the electrical industries. Even in the areas of mass production and automatic processes, where America was believed to be technically superior, the achievement had been won "almost wholly within industry."

Wickenden took a large view in his criticisms of electrical engineering education. He believed the problems stemmed in part from a fragmented national educational establishment that generally lacked a unified purpose. Still, the lapses in the faculties themselves were serious. In spite of the ability and distinction of a considerable number of teachers, "wide observations abroad" led Wickenden to conclude that most teachers in America were "inadequately prepared either by scientific training, professional experience or broad personal culture." College engineering faculties, in short, lacked "a strong spirit of inquiry and creative effort." He listed several basic reasons:

> There are still relatively few men of high research capabilities in our professorial chairs; industry has requisitioned many of the most fertile for her own fast growing research establishment; and until quite recently there has been far greater incentive to textbook writing and to incidental practice than to research.

Wickenden's critique was not launched as an attack, but as a diagnosis given with deep concern. And yet his criticisms, without apparently attempting to do so, countered almost every point made through the previous forty years that emphasized engineering practice and business methods in educating the electrical engineer. This language was more restrained than his 1930 words to a banquet crowd in Cleveland that honored him as the new president of the Case School of Applied Science: "We are flooded with handbook engineers; we need thinkers along new lines," he declared. Recognizing the problems as of long duration, he drew his prescription from the equally long engineering commitment to the ideal of cooperation: "In the absence of any national educational authority, we should strive to [bring together] the various groups concerned—schools, colleges, professional societies, industries—and put to the test the American capacity for self-government through group cooperation."

Cooperation was not just an ideal to the engineers. However, it worked in practice only insofar as there were mutual interests to allow cooperation. Around the question of education, however, where cooperation would be required to develop an electrical engineering education that was appropriate to the diverse world of engineering work, there were only conflicting interests. And yet, what failed in the educational realm did not fail in the world of standards. There, after thirty years of building a standards program, the AIEE engineers would be able to declare on the eve of World War I that their standardization efforts provided an ideal model of the cooperative spirit in action.

The industrial connection: Standardizing engineering practice, the early years

Most members who looked back during the AIEE's fiftieth anniversary year of 1934 either forgot the Institute's long history of working with standards or muddled it. For example, Comfort Adams, who had joined in the 1890's when he was a young college instructor, remembered the first standards committee as beginning in 1907, just a few years before he assumed the chairmanship. On the other hand, Westinghouse engineer Charles E. Skinner, like most members who had written on the Institute's history, recalled the society's standards work as beginning in 1898, when the first committee met to standardize apparatus. Yet, Arthur Kennelly, whose standards involvement at the time of the fiftieth anniversary was the longest, dated the beginning a decade earlier. Kennelly perceived three stages in the Institute's standards work during the first half century. According to Kennelly, the AIEE standards activity began in 1890, with the goal of establishing standardized "units, standard definitions, and nomenclature." The year 1898 marked the second stage, when the society focused on "projects relating to applied science, engineering, and technology." Then, following the First World War, the Institute's standards activity turned to the problems of "production and manufacture" [26].

Kennelly's recollection provided a perceptive periodization; yet again, the starting date reflected his history with the society more than the actual story of AIEE standards work. His stages, nonetheless, provided an ideal and, thus, a clarifying schema for viewing this critical aspect of the Institute's technical mission. Beginning with uniform terminology and consistent definitions, the journey the AIEE power engineers took through the standards field recapitulated engineering involvement whenever a new technical field was opened: moving from basic definition to creative development to broad application. And yet, the AIEE's pioneer development of electrical standards was an historical event as well as a patterned response to technical development in a mature industrial order. And so this standards work, undertaken at a particular moment, held special meaning for the first generation of professional electrical engineers. As Steinmetz and others would argue, the standards process suggested a social standard as well as a technical one. It embodied the early electrical engineer's cherished social value: coordinated activity.

The path to that coordinated activity, however, was far rougher than Kennelly's neat stages suggested. Several attempts came before the solid steps of 1890 and 1898. Within a month of its founding, after the Institute's organizing committee submitted the society's first formal report, a dozen standing committees were organized around the fundamental areas of electrical technology and were headed by the era's leading technologists: On

Dynamo-Electric Machines; Edward Weston. On Telephones; Alexander Graham Bell. On Telegraphs; George Hamilton. On Arc Lamps; Edwin Houston. On Incandescent Lamps; Thomas Edison. On Prime Motors and Transmission of Power; Weston. On Electric Railways and Signals; Stephen D. Field. On Underground, Submarine, & Cable Work; Hamilton. On Electro-Chemistry & Metallurgy; Nathaniel Keith. On Batteries, Voltaic; Charles D. Haskins. On Galvanometers and Measurements; Haskins. On Batteries, Secondary; Houston. This list documents not only the rich technical diversity of the early AIEE but also the expansive ambition of its founders. They had formed a professional engineering society determined to influence the technical development of their field. Thus at that same historic meeting, acting to provide a base for the new society's technical work, Institute directors established an administrative structure, appointing committees on publications, on finances and office management, on the library, and on meetings and exhibitions. Entering the larger political and economic arenas proved difficult for a new society representing the new engineering fields. A year later, only the four administrative committees existed [27].

The founders had overextended themselves. Several of the technical committee heads were men with demanding entrepreneurial and managerial involvements: Weston and Edison especially. Bell had already begun drifting into other pursuits; and Field, a lawyer and nephew of the builder of the Atlantic cable, helped advance telegraphic interests as an entrepreneur, not as a technologist. Only Houston, and, to an extent, Hamilton and Keith, continued as active members of the Institute. In fact, none of the three were technical leaders — the AIEE group that would necessarily direct the society's work with standards. Similar efforts at devising a technical role beyond providing technical information to the membership continued to mark the society's efforts. However, each step became surer. The next step came, in 1885, when Navy Captain O. E. Michaelis asked Institute members to consider acting on the motion, made at the 1884 National Conference of Electricians, that sought to establish standards for "electrical and other units." The desire to standardize names for electrical units cut across the nation's scientific and technical communities. Its natural beginning was at such meetings as the Conference of Electricians, at which, besides the presence of leading American physicists, Michaelis reminded his audience, the Institute was "so largely represented." He called for an independent Institute committee to investigate, as well, the area of "structural materials" [28].

Though no action was taken, Michaelis' assumption that the Institute had a role to play and his articulation of a framework for that involvement led directly to later efforts in the areas of both units and materials. Four years later, Columbia professor and manufacturer Francis Crocker proposed that the "American names for electrical units" be adopted at the next general meeting of members. Crocker successfully moved for the appointment of a standing Committee on Units and Standards, with Arthur Kennelly as chairman. The committee acted quickly, preparing a standards report within a few months. By 1893, Kennelly's committee had settled on the gilbert, weber, oersted, gauss, and henry as names of units. The Council approved the report and authorized the Committee on Units and Standards to meet with "other scientific societies . . . with a view to unity of action" [29].

Involvement in this issue of naming electrical units pulled the AIEE into a larger professional world, extending to other nations as well as to other professional societies. In 1889, President Elihu Thomson had appointed a delegation to attend an electrical congress two years later in Frankfort, Germany. The delegates' assignment was to promote the name "henry" as the practical unit of induction. They also urged adoption of a name and value for a practical unit of the intensity of magnetism and, as an aid to commerce, sought a normal value for the resistance of copper. With a rising sense of purpose and importance, the Institute moved to organize an international electrical congress to meet in August 1893 at the Columbian Exposition in Chicago in order to push for a resolution of these issues. After naming a committee of its most distinguished members to plan that affair, however, its actions were superseded by the organizers of the Chicago fair. The Chicago group proceeded to arrange the congress without granting a special role to the AIEE. But the society was committed to its drive for a standard nomenclature, and swallowing its pride, accepted a lesser part at the Chicago meetings. It was an effective move, for the efforts of the Institute and the American electrical community at Chicago won provisional adoption of the gilbert for magnetomotive force, the weber for flux, the oersted for reluctance, and the gauss for flux density. Within a year, an AIEE memorial to the U.S. Congress aided in getting that body to adopt these units [30].

The second area of interest that Michaelis called to the society's attention, in 1885, concerned the twin to the scientific interest, the industrial connection, which stemmed from the engineer's intimate relations with the nation's commercial life. Since Kennelly did not arrive in America and join the AIEE until 1888, he missed these early attempts to move into industrial standards. He also missed the tension present during the early discussions, as the engineers sought their own identity within the industrial context. For that was the arena they entered when the Institute took up the need for a standard wire gauge, the "structural materials" of which Michaelis spoke. As the mechanical engineers had initiated its standards work, after the Civil War, with a drive to standardize screw threads, so electrical engineers sought to standardize the gauges of wire. Since technical uniformity meant lowered costs, large industrial firms naturally desired to standardize materials, design, and apparatus. Though the issue concerned engineers as professionals and as engineering workers, the question of distinctive engineering values emerged when two newly organized industrial associations, the National Telephone Exchange Association and the National Electric Light Association (NELA), sought the AIEE's support for a standard wire gauge that they had already adopted. In October of 1885, the Telephone Association adopted as a standard the New English Board of Trade Standard Wire Gauge. The year-old NELA followed their example in February. At that time, the secretary of the Telephone Association wrote to Institute secretary Ralph Pope in order to gain the Institute's support to "bring about its universal use as a standard wire gauge in the United States" [31].

The Telephone Association had informed Secretary Pope about their act in the fall. So Pope had attended the conferences of both the Telephone Association and NELA and was already committed to the Board of Trade gauge in May when he asked for its immediate adoption at the AIEE's annual convention. But the members resisted his counsel of quick action, which led to a confrontation between the secretary and his brother, Franklin Pope, who, only minutes before the motion, had been elected AIEE president. The new president recommended caution, since it was a "matter of a good deal of practical and commercial importance." He advised, instead, the appointment of a committee. It was not, he said, "a matter that should be decided off-hand."

But Secretary Pope was confident that "the matter was very carefully gone over" at the meetings of the telephone and electric power associations. He feared the appointment of an Institute committee would delay action, and as a result, "we will come in at the tail end" after it "has been practically adopted." With the exception of the secretary, however, none of the engineers present wanted Institute action without at least an investigation of the literature on the subject and an examination of the year-old report of the Telephone Exchange Association. Therefore, Franklin Pope appointed the Institute's first standards committee. After some haggling, George M. Phelps, Jr., Cyprien Mailloux, and O. E. Michaelis were named to the committee. This decision to enter formally upon standards work was received with a good deal of enthusiasm. Mailloux wanted to use the monthly meetings to spread "the light" and keep "the Institute awake" about this important issue. And though Ralph Pope again urged immediate adoption, suggesting that the committee report back after the noon recess, as NELA had done, his brother's view that a committee report "at the next general meeting will be in season" prevailed [32].

Though Mailloux had argued for adoption in the belief that the "grand millennium will come when we designate wire by thousandths of an inch as they do in Europe," the engineer in him was unable simply to read the literature and accept the proposed standard. Thus, before the committee reported to the Council the next November, Mailloux "had made some investigations into the subject, and plotted some curves from the various gauges with a view to proposing a gauge based on the sectional area of wires." His undertaking to turn the work of the committee into a research project pointed to no quick adoption, and, indeed, he admitted that he "had arrived at no definite result." In March, 1887, the committee promised a formal report for the annual meeting in May. But no report was forthcoming, and the Institute failed to approve a standard wire gauge. When at the end of 1889 a second standards committee was named "to formulate . . . a standard wiring table for lighting and power purposes," the action that followed marked a quickening of the Institute's commitment to standards work. Crocker took the chair of a committee containing his manufacturing and engineering partner Schuyler S. Wheeler, engineering designers Arthur Kennelly and William Stanley, and Johns Hopkins professor, Louis Duncan. Within six months, a preliminary report went to the Council and was soon printed in the *Transactions.* Two months later, a final report was submitted for adoption.

The fate of that report suggested, however, that the society still had not determined whether their role was simply to inform, as was the practice of the American Society of Mechanical Engineers, or to specify actual standards. For instead of adopting the report, which, as Mailloux argued was needed so that "the practitioners may know what to adopt pending further investigation," the Council decided, after a lengthy discussion, simply to "receive" it. Several years later, when a new Institute standards committee was appointed, the Standard Wiring Table Committee was dismissed and its report consigned to the new committee [33].

Despite the failure to adopt a standard wiring table, this early effort at standard-setting prepared the Institute for decisive action in 1898, when the New York Electrical Society asked the AIEE to consider a "plan for standardizing certain apparatus." The AIEE soon announced a topical discussion on the subject of "generators, motors, and transformers," with meetings to take place simultaneously in Chicago and New York [34]. The main debate was in New York, however, where representatives of the major electrical interests gathered. They included engineering researchers and designers from large and small manufacturing companies; a managerial spokesman for electric power interests; an officer from one of the giant manufacturing firms; consulting engineers who advised users of machinery — especially the builders of central stations and isolated plants; and college professors who taught future engineers and often consulted, as well. Besides the Institute's president, Francis Crocker, who presided, those present included John W. Lieb, Jr., high in the managerial ranks of the Edison Illuminating Company of New York; Edwin W. Rice, now technical director at General Electric; Cary T. Hutchinson, an MIT professor; and Gano S. Dunn, an engineer at Crocker's manufacturing firm. With them were Steinmetz, Kennelly, Mailloux, and several others. It had become clear that the industrial standards movement was ripe for development in the electrical field when the movement's turn from nomenclature to electrical apparatus attracted engineers from across the spectrum.

Crocker appropriately guided the attempt to make a new departure in standards work. After a decade of leadership in the Institute's standards program, he was a natural choice. When he first undertook standards work in 1888, Crocker was five years out of college and already a professor and a manufacturer. After graduating from Columbia's School of Mines in 1883, he formed a partnership to manufacture electric motors. Then, the year he joined Columbia as a professor, he and Schuyler S. Wheeler organized the Crocker-Wheeler Company and made it an important producer of electrical apparatus, especially dynamos. Hence, Crocker was well placed to moderate the standards discussion in 1898, for he was a manufacturer without the biases of the big companies and a professor with the experience of an engineer-entrepreneur.

The meeting rested on the assumption that all present agreed on the need to standardize electrical apparatus. Furthermore, as Crocker pointed out, ample precedent existed for such an undertaking. The American Society of Mechanical Engineers had published an "elaborate report" from its Committee on Standard Methods of Steam Boiler Trials, and organizations such as the

AMERICAN INSTITUTE OF ELECTRICAL ENGINEERS.

HAVEMEYER BUILDING, 26 CORTLANDT STREET,

NEW YORK, N. Y. JUNE, 1899.

[COPYRIGHT 1898.]

REPORT OF THE COMMITTEE ON STANDARDIZATION.

[Accepted by the INSTITUTE, June 26th, 1899.]

———o———

To The Council of The AMERICAN INSTITUTE OF ELECTRICAL ENGINEERS.

Gentlemen :

Your committee on Standardization begs to submit the following report, covering such subjects as have been deemed of pressing and immediate importance, and which are of such a nature that general agreement may be expected upon them.

While it is the opinion of the committee that many other matters might advantageously have been considered, as, for example, standard methods of testing ; yet it has been deemed inexpedient to attempt to cover in a single report more than is here submitted.

Yours respectfully,

FRANCIS B. CROCKER, *Chairman.*
CARY T. HUTCHINSON.
A. E. KENNELLY.
J. W. LIEB, JR.
CHARLES P. STEINMETZ.
LEWIS B. STILLWELL.
ELIHU THOMSON.

255

The year 1898 marked the point at which the AIEE turned from focusing on measurements and nomenclature in its standards work to concentrating on industrial standards.

American Society of Civil Engineers and sister societies in England had taken similar steps. "It is very common on both sides of the Atlantic," Crocker assured the members, so "we should be doing nothing radical or unusual." The tasks before them, then, were first to decide the "feasibility" of taking up standards and then to develop a "policy" to guide that work. Determining feasibility meant deciding which technical areas were capable of standardization. Edwin Rice, who opened the meeting, thought the Institute should confine its work to defining and determining methods for testing certain characteristics of apparatus. All agreed that standardizing actual apparatus was not a proper task. As one member put it, the apparatus "will eventually standardize itself," since "commercial conditions" dictate that standard designs cost less than special construction. And though some agreed with Rice that methods of testing could be standardized, the consensus was that specifying "conditions of performance and defining terms clearly was preferable to attempting to outline standard methods of testing." Steinmetz, too, thought the committee should first decide on the meaning of terms used in electrical engineering. Because engineers used terms like "efficiency" in different ways, precise definitions were highly desirable. Standard methods of testing, "now one very great difficulty," should also be a goal, even if the problems involved would not soon be solved.

Crocker's second policy matter turned out to be the most heated area of discussion as well as the issue most easily decided. It concerned, as Kennelly put it, "how the committee should be made up." MIT professor Cary Hutchinson admitted to having "stirred up the meeting" when he asserted that electrical manufacturing companies should not be directly represented. Since all companies would likely demand a position, he argued, the committee would become too cumbersome. To have one represented and not others "would prejudice the work at the outset" and limit access to information from the shops of rival firms. Crocker seized the opportunity to invite discussion on this "most important question" and, in doing so, revealed a decisive majority behind the obvious necessity of admitting representatives from the manufacturers. Kennelly thought that a committee without manufacturers would be "like playing Hamlet with Hamlet left out." Gano Dunn said that it would be as wrong to omit manufacturers as it would be to exclude consulting engineers. Still, Hutchinson insisted that all necessary information could be gathered without company representatives on the committee. Steinmetz confessed to being "stirred up" by the discussion, telling Hutchinson that the "question should not have been raised at all." Like others there, he thought the presence of engineers of high professional standing on the committee would banish conflicts between rival manufacturers:

> If the Institute intends to produce something of lasting value, which will be accepted and adopted by the whole continent, then the committee doing the work must be composed of men of such standing and reputation that, regardless of whether they are connected with manufacturing concerns or not, there can be no question that they will be impartial and not influenced by the fact that they are connected with this or that company.

The men of standing named to the first committee were Francis B. Crocker, as chairman; Professor Cary Hutchinson; GE engineer Charles Steinmetz; Elihu Thomson; independent consultant Arthur Kennelly; power company representative John Lieb; and Lewis B. Stillwell, who had been a Westinghouse engineer for seven years before leaving in 1897 to join the Niagara Falls Power Company. In short, the committee members had experience in all the areas necessary to coordinate action in standardizing apparatus.

From a series of special-subject committees, grown increasingly dependable and prompt, the Institute had launched a full-blown effort to enter the heart of standards work: that of the machinery itself. Acting with the efficiency it had hoped to bring to the electrical industry, the committee completed its report within eighteen months. Its twelve pages followed the cautious technical path laid out in the discussion. The committee avoided the task of standardizing the actual apparatus, leaving that, as Kennelly said, to "the process of evolution among business interests." Omitted also were standard methods of testing. Though definitions and conditions for testing such matters as temperature, insulation, and rating were included, the bulk of the report dealt with the efficiency of apparatus, that is, with "the ratio of its net-power output to its gross-power input." Methods for measuring efficiency and probable sources of energy loss were given for a variety of basic electrical machines as well as for transmission lines. Cautionary asides were included in such matters as the need to distinguish between systems, or "plant efficiency," and "machine efficiency" [35].

Formally adopting the report, the AIEE fully entered the work of technical standards. During its first two decades, the Institute had thus established relations with the national and international electrical communities, aided in achieving a standards agency of the federal government, and now had joined with business firms to begin to rationalize industrial design and manufacture. That it came first in the power engineering field followed not only from the field's centrality in the manufacturing companies, but also because it was the most competitive field. This was important since, just as electrical engineering educators had turned for its curricular content to the ideal of "best engineering practice," so did standards workers draw their data from practice.

By 1900, then, the Institute had committed its efforts to the work of establishing industrial standards. Five years after the 1898 discussion, Charles Scott made a critical contribution to that commitment when he created the technical committees. These bodies were to be the Institute's response to the growing specialization of the profession. They were designed to bring precision to the Institute's technical work, to more accurately reflect actual practice. This was Charles Scott's intention in 1902 when he inaugurated the first technical committee. Though his act partially fulfilled that abortive attempt of 1884 to cover all technical fields, even twenty years later, such a move could not be taken at once. Each specialized committee would come in time, and, then, only if the field possessed a constituency within the Institute.

As Scott explained his innovation in a letter to Steinmetz in 1902, technical committees were necessary for continued Institute growth, as they would help the members keep "in touch with the latest engineering practice in new lines." Information would flow from the committees as papers at meetings, as special reports to the society, or as reports to the Standardization Committee. Their work would thus present to members "the standards of engineering practice and efficiency of electrical work." Following his perception that "electrical engineering is crystallizing," Scott established the technical committees to advance that process. Two were initially started on high-tension transmission and on engineering data. The Transmission Committee was the critical one, for as Scott explained, the area of high-tension transmission, while "developing quite rapidly," was still "not definitely crystallized." To achieve this technical goal, however, the committee's makeup was important, and so Scott appointed a "consulting engineer [as] chairman, and representatives from several large manufacturing companies and a western university." Scott thought it vital to continued technical innovation for the Institute to attend to these specialized fields. Adding technical committees, in fact, perpetuated the Institute's objective of bringing together the achievements of a "diversified" profession so that it would "constitute a single total of accomplishment." This, he explained, adds up to that "which we designate as progress" [36].

Scott's aim, at the simplest level, had been to establish specialized committees to keep the Standardization Committee current. Over time, then, the technical committees reflected the disciplinary contours of the electrical field, as some remained through the years while others faded in and out. All, in part, served to chart the ebb and flow of electrical concerns within the Institute. Until 1909, the Transmission and Engineering Data Committees were the only two. In 1909 and 1911, committees on Industrial Power, Electro-Chemicals, and Power Stations were formed; in 1910, committees were organized on Electric Lighting and Telephony and Telephone. By the 1930's, Scott's creation had been replicated many times. At first, the Standardization Committee added specialized subcommittees, but the technical committees soon replaced them as "the originator of new standards" [37].

The AIEE and the world of standards

The organizational structure for standards work had to change in order to keep up with technical advances. The technical committees provided the Institute with a flexible posture before that change, a posture that would be sorely needed as new chapters in the electrical revolution were written. In the matter of standards alone, the society's decisive move, near the turn of the century, demonstrated the growing strength and sureness of the organized leadership of an electrical profession that was equally dynamic and expansive. That leadership was enhanced by the society's simultaneous entrance into a world of standards beyond the Institute committees and specialized electrical engineering issues. On the one hand, the move involved a linear extension into larger arenas—into participation on national and international bodies—and the advance of cooperative activities already undertaken. On the other hand, the AIEE had entered a world of standards fraught with complexity from both the raw competition among consultants in the fast-growing power industry, and the lengthy campaign of a corporate executive, who wished to instill in the engineers an appreciation for the savings in costs that would result from a standardized design.

Recognition of this latter interest had been clear to most engineers at the '98 discussion, aware as they were that the standardization of apparatus was an industrial matter. Industrial standards were naturally of concern to the engineers of the AIEE, Kennelly explained, because "the proper province of an institution such as ours" was to assist "the business which it is our pleasure and advantage to promote." And in his opening remarks to the '98 discussion, Edwin Rice carefully explained the commercial importance of engineering standards: "Standard lines" were already being "arbitrarily determined" by manufacturing companies acting out of their experience in the marketplace and serving their own ends. "Competition" undercut this method of reducing costs in design and production time." It tended "to increase the number of standard sizes" and create a confusing variety of dynamos, motors, and transformers as "the industry . . . added enormously to the variety of apparatus in use." Rice was not criticizing so much as lamenting the industry's dilemma. For even with the confusion, the attempts at standardization had led to "phenomenally low costs" and "great improvement in . . . quality." In short, the benefits of standards encouraged large companies "to organize, elaborate and perfect methods of manufacture." These efforts, Rice concluded, were preferable to AIEE standards for apparatus "which the rapid evolution of the business would soon render useless" [38].

Like Rice, John Lieb did not think such work on standards required the Institute's consideration, since standardizing apparatus followed from the "general tendency of American manufacturing methods . . . to secure interchangeability . . . of parts." As a power-industry representative, Lieb was especially anxious to cut machinery costs for the central stations. Lieb, too, had imbibed his ideals with Edison and Samuel Insull at Pearl Street and had remained with the company that grew out of that pioneer central station. During the nineties, Lieb often traveled in Italy, consulting on electrification projects for the Edison Electric Illuminating companies. By 1898, he was fully launched on the career that, a quarter-century later, won him the AIEE's Edison medal "for the development and operation of electric central stations." Lieb spoke strongly against practices that contributed to the costliness of specific designs, with "each construction requiring special patterns and special development." The issue of standardized design continued an old theme. It had been part of a search for technical efficiency in tandem with economic efficiency. It had informed the combination of technical and marketing sensibilities that underlay Edison's genius. In a competitive atmosphere, costs were paramount. For Edison, the competition over costs had been with the gas companies; then, it became an element in the rivalry between the giant electrical manufacturers. And now there was the growing contest between the urban central-station owners, which also involved the independent consultants, who designed and built central stations and isolated plants.

Standardization meant economic savings, a concern of both engineers and managers. In his presidential address, Kennelly urged engineers to heed costs, warning against the recent tendency of manufacturers to favor "special machinery of independent design" over standardized apparatus. Special design interfered with "natural shop methods of standardizing and cheapening production." Because electrification depended on competitive rates, "it needs no homily," Kennelly concluded, "to drive home the conviction that the unnecessary introduction of special machinery is a puncture in the tire of progress" [39].

As a matter of fact, the strongest and most persistent calls for an end to special design came from the power industry, and especially from Chicago Edison chief Samuel Insull. Insull was a national leader of the central station movement and assumed early the position of delivering the message of standard design to the AIEE. Soon after learning of the electrical engineers' 1898 action, he used the opportunity of his NELA presidency to encourage their collective act. Insull commended the AIEE for recognizing the "paramount" need for special machinery to give way to standard designs. Standardization served both manufacturer and user, he argued, giving the United States its competitive edge over Great Britain: "In one case, the machinery is really manufactured; in the other case, the builder runs a jobbing shop." All of his arguments reinforced his basic economic point that "constant duplication of parts, resulting in constant duplication of a given piece of machinery, means . . . constant reduction of cost." Whereas Insull approved the AIEE's initiative, he strongly condemned the acts of some individual electrical engineers for their special designs. He urged his NELA audience of operating company executives to resist the efforts of "the electrical engineer . . . to draw up specifications" that tend "toward the specializing of apparatus" and interfere "with rapid manufacture and low cost of the product." Insull did not censure the engineer who "adapts his requirements to the standard apparatus"; rather, he criticized "designing engineers" who wanted machines they could "point to as their own design" [40].

The tension injected into engineering discussions by the campaign to standardize design paled beside the emotions that could be raised in asking competing power engineering consultants to share their knowledge on technical committees. Scott quickly found this out as he set up the Transmission Committee. He had no trouble with the initial appointments. Ralph D. Mershon, who had left Westinghouse recently to embark on a consulting career, had suggested the topic and was named chairman. Two other men were quickly lined up from Westinghouse and General Electric. However, when Scott moved to achieve a geographical balance by appointing a West Coast engineer, he found how valuable some engineers held a knowledge of best practice to be [41].

After a first choice declined, Scott appointed R. S. Masson, a consulting engineer who maintained offices in Los Angeles and San Francisco. Scott thought Masson a good choice, "as he is on the ground and could work at short range." A month after he had appointed the new member, however, Masson "startled" Scott by refusing the post. Scott wrote to Masson again to explain the purpose of the committee, insisting that Masson's distance from New York made him important to the committee. With Scott thus resisting his resignation, Masson wrote an "entirely personal" letter to tell Scott what he "could not express" in the formal resignation to Secretary Pope. Though the California engineer was "uncertain what an outsider would think of the views expressed herein," Masson believed he would be materially damaged if he supplied the requested information on "the regulation and operation of a transmission plant including everything connected with the transmission business." He described the fierce competition among consultants to the California power companies that had led some men to pretend to be gathering information for purposes of study, yet "when they had enough of it collected, . . . they discontinued asking and probably refused to give any information back." Though he claimed to personally prefer the open exchange of information, the Committee wanted "absolutely everything which I have available for the production of bread and butter." In fact, several of his clients were unwilling for him to participate, since "they had employed me with the intention of securing engineering which would place them somewhat ahead of those who do not have engineers." Indeed, these attitudes were the "only reason for the existence of independent engineers," he informed the AIEE president. He concluded that "self preservation must ever be one of the underlying principles of life . . . . It would be a death dealing blow to equip my competitors" with the information wanted [42].

Scott agreed that an engineer who told all he knew was "freely dispersing his capital in trade." But that was not the information wanted: "standardizing transmission work" required making known "established ways of working." Good practice in spacing and transposing power lines and placing telephone lines was common knowledge: the committee had only to gather "reliable actual data on present practice." Scott argued that it was possible for an engineer to be loyal to his employer and to share ideas and information with professional colleagues — as he himself had done in papers before the AIEE. He organized his data to "lead to a natural conclusion," without judging, for example, "whether A.C. or D.C. is best for a particular case." Scott asked Masson to reconsider once more before telegraphing his answer. Nonetheless, Masson ended the exchange with a simple message: "I will refuse you nothing but repeat request to be released" [43].

The other dimension to the AIEE's entry into a larger standards arena proved more benign, beginning when the Institute's standards work evolved into a source of continuous Institute contact with other engineering societies and public and private agencies. Professional electrical engineers had entered, in effect, a world of standards in the process of nationalization. It was an expanding world, as well. The family of American engineering societies continued to grow explosively through the first decades of the twentieth century. During this same period, moreover, new federal agencies appeared and greatly expanded technical functions dating from before the Civil War. Nor was the AIEE passive in this development. In 1900, an ad hoc committee joined in the general drive to get Congress to establish a "Standardizing Laboratory." Congress responded by holding hearings late that year and, in March, 1901, established the National Bureau of Standards. Its first director was Samuel Wesley Stratton, a physicist who headed the Weights and Measures Office in the U.S. Coast and Geodetic Survey. Stratton had pushed for a bureau from within by channeling testimonials from scientists, engineers, and industrial executives to congressional supporters [44]. The overlapping work of the federal bureau and the Institute was recognized when Institute officials appointed Stratton to the Standardization Committee.

The two agencies, one a professional body, the other a public agency, intersected at common points because each sought the same objectives. An early account of the Bureau explained its purpose as determining "physical constants and the properties of materials, when such data are of great importance to scientific or manufacturing interests." E. B. Rosa, the first director of the electrical branch, described his division as a "National Physical Laboratory"; he used the word "physical in a liberal sense," he explained, to include "chemistry and engineering" as well as physics. In 1912, Stratton described the context of the Bureau's work in words readily applicable to

Institute purposes. The agency's involvement in standardizing "industrial processes and products," Stratton wrote, "made it imperative that the Bureau keep in close touch with the advancing needs for such work" [45].

Similar goals prevailed within the vastly larger world of international standards. An international body had been initiated in St. Louis in 1904 and was formally organized in 1905 and 1906 as the International Electrotechnical Commission (IEC), with British engineer C. le Maistre as general secretary. In June, 1906, Crocker, Kennelly, and Mailloux represented the Institute at meetings in London of what was then called the International Electrical Standardization Commission. At the end of the year, their report was accepted by the Board of Directors. Six months later, the Board appointed twenty members to serve at the local IEC and to report to the National Bureau of Standards. The IEC continued as the focal point for international standards work through the 1930's. At a conference held at the end of World War I on "industries in readjustment," Comfort A. Adams observed that international standardization had led to "a common industrial language" and, thus, to the removal of barriers that separated nations and caused misunderstandings. The IEC was responsible for the movement's success, he said, "to the great advantage of our foreign commerce" [46].

Adams played a central role in that success. He represented the Institute in 1917 when a group of American engineering societies began to plan a national standards organization. As an active member of the Standards Committee, he had already begun to work with the other engineering societies. Now, a more formal alliance was being formed as representatives of the nation's civil, mining, mechanical, and electrical engineering societies met with the American Society for Testing Materials to decide on a "plan of organization." In 1918, these groups established the American Engineering Standards Committee (AESC)—later the American Standards Association (ASA). The AESC coordinated the overlapping standardizing work of the American societies. Also, as Adams explained, it served as "an authoritative national body" to meet with other national groups in the drive for international standards. The tightening network within which the Institute carried out its standards work was understood, as was the rationale for the AIEE itself, to be the fruit of cooperation. Institute spokesmen had frequently explained their professional activities as a reflection of the cooperative spirit. Although the idea always possessed a practical side, standards work made it more concrete. The cooperative idea was especially pronounced in 1916 when, as war raged in Europe, IEC general secretary le Maistre met with engineering groups throughout America to discuss international standards. As a representative of the IEC, he had already visited Boston and had attended a Bureau of Standards conference in Chicago. Now as he spoke at the annual convention in Cleveland, le Maistre praised American engineers for having made "the cooperative spirit so alive." He credited the society's achievements with having flowed from the 1898 Standards Committee, an act, he said, that reinforced such factors as competition, demands by labor for higher pay, and capital's desire for "a better return" to compel "the electrical, in common with the whole engineering industry, to introduce modern order and system into all its methods of production." The AIEE's active involvement, le Maistre believed, had forced "individualistic methods . . . to give way to co-ordination and collective effort" [47].

Joined during the discussion by veteran AIEE standards workers Steinmetz and Adams, Institute members echoed the British engineer's praise for the world's tendency to collective effort. When the drive for international standards began, Steinmetz had doubted its "feasibility," believing the difficulties of "getting . . . unanimous agreement . . . between all nations" to be insurmountable. Yet Steinmetz had come to believe that in spite of the "state of suspended animation" induced by the war, "the world is being impulsively driven into shape for co-operation, and all standardization is based on co-operation." The end of the war would find the nations greatly changed, and the "old idea of every one for himself," Steinmetz predicted, will be replaced by a belief in "the necessity of co-operation for the mutual welfare." So successful had been the movement that Steinmetz felt he could now announce his retirement from active participation. It was natural, he said, that "the younger generation, our pupils, [should take] over the work." Comfort Adams was the GE engineer's star pupil, his active involvement having begun at Steinmetz's urging over a half-decade earlier. Adams found in the standardization process itself the germ of the growing acceptance of co-operation. At the national meetings he had attended, Adams had seen the experience of achieving agreements out of "apparent conflicts of interest" lead to "mutual education." He believed a similar process would, in turn, infuse the international world of standardization where "the narrow minded partisan must in the long run give way to the broader co-operative spirit" [48].

Steinmetz and the cooperative nation

The Institute's response to technical standards and educational issues had solidified by the time of America's entry into the world war. In the three and a half decades after 1884, the focus of standards work had shifted from cooperation with scientists and alliance with the American engineering community to concentration on industrial standards at the national level. From there, the impetus of the world war had carried American standards workers into the international industrial order. For its educational concerns, the AIEE had not so much required committees — though an educational committee was created in 1906 — as it entailed using the existing meetings and papers programs as a platform for the exchange of ideas. Electrical engineers had thus aired their views on the teaching of science and theory while a consensus gathered around curricula that reflected best engineering practice. As with standards, this conclusion led to a general emphasis on practice in engineering training and specifically to cooperative programs between industrial firms and university departments.

Yet a larger question remained: how to move from engineering cooperation in education and standards to a cooperative national order? By no means did all engineers take their engineering concerns into the national arena, and most who did generally couched their queries around the notion of "status." Nevertheless, two prominent electrical engineers, during these early years, sought the place of engineering in terms of cooperation, an ideal firmly entrenched in the engineering mind of that era. The two engineers were John J. Carty (AT&T's chief engineer until 1919 and a leading architect of the Bell system) and Charles Steinmetz (who, besides being a distinguished GE research engineer, was a dedicated socialist). Neither turned for an answer to an actual "cooperative spirit"—Comfort Adams' phrase—perceiving more precisely the role of the business corporation as the most active agent for a national order. Therefore, during the years between 1910 and World War I, first the self-trained Carty and then the highly educated Steinmetz turned their considerable intelligences to an analysis of the American corporation and the place and potential of the bureaucratic phenomenon in the life of the United States.

Viewing the corporate organization from the front office gave an immediacy to Carty's thoughts. Unlike Steinmetz's disdain for administrative duties, Carty preferred—as he wrote in an application for a Signal Corps assignment in the wartime Army—"engineering administration." He had joined Bell at the end of the eighties, early contributing as an inventor before spending nearly two decades as chief engineer to the New York Telephone Company. Then in 1907, when Theodore Vail returned to head the parent company, Carty became chief engineer at AT&T. Vail wanted his help in reorganizing the firm and fulfilling his dream of a truly national system—to do, in short, what the company's directors had refused him twenty years before. But even though the Bell companies had been founded over thirty years before, when Carty joined AT&T's headquarters staff in 1910, there existed no central engineering office from which to direct the technical development of the system. This was in spite of a large research and development staff. In addition to American Bell's department of research in Boston, the Western Electric Company—Bell's manufacturing subsidiary—had research groups in Chicago and in New York. Carty first merged the Boston research personnel and laboratory with the research staff in New York and transferred to New York a large contingent from the Chicago laboratory. He sought the critical mass necessary to achieving technical breakthroughs by which to make long-distance telephonic communication practical [49].

This work had taken him to the West Coast in 1909, and it still engaged him a year later when Dugald Jackson wrote from MIT for Carty's ideas on the organizational structure of the corporation. Carty enthusiastically responded. He had "studied the matter a great deal and [was] exceedingly anxious" to find the time to prepare a paper on the topic. He sent a pamphlet that the company had recently distributed to the associated companies. Though not entirely satisfactory, Carty told Jackson, it "will give a fair idea of the functional organization, so called." He also wrote a six-page letter, to which he gave the title "Corporate Organization"; upon reflection, he added five pages that offered the term "unifunctional organization" to reflect more accurately the evolving organizational structure at AT&T and the Bell companies. Though he never found the time to write his paper, he returned six years later to the basic theme of what he then called "modern organization." Carty was prompted, this time, by an invitation from the AIEE to contribute a paper to a special meeting in New York on "The Status of the Engineer" [50]. Having fairly satisfied himself in 1910 with an analysis of the internal workings of the corporation as a coordinated network of specialized functions, Carty turned in his 1916 paper to the national arena. His straightforward acceptance of the role of the engineer in the making of a national order led him to welcome the spread of "large organizations [to] all parts of the world." As the corporate form became more widespread, he found the "same tendencies . . . at work" in "communities and nations" as in corporate organizations.

Since Steinmetz viewed the rise of the corporation as both a structural entity and a complex historical event, he viewed its programs more critically than did Carty. Similarly, Steinmetz viewed the role of engineering more concretely. Though he believed engineering methods were applicable to the problem of national organization, how that was so required analysis more than assertion. To his way of thinking, engineering represented only one aspect of the nation's social and political economic order. Engineering successes could, at best, serve as a model for change. Indeed, establishing the place of engineering in the national order was a secondary if necessary consideration as he explored the health and prospects of America during the years before the war. Steinmetz worked out this larger vision in his ambitious work of 1916 called *America and the New Epoch;* yet he also published essays in magazines as

Charles Proteus Steinmetz, an engineering scientist with a vision of a new epoch for America.

various as *GE Review* and *Harper's Monthly Magazine.* Steinmetz's social writings were unprecedented, not just because, as he explained, of his "unusual opportunity [to observe] all sides of the politico-industrial structure of to-day," but also because no engineer had yet attempted to examine the fundamental framework of American society. He pursued what Adams called the "philosophic aspect." But while Adams referred to standardization alone, Steinmetz carried the exploration to levels of social activity that subsumed the technical. The result was a carefully laid path to the "co-operative organization of the nation" [51]. With it came a sharp-eyed assessment of the place of engineering on the eve of World War I.

Steinmetz was well equipped to assess the engineering role for a number of reasons, but three were paramount: the freedom he enjoyed at GE, the rigorous education he received in Germany, and the special brand of socialism he imbibed in Germany during his student days in the 1880's. His independence as a corporate engineer was singular among contemporaries. The rebuff by Bell officers in the 1890's of John Stone Stone's request for freedom from departmental affiliation and Bernard Behrend's battle losses at Allis-Chalmers a decade later were more familiar outcomes. Steinmetz's European education, on the other hand, was the kind Wickenden later found so often lacking in engineers educated in America. Only from such a solid base could Steinmetz have launched his investigation.

The crucial element in Steinmetz's intellectual makeup, however, was his socialism. He became a socialist during the 1880's when, in response to repressive measures by the German government, Germany's Socialist Labor Party adopted views closer to the social theorist Ferdinand Lassalle than to Karl Marx. Lassallean socialism taught that radical change could be achieved without revolution, that the iron law of wages could be abolished by legal means through the "cooperative control of collective labor." During the period that Steinmetz joined the socialist cause, many other Germans were doing so as well: the Socialist vote rose from just over 300,000 in 1881 to nearly one and a half million in 1890. This nonrevolutionary brand of socialism assumed that capitalism was developing a flexibility that would avoid a major economic crisis. The increasing tendency to economic concentration, moreover, would be accompanied by a growing trend toward an equal distribution of wealth. The leading German socialist of the late nineteenth century, Eduard Bernstein, perceived "the common interest" gaining in power "as opposed to private interest." A rational and equitable national order, rather than being inhibited by the capitalist state, as the Marxists believed, was instead stymied by the stubborn refusal of powerful private interests to adopt the cooperative spirit. This, in essence, was the socialist position Steinmetz brought to his growing political activism and to his analysis of the American order in the middle of the 1910's [52].

In spite of Steinmetz's excursion into the realm of history and social thought, the engineering scientist insisted on the rigor of his analysis and, thus, the objective validity of his conclusions. Their correctness rested, in turn, on his historical approach. "Historical facts," Steinmetz wrote, as "in any physical or engineering problem [lead to] conclusions which follow from the premises." The facts he looked to, moreover, were not those of "war and revolution, conquest and defeat." The Marxist in him saw such events as only "outward appearances." Steinmetz looked, rather, to the "history of industry, arts, and commerce," since "the true history of the human race . . . is made on the fields and farms, in the factories and workshops, in the business houses and shipping-offices" [53].

The cornerstone of Steinmetz's vision of the new epoch was the transitional role of the corporation. From this early belief, Steinmetz had become a staunch supporter of big business, a position he maintained throughout his life in America. His stern rebuff of the professor at the '98 standards discussion was characteristic—telling him that the idea of excluding corporate engineers from the Institute's technical work was not just wrong, it was irrelevant. At a meeting of the Institute, some years later, Steinmetz defended Samuel Insull's preference for the centralization of power production when Lewis Stillwell and other independent consultants to isolated power plant interests criticized an address by the Chicago power executive. In the preface to *America and the New Epoch,* then, Steinmetz insisted that the industrial corporation was not "the greedy monster of popular misconception." No matter how "crude and undeveloped . . . its social activities," the corporation would lead to the centralized society. "In some respects," he wrote, "the corporation may be considered as the first step toward socialism." More precisely, he believed the corporation had the possibility of fulfilling the aims implicit in the cooperative spirit. Steinmetz perceived the great historical movement of his time to be the shift "from competition to co-operation." Individual efforts had rarely achieved any "great work." Usually the private corporation or "the public corporation—municipality, State, or nation—had to step in." But in America, the wasteful competition of the nineties had disappeared in the industries that had consolidated. Still, the reign of the principles of the "co-operative era"—"control of production; control of prices; interchange of information"—had not lasted. The protests of a "misguided public" had led to "the interference of the Government." Though corporations grew more peaceful, industrial America had "gone backward." Steinmetz admitted that corporations, in the past, had often been crude and inefficient, disregardful of both employees and public, as if they had "no obligations at all." In 1916, however, he believed they were becoming more responsible. Corporations were beginning to recognize, more often, that "the social (and educational, as part thereof) relations with the employees and the general public" were a legitimate part of corporate activity [54].

Steinmetz knew there would be potential resistance to his analysis—Insull and other corporate leaders had already encountered and spoken of the widespread public mistrust of corporations. Thus, determined to have his conclusions fairly considered, Steinmetz refused to act as apologist for the corporation. He denied that the "wide-spread hostility" was the work of "demagogues," as some charged. Instead, he searched for the "structural elements . . . from which a continuous, competent, and responsible government could develop by evolution." It was at heart an engineering approach. Steinmetz wanted a government that sought "the efficient industrial cooperation of all citizens in the interest of all, under democratic principles." Efficiency was the means, the elimination of waste, the goal. It was the standards movement writ large. A "responsible government" would ban overproduction and the practice of "creating a demand where it does not exist." Goods could be marketed by a single organization so that duplication of sales force and misrepresentation of products would end. Ways of ending waste included such measures as forbidding a railroad to compete with a waterway when the latter offered more economical transportation, not allowing cattle to graze on land better fitted for wheat, and protecting national sources for hydropower from the "reckless deforestation of the head waters." Efficiency, finally, relied on cooperation throughout the society, "between all producers, from the unskilled laborer to the master mind which directs a huge industrial organization" [55].

The one ray of hope Steinmetz saw as he tried to envision a new era for America came again from the world of engineering, specifically from among the activities of organized engineering. Only there were to be found positive signs of national organization, of the rise of the "concentral method" of government. Engineering achievements could not alone deliver the new order. Since the largest national societies limited their work to "certain definite fields of activity," they could be counted on mostly for "assistance and co-operation in the industrial reorganization of the nation." However, national engineering societies had done "much successful industrial organizing work," especially in the area of "engineering standardization." That work had begun with America's national engineering societies and spread abroad from that base. Now the cooperative spirit engendered by the work of standardization was "beginning to reach our Government" [56].

Steinmetz was not sanguine about the chances for such radical change. Still, he worked for change in the areas he knew. Besides his work in standards, as Steinmetz became politically active after 1910, he turned also to educational reforms, serving as president of the National Association of Corporate Schools and being elected on the Socialist ticket as a member of the Schenectady school board. From this experience, he learned, first hand, that America, unlike Germany, had not achieved "a universal system of industrial education . . . leading to the highest fields of engineering" no more than the country had developed a "co-operative organization of the nation" [57].

He believed, though, that if the country were to become an efficient and rational cooperative society, it would do so out of the spirit of engineering. The AIEE had exhibited, in part, this larger vision of Steinmetz, functioning cooperatively in the setting of standards. This had come through work carried out in alliance with industrial and governmental interests. Here Steinmetz found his strongest model for a cooperative America. Indeed, for many of the Institute's early twentieth-century spokesmen, the promise of engineering became the promise of America as well. Yet it was to be within the AIEE itself—which Scott and others understood to have been constructed on the bedrock of cooperation—that the cooperative ideal would prove most deficient. For beginning around 1905, Institute members increasingly found themselves embroiled in controversy, at first over the need to regulate the business practices of consulting engineers and then over the old issue of constitutional membership criteria, with both having implications for the place of professional standards in electrical engineering. No less than the cooperative achievements in standards, these conflicts, too, would shape the character of the AIEE.

4/THE PROFESSIONAL STANDING OF ELECTRICAL ENGINEERING

Our organization is powerful, is of a very high standing; it is up to us and it is within our power either to increase the standing of the electrical engineering profession, to put a ban on everything we consider improper, to raise the code of ethics of the electrical engineering profession, or to let matters slide and trust to Providence whether our standing shall rise and fall. I believe we should not do that.

Charles Steinmetz, 1906 [1]

Insull's dinner

It was a stately affair, having the air of a group of diplomats relaxing over a grand meal after several days of long meetings and listening to colleagues proclaim their mutual achievements. Taking place late in September, 1911, the occasion marked the departure from America of a half-dozen distinguished British electrical engineers. Power company executive Samuel Insull was honoring them with a dinner at Delmonico's restaurant in New York, "the de Ferranti Dinner," he called it in his published version of the evening. More accurately, it might have been called Insull's AIEE dinner [2].

The ostensible honorees were major contributors to the design and economics of large-scale central stations. This was especially true of guest of honor Sabastien Z. de Ferranti, who in the 1880's had built the first large generators of 1200 kilowatts at Deptford, England. As Charles Steinmetz told the guests, de Ferranti was one of the "pioneers who had in those bygone ages done the work we are just beginning to do now." He meant such technical feats as creating generators of many-thousand horsepower and transmitting 10,000 volts of electricity through underground cables. Though Insull, too, prized these achievements, he had other purposes, which appeared in the full guest list and addresses made that evening. In planning the dinner, Insull took advantage of a luncheon being given by the AIEE to honor de Ferranti [3]. The AIEE had invited many of its most illustrious members, a group representing various engineering roles but bound by field and a special relationship to the Institute.

Having been an Institute member since the 1880's, Insull knew that the engineers at his dinner represented the leadership of both the Institute and the profession. Since 1892, fifteen had served as officers, nine of them as president. Besides Steinmetz, Francis Crocker and his partner, Schuyler Wheeler, were there. Others present were Charles Scott, now head of electrical engineering at Yale; Henry G. Stott, superintendent of motive power for

New York's rapid transit system; and Dugald Jackson, who just a month before had passed the presidency to Crocker-Wheeler engineer Gano Dunn, also in attendance. Insull's chief engineer, Louis Ferguson, attended along with John Lieb, associate general manager of New York Edison. In addition, two future AIEE presidents attended: Michael Pupin, of Columbia University, and Edwin Rice, only two years from the presidency of GE as well. Counting the five British engineers and several more Americans, almost one-half of the forty-six men at Insull's dinner were electrical engineers.

But they stood for something more precise than electrical engineering: except for Pupin, all worked in the field of power engineering. It would have been difficult to gather the leadership of the Institute and have it be otherwise. Engineers working in the power field were especially influential within the AIEE during these years. Between the 1901 presidency of Carl Hering, a University of Pennsylvania professor whose field was electrochemistry, and that of Bell chief engineer John Carty in 1915, every president of the Institute, save one, made a substantial career in power engineering. Five were connected with manufacturing companies and seven worked either for the utilities or as consultants in designing and constructing central or isolated plants. In contrast, during the decade following 1915, only two presidents had careers in power engineering.

In his relations with the Institute's engineers, Insull had always sought to impress upon them the importance of financial and industrial concerns. His early promotion of standards had demonstrated this, as did the names of the remaining guests, most of whom were executives from GE, Westinghouse, and the operating companies. Also present was Sidney Z. Mitchell, who headed the Electric Bond and Share Company, a holding company whose expanding power network would soon extend from Indiana to the Atlantic Coast [4]. With him were such men as the president of the Electric Storage Battery Company of Philadelphia and a vice president of the National Conduit and Cable Company, representing an area of the industry that mushroomed with the spread of electrification. Leaders in technical publishing were present, including James H. McGraw of McGraw Publishing Company. Insull had also invited several bankers, among them his friend Frank A. Vanderlip, president of the National City Bank of New York. Like Insull, he had high standing in the national corporate community.

Insull introduced the British engineers and the speakers and, in his own talk, set the agenda by summarizing his key ideas regarding the industry. Urging, as always, the necessity of building large central stations, he praised de Ferranti's achievement of a quarter-century earlier. From the achievements of another English guest, Arthur Wright, Insull explained that the industry had learned how to sell large amounts of energy to many customers, large and small, and to do so profitably [5]. These two issues — massing production and distribution and expanding sales — were two of the principal messages Insull constantly reiterated in his talks to the industry. De Ferranti followed Insull's chart of growth and expansion with an engineer's vision of continued technical innovation. He described electricity as a means of "conserving our natural resources" and "the greatest labor saver. . . ever. . . invented." From this fact had come his great interest "in the larger uses of electricity, . . . in what I . . . call the universal application of electricity to almost all purposes." Insull reacted diplomatically, yet characteristically, to the engineer's vision. He too favored growth, but pointed out that de Ferranti did not indicate where the capital would come from for the generating stations and the distribution systems that was needed to fill out his picture. Given this dilemma, he found it "natural for us to look for a banker" and introduced Vanderlip.

Vanderlip was overtly upset by the British engineer's expansive vision of progress. As a banker, he played a critical role in Insull's industry, whose expanding systems constantly needed fresh capital. Vanderlip and Insull were also committed to economic concentration and so shared membership in the National Civic Federation (NCF), a powerful business group founded in 1900 to defend and extend the new corporate order in America. The AIEE had connections to the corporate group beyond that of individual members. In 1910, for example, at the invitation of the NCF, the Board sent delegates to the Federation's annual conference in Washington [6]. In his remarks to the dinner gathering, however, there was a sharpness to Vanderlip's remarks as he assumed Insull's usual role in cautioning the engineers to moderate their penchant for novelty. Vanderlip bluntly qualified de Ferranti's call for continued improvements in the process of converting the latent energy of coal into electricity. Having to discard machinery after only a brief service "rather frightens the capitalist," he said, and especially so when he sees "the danger of competition coming in with newer forms of machinery, endangering old investments." The engineer's work of getting "nearer to the point of efficiency," though necessary, must retain "the confidence of capital."

Whereas Vanderlip echoed Insull's fear of excessive innovation, the final speaker, Steinmetz, spoke for the aims of both sides. Insull might have known Steinmetz would be attracted to de Ferranti's idea of universal electrification. This was implied when he introduced the GE engineer: no matter the "accumulation of talent, electrical talent, scientific talent" around the table, Insull declared, Steinmetz stood above all. As a research engineer, then, Steinmetz praised American engineering creativity for surpassing de Ferranti's achievements by moving from counting in kilowatts to hundreds of megawatts. But Steinmetz admitted that more than the "technical side of distribution of electric power . . . must be considered"; there was "the side of selling." He praised the pioneers in that field, by whose labors "we will all see . . . electrical energy running the world."

In spite of important differences between businessmen and engineering innovators like de Ferranti and Steinmetz, a fundamental belief united the men of industry who sat down at Insull's table. They believed not only in electrification but also in its realization through the central station concept. Insull had brought to his table, in short, engineering and business leaders whom he could reasonably expect to support the centralization of electric power. Not all engineers accepted the logic of his principle of massing production and distribution. When Insull read a paper on railroad electrification at the AIEE conference in Chicago in 1912, opposition came from ex-president Lewis Stillwell and other consulting engineers whose livelihood partly relied on the continued existence of isolated plants and who argued for a mixed system. Nevertheless they presented no serious obstacles to Insull's campaign [7]. For although other industry leaders could have possibly gathered as distinguished a group of AIEE engineers, Insull was uniquely able to do so. It was an unexpected result of the revolution brought to the electrical community by power engineers. With it came, after 1900, the rise of the operating companies and an opportunity for engineering representatives of the electric utilities—Dugald Jackson called them "public service corporations"—to move into leadership positions within the society.

Yet the central station fraternity never took control of the Institute. This was so even though electrical engineering was, in itself, a broad disciplinary grouping which had always included such fields as telegraphy and telephony, electrochemistry, and the newer area of wireless telegraphy. For, in spite of the isolation that the growing concentration of power engineers brought to the Institute within the discipline, the AIEE contained contending groups. The diversity of types existed within the power engineering field itself: Eminent electrical scientists, highly respected and wealthy consulting engineers, professors with national reputations, and corporate engineering researchers and designers—each of these groups represented distinctive sets of values and entertained particular notions about professional standards. Thus the central station engineers and the executive officers did not operate in a vacuum, a fact that became abundantly clear as, first, the drive for a code of ethics and, then, the constitutional issue of membership criteria roiled the Institute's waters during the ten years after 1905. In the first case, central station engineers lined up with engineers from industrial areas, such as manufacturing and transportation, to ensure that the code stamped out some of what Charles Scott called, the "chicanery" of the consulting engineers. The controversy, in short, arrayed the groups represented at Insull's dinner against a group that was conspicuously absent. In the second case, the issue of membership grades came up once more as some members launched a six-year campaign for an intermediate membership grade for executive officers and managers without engineering credentials.

Yet, in the middle of this turbulant decade, the question of professional standing arose. It infused the issues out of which the ethics code came, stood at the heart of the constitutional changes made in 1912, and was the major question in the public battle that erupted the following year, finally carrying contending groups of national electrical engineering leaders to the New York State Supreme Court. But the more permanent results of this period, during which the disciplinary reach of the Institute narrowed and its early professionalism was shaken, was the rise of an electronics branch of the discipline and the founding of a new engineering society: the Institute of Radio Engineers.

Engineers and "systematization" in the power industry

The influence of the central station fraternity within the Institute drew considerably on the power of the idea of centralization itself. For if Insull's

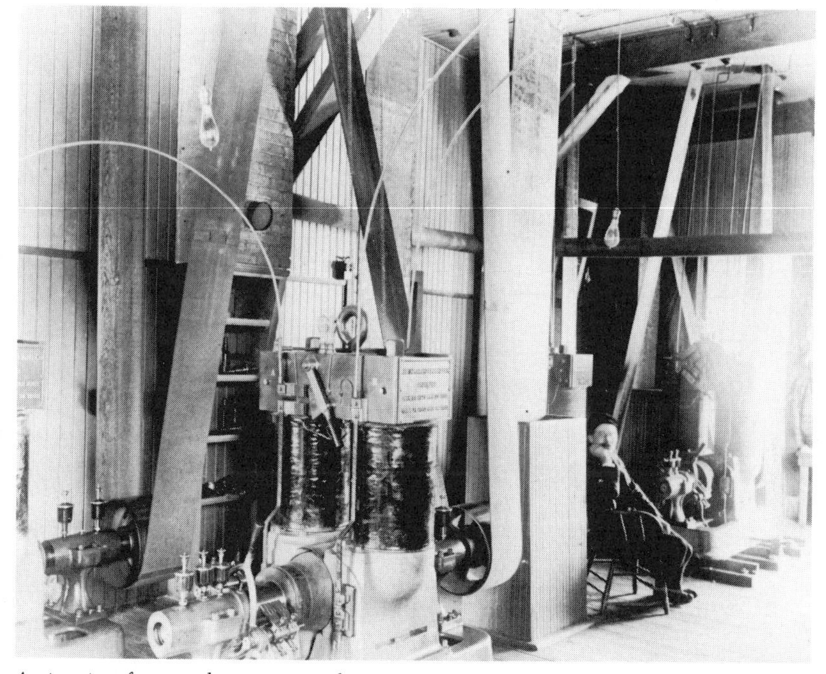

An interior of a central station around 1910: stations grew bigger according to Samuel Insull's strategy of massing the production and distribution of energy.

to indicate whether engineers belonged with "superintendents and managers" or with "clerks, stenographers, and other salaried employees" [16].

The slight presence of central station engineers within the electric power industry extended to the appreciation of engineering skills held by the industry's leading promoter. In a speech to the National Electric Light Association, in 1910, Insull reflected on the engineer's place in "twenty-five years of central-station commercial development." It had not been difficult, he said, for central station managers to hire "first-class operating assistance" or "engineers of constructive capacity" to design and build plants and distribution systems. More difficult was the problem of finding engineers with knowledge of the commercial end of the business. Insull thus advised young engineers engaged in the technical side to become familiar with the commercial aspects. If they learned how to sell the product and to extend the categories of consumers, "they will stand a chance of achieving distinction and profit far greater than most of them can achieve in the operating and purely engineering side of the business" [17].

"Commercial engineering"

Once again, Insull's ideas about engineering gained influence from their relation to broader currents in engineering culture. By 1910, his advocacy of commercial engineering reflected, to a certain extent, the wide acceptance of a new, more inclusive definition of engineering work. Increasingly, electrical engineering work included the skills of planners, managers, and administrators, as well as fields that fundamentally involved the application of engineering methods. A writer for the *New York Evening Post* called it "commercialized engineering," but electrical engineering spokesmen favored the phrase "commercial engineering" [18]. Though the growth of the phenomenon was contemporary with the rise of "scientific management," its origins lay in the dilution of the idea of engineering itself. Scientific management applied the engineering method specifically to the production process; commercial engineering, however, came out of the front office, appearing in guises as various as public relations managers and salesmen. Engineers had earlier argued that the engineer could be a businessman, now many insisted that managing a business was really engineering work.

The shifting perceptions of engineering became strikingly clear in the uses Institute leaders made of British engineer Thomas Tredgold's classic definition of engineering. Tredgold's early nineteenth-century definition was used as a touchstone as Institute engineers attempted to define their changing perception of engineering work. The contrasting use Charles Scott made of Tredgold's words in his presidential address in 1903 places in relief the sharp break from tradition that took place. Scott defined the engineer, after Tredgold, as "one skilled in the application of the materials and forces of nature to the use of man." Scott's point was technical and was meant to distinguish the electrical engineer as one dealing with "forces, with energy in its moving, kinetic form." Speaking five years later, Henry Stott too began by quoting Tredgold's definition, but he used it as a foil for a different conception of engineering, wondering if the older definition were "broad enough?" and if Tredgold's words "hold good to-day?" [19]

Engineers fell into two "classes" as Stott saw it. Some were "restricted to a specific vocation, such as electrical, steam, hydraulic, pneumatic, or sanitary engineering"; others took their "place not only as an engineer but also as a public-spirited citizen and leader." Stott lamented the growing specialization of electrical engineering that followed "the consolidations of manufacturing and other interests into a few large concerns." Industrial concentration served as "a factory" by producing narrow engineering specialities that had blinded the engineer to "the entire sphere covered by his company." Engineers could overcome their blindness, however, by broadening their field through study, experience, and the engineer's "natural adaptability for administrative work." Stott modified Tredgold's definition, therefore, to one more accurately described as "the position of the engineer to-day." Engineering had become "the art of organizing and directing men, and of controlling the forces and materials of nature for the benefit of the human race."

Up to seven years after Henry Stott's presidency of 1908, five presidents alone drew on Tredgold's assertion to promote commercial engineering within the AIEE. By 1912, Stott's modified version of Tredgold had become the original one in the mind of Institute president Gano Dunn. He began his presidential address on the new engineer with what he apparently thought was the Tredgold definition. His source, he said, was the wall of a great engineering library. He quoted instead the modified version Stott had ended his address with four years earlier. Thus, beginning where Stott had ended, Dunn concluded that engineering was so "broad and all-embracing" that 150 years hence there would be no profession "in respect to a large part of what we now call engineering" [20]. Dunn's view drastically de-emphasized the traditional technical content. Engineering became "educated common sense," serving as a handmaid to science, "doing the practical chores of life," concerned with the useful and with "costs and expediency." Electrical engineering carried the tendency to breadth the furthest. Its "forms of thought" being "so illuminating that engineering does not bound them." From concern with single-phase motors and coronas, Dunn concluded that electrical engineering had become a "way of thinking."

Dunn diluted the content of electrical engineering in spite of his own experience. His views of 1912 would have been more understandable if, in 1891, he had simply left his first job with Western Union for the Crocker-Wheeler company. But during his five years with Western Union, he had won a bachelor of science degree from the City College of New York and an electrical engineering degree from Columbia. Toward the end of the century, Dunn earned an M.S. from City College. While advancing his technical knowledge, Dunn also rose in the business world. In 1898, he moved into a managerial position at Crocker-Wheeler. He stayed as vice president and engineer until the year before his 1912 AIEE presidency, when he accepted a vice presidency with an engineering and construction firm, the J. G. White Company. Thus, after two decades of work with generators, electric motors, electromechanical controls, and systems operation, Dunn began at J. G. White to direct the design and construction of electric power systems for waterpower installations, central stations, and railroads. But Dunn was not to be fixed as engineer or executive, for he continued to complement his executive experience with technical studies. In the early 1910's, he returned to Columbia to study under mathematical physicist Michael Pupin, alongside young electrical engineers like Edwin H. Armstrong and Alfred N. Goldsmith, both of whom would play influential roles in radio engineering.

That these technical influences on Dunn gave way before the ideology of commercial engineering suggests the force and pervasiveness of that intellectual movement. Stott, therefore, had not initiated it with his 1908 revision of Tredgold. Several years earlier, in a list of the characteristics of an engineer—included in a paper to the Philadelphia section—Charles Day had enumerated, in addition to technical knowledge, such qualities as intelligence, tact, grit, and good health. Day had taken these traits from Frederick Taylor's description of the "well-rounded man" in his classic paper of 1903, "Shop Management." In reaching into the mechanical engineering field for a broader conception of the engineer, Day tapped the earliest American source of engineering as "organizing and directing men." Even Taylor, the founder of scientific management, first worked out his ideas in the 1880's while employed as a mechanical engineer with William Seller's Midvale Steel Company. Taylor's extension of the engineering field was part of a larger movement, however, one that received clear expression in 1886 when a leading mechanical engineer read a paper, which became "a kind of manifesto of the movement," to the American Society of Mechanical Engineering on "the engineer as economist" [21]. Though scientific management proper found a hearing within the electrical industry, the more general idea of the engineering approach being broadly applicable to managerial tasks won most attention within the AIEE. Since commercial engineering and its variants concerned the nature of engineering, the ideology struck especially close to the question of engineering professionalism. The signs of this influence were numerous, ranging from the number of papers given on the problems of central station management to the skein of presidencies held by power company directors and ideologues: John Lieb (1905) and Louis Ferguson (1909) are examples of company directors and Dugald Jackson (1911) of ideologue.

Yet of them all, Ferguson most explicitly stated the ideology of commercial engineering. He gave his presidential address on Insull's favorite theme, "Centralization of Power Supply" or, as Ferguson the engineer put it, "the importance of a good load-factor." His address revealed a number of purposes. Technically, he called for an end to "independently produced power" and the combining of single systems into large networks. He also advocated a political objective, urging that the power industry be extended "monopoly privileges with proper regulation" [22]. But above all, he advocated the universality of "the commercial electrical engineer." Technically minded engineers — labeled by Ferguson, the "progressive. . . electrical engineer,"— still were needed to solve problems in the field. However, commercial engineers were more representative since this group "really includes nearly all of us."

Ferguson's assertion that the purposes of the power industry best represented the profession was bold, given the protests made to his nomination. In a spirit reminiscent of his alarm at the rumored nomination of the physicist William Anthony, Cyprien Mailloux again took the lead in questioning the presidential qualifications of a candidate. In Anthony's case, he wondered if a man without practical interests was proper for the presidency; now, in the case of Ferguson, he worried about a candidate in whom commercial interests crowded out professional standards. Like Gano Dunn, Mailloux had returned to Columbia University to study with the young radio engineers and had, indeed, spent most of his sixth decade in school, first to get an undergraduate degree at Brooklyn Polytechnic, then in doctoral studies at the university.

When Dugald Jackson defended Ferguson after Mailloux had distributed a circular letter criticizing the Chicago engineering executive, Mailloux defended his sense of engineering values. Jackson described Ferguson as "a representative of American Central Station Engineering" and a gentleman. Besides, Jackson argued, Ferguson had done much to promote Institute affairs in Chicago. For his case, Mailloux proffered supporting letters from many other Institute engineers who believed something more was needed than a "good fellow . . . to be president of the A.I.E.E." If it was a matter of being president of the Association of Edison Illuminating Companies or NELA, Mailloux said, Ferguson would "grace and honor the position." Mailloux's discomfort with the elevation of commercial engineering within the Institute was shared by other influential members. When advising Jackson two years later on appointments to a committee on ethics, Steinmetz recommended that Jackson appoint members who "do engineering work, but not . . . men who have more or less left engineering to accept administrative positions" [23].

John Carty's critique of commercial engineering was even more pointed in his lengthy letter to Jackson in 1910 on "corporate organization." And a half-decade later, at the special AIEE conference on the "Status of the Engineer," Carty spoke sharply against the dilution of electrical engineering [24]. Carty vigorously defended engineering work, disdaining the use of the term "commercial engineering . . . [which] often contains no engineering whatever." He believed "efficiency engineering" to be nothing "but a meddlesome interference with factory management" and thought that such branches as "public relations engineering" were "absurdities." He looked to the "great day" when scientific and common-sense management became "plain management." At the same conference, Henry Stott joined Carty and reversed his earlier widening of the field, defending, instead, engineers who did "actual engineering." Stott now found Tredgold's nearly century-old concept of engineering as "the art of directing the great sources of power in nature" to be a "powerful definition." Carty explicitly advised the engineer to avoid encroaching upon other professions and "call things by their right names," recognizing "that within his own province there lies a career exceeded in usefulness by none of the other learned professions."

Yet the criticism of commercial engineering within the AIEE could not alter the inroads the idea had made in the way a great number of electrical engineers thought about themselves and their field. At the 1915 conference, all the speakers except Carty and Stott supported the broader definition. Two executive engineers — GE's president, Edwin Rice, and the head of Stevens Institute of Technology, Alexander C. Humphreys — looked beyond the commercial sphere, prodding the engineer to enter the political world of elected office and to seek positions on the growing number of regulatory commissions. Another speaker praised scientific management — "first developed" by an engineer — as having contributed, "to the wonderful advance in the handling of workmen" and held up Tredgold's definition as being "far too restrictive and must hereafter include men as well as materials." College president Humphreys went so far as to criticize a speaker's call for a more rigorous college education, insisting instead that the training of the engineer emphasize "the school of experience."

Taking the conference on "the status of the engineer" as a measure, even with those who demurred, the broad definition of engineering stood firmly in place by the eve of the First World War. After the war, commercial engineering would recede both as a label and as an overt ideology. But the place of engineering management remained both as courses in college departments and as a category of professional engineering. In itself, commercial engineering never roused controversies within the society. Even during the years of its ascendance, dispute over it was muted and was more often reflected in statements made in its defense. An example was when Jackson urged engineers to defend the utilities from their critics and then denied that this would "lead engineers out of a professional spirit and into commercialism" [25]. However, more than in ideological assertions, the basic question of professional standing, which was raised by the dilution of the discipline, found expression in concrete events like the move which began in 1906 by some Institute factions to adopt a code of ethics.

The ethics code of 1912: Engineers versus engineers

The members who helped write an ethics code thought from the first that, like technical standards, it would represent an amalgam of opinions from representative sources and would be achieved through consensus. So when Schuyler Wheeler wrote Dugald Jackson, in the spring of 1910, to revive the issue, he argued that "It is just as desirable to standardize in this direction as in apparatus." Jackson was to be the new AIEE president, and Wheeler wanted the Code of Ethics he had initiated in 1906 to be finally adopted. Steinmetz's assumptions were similar. He advised Jackson that the committee should represent all "sides of the industry," including consulting, operating, and designing engineers, plus a professor with a consulting practice. He also suggested that a committee of leading men be appointed to smooth the path to consensus. This was in essence what took place. Jackson appointed a representative group of engineers—with continuity provided by Wheeler, Steinmetz, and Harold W. Buck from the original committee. In 1911, Jackson appointed an eighteen-member advisory group of "men prominent in the electrical field." But an ethics code was not to be achieved in a manner akin to the standards process. Though Wheeler called for a broad-based code to improve the "moral standard or the ethics of the profession," the document was mostly a set of guidelines to regulate the business practice of consulting engineers [26].

Controlling the consultants was certainly not the whole of the code's aim. When the process was completed in March, 1912, the Board of Directors had adopted a code that it hoped would establish "a standard that shall be educational as well as admonitory." But most of what remained concerned the consulting engineers, as their reaction would confirm. For after the initial document had passed through six years of revision, sections dealing with other issues were often excised or moderated. But just as the strictures for consultants remained virtually unchanged and thus indicative of the central purposes of the code writers, it is also revealing to observe the issues that were muted or simply cut. To understand the import of the engineers' attempt to give themselves what the "older professions"—clergy, physicians, and lawyers—already had, an examination of what was discarded as well as what was kept becomes necessary [27].

The first committee—Wheeler, Steinmetz, and Buck, who had just left GE after eleven years to become chief electrical engineer for the Niagara Falls Power Company—submitted an initial version of the code in June 1907. Wheeler reported that the committee had been "most harmonious in all of its

CODE OF PRINCIPLES OF PROFESSIONAL CONDUCT

OF THE

AMERICAN INSTITUTE OF ELECTRICAL ENGINEERS

ADOPTED BY THE BOARD OF DIRECTORS, March 8, 1912.

A. General Principles.
B. The Engineer's Relations to Client or Employer.
C. Ownership of Engineering Records and Data.
D. The Engineer's Relations to the Public.
E. The Engineer's Relations to the Engineering Fraternity.
F. Amendments.

While the following principles express, generally, the engineer's relations to client, employer, the public, and the engineering fraternity, it is not presumed that they define all of the engineer's duties and obligations.

A. GENERAL PRINCIPLES

1. In all of his relations the engineer should be guided by the highest principles of honor.

2. It is the duty of the engineer to satisfy himself to the best of his ability that the enterprises with which he becomes identified are of legitimate character. If after becoming associated with an enterprise he finds it to be of questionable character, he should sever his connection with it as soon as practicable.

B. THE ENGINEER'S RELATIONS TO CLIENT OR EMPLOYER

3. The engineer should consider the protection of a client's or employer's interests his first professional obligation, and therefore should avoid every act contrary to this duty. If any other considerations, such as professional obligations or restrictions, interfere with his meeting the legitimate expectation of a client or employer, the engineer should inform him of the situation.

4. An engineer can not honorably accept compensation, financial or otherwise, from more than one interested party, without the consent of all parties. The engineer, whether consulting, designing installing or operating, must not accept commissions, directly or indirectly, from parties dealing with his client or employer.

5. An engineer called upon to decide on the use of inventions, apparatus, or anything in which he has a financial interest, should make his status in the matter clearly understood before engagement.

6. An engineer in independent practise may be employed by more than one party, when the interests of the several parties do not conflict; and it should be understood that he is not expected to devote his entire time to the work of one, but is free to carry out other engagements. A consulting

2227

After six years of constitutional maneuvering and paring sentences from a proposed "Code of Ethics," in 1912 the AIEE's leadership accepted a "Code of Principles."

conclusions," not disagreeing "over a single feature." Yet immediately upon the printing of a preliminary version in June, disagreements from other members led to rejected phrasings and deleted paragraphs, and then to a second version, the "Proposed Code" of August. This version went to the edge of adoption before, as Wheeler later wrote, it was "negatived on technical grounds." Though the main parts were compatible with ideas Wheeler had sketched in his 1906 presidential address on "Engineering Honor," his ideas had been significantly expanded. The first of five general principles made the most explicit reference to ethics, advising the engineer to pursue the same "ethical principles" in professional life as in "the social relations of every-day life." The remaining items related to matters of business practice. One warned the engineer to guard against the ill use of his name; another, "to incline toward and not away from standards of all kind, since standardization is peculiarly essential to the general progress of the profession." This rule applied to "construction" or design, measurement, and nomenclature as well as to "conduct, or ethics." A final note to design engineers advised—much as an Insull speech might—that "even the tendency to give individuality by providing special construction may usually be avoided with advantage." Though in agreement with this view, Jackson remarked in a detailed commentary that he thought standardization and standard design "a matter of business practice rather than of engineering ethics" [28]. The paragraph was removed.

The issues lying outside the realm of the consulting engineer drew the most criticism. These issues dealt with the engineer-employee, with what later generations called whistle-blowing, and with the question of professional standards. In the first instance, the August 1907 code briefly and decisively adjudicated the rights of the "engineer employed under permanent engagement." "Designs, data, records and notes" made during his employment on work issuing from his position, the code declared, "are his employer's property." This was criticized by Henry Stott, who succeeded Wheeler as president. An immigrant engineer from Scotland who had become an expert in underground cables and thus chief electrical engineer for New York City's transit company, Stott rejected the denial of the ownership of his records to the engineer "under permanent engagement." As the code gave possession to the consulting engineer, he said, "I do not see the fine point in that distinction" [29]. Nonetheless, the provision passed with little change into the adopted version.

All versions contained a whistle-blowing provision, that is, advice to the engineer in the face of dangerous irresponsibility by the client or the employer. However, the advice given changed significantly from version to version. The June 1906 code was unequivocal: "Operating engineers" were responsible for "defects in apparatus or dangerous conditions of operation." Besides bringing them to the employer's attention, the engineer was to take corrective action. If the dangers were not removed, "they should withdraw." Stott thought this advice "academic," asking, during a discussion, if there was "in this room any operating engineer who would do a thing like that? I for one would not" [30]. He believed the engineer should rather "stand by the apparatus and his employer until such time as the defects can be remedied." In the August version, then, the engineer was advised simply to bring the conditions to the attention of the employer and "insist upon the removal of the causes of danger."

The last section of the early code was removed on the eve of adoption of the 1912 code. It concerned the engineer's relations to "the Standards of his Profession." Here, the code defined and distinguished among three categories of electrical workers: "Electrician" applied to those trained for certain classes of work, including tasks such as installing electric lights and signaling systems, and operating small electric plants. "Electrical engineer" referred to engineering-school graduates or to "such men as possess an equivalent knowledge of electrical engineering." Finally, engineers could use the title "consulting electrical engineer" only by possessing "such knowledge and experience" as entitled them to full membership in the AIEE. Further changes would occur before the final code of 1912. The general principle applying the ethics of "social relations" to engineering work were excised and the whistle-blowing paragraph was revised once more to instruct the engineer, not to insist upon the removal of dangers, but to bring them "to the attention" of the employer. This was supplemented in 1912, however, by a paragraph in another section that advised the engineer not to "permit non-technical persons to overrule his engineering judgments on purely engineering grounds."

But these concerns were peripheral to the main goals of the code writers, as confirmed by the space given to the nearly unchanged sections on "the engineer's relations to client and employer" and on "the ownership of engineering records" [31]. Of twenty-two paragraphs organized under five topics in the 1912 code, three-fifths filled these two sections. Like the presidential address of 1906 that inspired it, the 1912 code concentrated on business practice, devoting only a few lines to ethical principles and two final, brief sections to the "engineer's relations to the public" and to "the engineering fraternity." At the heart of the code, then, were the principles regulating the engineer's relations to clients and employers. In these, the code specified appropriate behavior for different situations, though throughout advocated that the engineer recognize that his "first professional obligation" was to protect the "client's or employer's interests." This required a number of basic steps. He should seek the consent of his primary client before accepting compensation "from more than one interested party." He should never accept a commission from a manufacturer whose products the engineer recommends to the client for purchase. And engineers should reveal their use of inventions and apparatus in which they have an interest. Yet the code upheld the freedom of the "engineer in independent practice" to "carry out other engagements," providing that consulting engineers inform their client in the case of possible conflicts. Eight paragraphs made up the section adjudicating ownership of the records of engineering work. They established conditions under which engineers owned their records and under which they did not. Federal patent law regulated the marketing of inventions and innovations, but nothing comparable existed for special designs of apparatus, installations, and systems. Thus, the code first recommended that an agreement regarding ownership always precede the work. It specified that the client retain ownership if the information came from the client. Similarly, designs furnished to an engineer or manufacturer by a client remained the client's property. Because such information usually contributed to the engineer's experience, he too could rightfully draw on it, even as he recognized that shared ownership required that the information "be considered confidential." In the reverse case, however, when a consulting engineer used his own plans—or designs of apparatus purchased from another party—the client had no rights to the design.

Unlike the paragraphs on the consultant's ownership of records, those specifying conditions of ownership for the employee-engineer referred to the employer's rights. But there were some areas in which the employee-engineer might own the fruits of his technical creativity. The worker's engineering accomplishments "outside of the field" for which he was employed belonged to the engineer. The 1912 code, however, qualified each case with the proviso that the engineer's ownership rights existed only "in the absence of an agreement to the contrary." A separate paragraph, moreover, designated that the employee-engineer's "designs, data, records and notes" resulting from work done for the employer was in any case "his employer's property."

In spite of these provisions, consulting engineers clearly understood the central purpose of the code. This was confirmed in 1910 by the actions of AIEE presidents Lewis Stillwell and Dugald Jackson. In March, Stillwell created a Committee on Engineering Relations to consider the "relations of Consulting Engineers to each other, to their clients and to manufacturing companies." The committee was charged with deciding the feasibility of setting "standard schedules of minimum charges for professional services." This put it squarely in the scope of an ethics code, so far as the consulting engineers were concerned, since the recently organized American Society of Consulting Engineers had adopted just such a code. Stillwell's opposition to the proposed code was so well known that Wheeler and Steinmetz delayed trying to revive "the ethics question" until a new administration took office. Jackson was noncommittal that spring, when Wheeler pressed him "to appoint a good strong Committee that is favorable to the Code." The question of a code, Jackson said, was "a more or less vague thing in my mind." Nevertheless, he recalled criticisms of the proposed code for dealing more with matters of "professional etiquette . . . than professional ethics." Wheeler prodded him by sending a copy of the long, paragraph-by-paragraph commentary Jackson had prepared on the 1907 code. The earlier committee, Wheeler told him, had adopted "most, if not all" of Jackson's suggested changes. After Wheeler wrote several times more, Jackson told him he wanted to see "the progress that is made by the Committee on Engineering Relations" before acting [32].

Several months into his presidency, Jackson informed Steinmetz that he wanted a code that, in part, forbade an engineer to bid "for contracts under his specification." But Steinmetz disagreed. He did not think the consulting engineers' committee had anything to do with ethics and, besides, an ethics committee should in no way be "related or connected with the present 'committee on engineering relations.'" That committee was, "brutally expressed: 'A committee of consulting engineers to fix minimum wages and eliminate scab labor.'" Jackson changed his tactics following this rebuff by turning instead to the Board of Managers, attempting during his last meeting in June to embody his ideas in resolutions instructing a new ethics committee. Though he had apparently dropped the notion of fixing bidding practices, both the preamble and the resolution creating the committee contained a new, explicit instruction that the code enhance the "professional status" of Institute members and guide electrical engineering practice—"particularly . . . the practice of consulting electrical engineers." Jackson's new committee included all the original members, with Stott and two other engineers, and Harold Buck as chairman. By December, the committee had prepared a rough draft of a new code to submit to the advisory group [33].

In spite of Jackson's capitulation to the opinions of Steinmetz and the others, all was not well. For the consulting engineers did not let the matter pass. Of four members of the advisory group who voted against the code, three were consulting engineers — Mershon, Mailloux, and Stillwell. As Jackson, the professor-consultant, had earlier sought, they wanted a document "like the codes of . . . societies of consulting engineers . . . sufficiently rigid" to enter the constitution and "to enable penalties or expulsion to be called for in case of violation." Yet the final code did what it had mainly sought to do from the beginning. It did not seriously undertake to deal with the employee-engineer's rights to ownership of his engineering records. This position was common to engineering codes, as engineer F. H. Newell observed a decade later: Engineering codes dealt with a "few men near the head of their profession" while ignoring the "great body of younger men" [34]. Nor did the code provide protection for engineer-employees faced with the dangerous irresponsibility of an employer.

Then what was the code? It was, of course, almost entirely an attempt to regulate the behavior of consulting engineers in the marketplace [35]. There was, in fact, a number of instances in which the behavior of consulting engineers had concerned both manufacturing engineers and others. In his inaugural address of 1903, Scott had spoken of the need "to place the profession . . . above suspicion of corruption and chicanery." He called "for high standards of dignity." Similarly, Wheeler had given an elaborate example of how the consulting engineer's situation lent itself to kickbacks and rebates. Later, at Insull's engineering dinner in 1911, when the Institute was again considering the code, banker Frank Vanderlip encouraged the engineers in their efforts with the argument that a code was "as important as any technical work which you are doing — the work of satisfying the public that you are being fair, that you are giving the public a square deal and the sort of square deal that the public demands." "Capital," he added, must feel "certain of that fair treatment" [36]. The consulting engineer was being bombarded from all sides, and the ethics code made a prominent weapon. Ironically, the final battle in the Institute, during these years, pitted a lone consultant against the Board as the protector of professional standards. For the president of the Institute from August 1912 to the summer of 1913 — the precise period of the constitutional struggle — was Ralph Mershon.

"Do not include questions of professional standing"

More concretely than the thinning of the ethics code or the currency of commercial engineering, the strength of nonengineering values became manifest in the attack on the Constitution of the AIEE. The target was professional standards; their weapon two paragraphs appended to the article on membership in the revised Constitution of 1912. For almost a year, these paragraphs, called the "Special Section," provided entry to the highest membership grade of the Institute—the newly created rank of Fellow—for a rush of executive officers and managers, conspicuous among whom was Samuel Insull.

Seven years earlier, John Lieb of the New York Commonwealth Edison left no doubt that it was the brand of engineering professionalism expressed in the 1901 Constitution that disturbed him. In his presidential address of 1905, he suggested reforms in several areas of the organization, but only one took root: his call for revision of the society's four-year-old membership criteria. Lieb criticized the 1901 membership rules for relegating businessmen and executives to "Associate" status, thus placing them "permanently in an inferior grade of membership with the beginners in professional service . . . " [37].

The Board of Examiners reacted quickly, defending the 1901 criteria. The chairman complained that members were endorsing transfers to higher grades without considering the professional qualifications of the applicant. The problem was a prominent one during the first decade of the century. When Jackson tried, in 1910, to appoint *Electrical World* editor William D. Weaver to the membership committee, Weaver declined, explaining: "I am not in full sympathy with the manner in which the Institute has indiscriminately added to its membership in recent years." Jackson also had criticized similar behavior by the Secretary's Office, which expanded the Associate grade at "the expense of the professional ideals of the Institute." Though the Examiners failed to act on either of the complaints, the issue, bared in Lieb's presidential address, commanded increasing support from AIEE members. In 1909, the Board of Directors established a Committee on Intermediate Grade of Membership, which two years later became the Additional Grade Committee. In choosing to add a new senior grade of "Fellow" rather than establish the intermediate rank of "Junior Member" or "Associate Member," the Board consciously followed the example of British engineering societies, which in turn had followed the precedent set by the Royal Society. By the end of 1911, the committee had prepared a set of constitutional amendments to establish the new grade, which in May, 1912, the members overwhelmingly approved [38].

For the most part, the revisions were unexceptional. As Institute president Gano Dunn explained in his February, 1912, communication to the membership: "no increase in the quality of experience" was required for the new Fellow grade over that previously required for Members. The changes allowing entry to the Fellow grade had rather to do with admitting persons outside the "strictly professional class." Yet this applied only to persons with ten years of experience, who had done original work in electrical science and had made special contributions by invention or publication, Dunn assured the members. These provisions, as the Board explained, broadened the Fellow grade, but did not "reduce its caliber." In the Member class —now an intermediate grade— a constitutional place was made for the "executive of an electrical enterprise of large scope," who had "responsible charge of the broader features of electrical engineering." But the rank of Member still included mostly professional engineers, Dunn pointed out. They will be the same type as Fellows, though generally with less experience. The Associate rank, rather, would remain the catch-all grade, providing a place for a part of the membership that had been "essential . . . in the Institute's success," including "lawyers, bankers, business men, authors, and friends of the art" [39].

These additions effected no fundamental changes in the AIEE's membership structure. Not even the 1901 Constitution had denied a place to executives. Yet their place among the Associates had become, as Lieb said, "humiliating." In actual fact, a new place for executives who directed electrical engineering work made technical sense. A man like Insull worked closely with technical matters — technical and financial matters being especially intimate in an emerging industry. Moreover, the structural changes he brought to the industry, and even his criticisms of special design, indicated a firm grasp of the technical character and potential of the field. His participation in the field put Insull far beyond the status of friend of the art. However, the reason went beyond an individual's personal characteristics. As the Board explained in 1912, there were over 7300 members in the Institute — the AIEE was the largest of the nation's engineering societies — and only 10 percent were Members. But the 90 percent who were Associates were "very largely . . . professional electrical engineers." Thus, one aim was to move a substantial number of those who possessed extended experience into higher classes, leaving only engineers with little experience in the Associate grade. Finally, the Board thought that three grades would give an "opportunity for increasing the standard of the highest grade" [40].

No one protested these changes to the constitution, and yet, within months of its passage, protests began in earnest. It was the Special Section, a temporary measure, that churned the generally calm surface of the Institute. These paragraphs allowed all persons who belonged to the Institute when the amendment was adopted in May, 1912, and who applied by the next May, the privilege of transferring outside the normal channels. Instead of going through the Examiners, applications would go directly to the secretary's office. This was special treatment indeed, so to provide a safeguard, transfers to the Fellow grade would require the names of "five Fellows or Members, who, upon inquiry, shall certify that he meets the requirements" defined in the constitution. For transfer to Member grade, four signatures were needed. Names then went to the Board of Directors for approval or disapproval in the normal way. The constitution had not been corrupted; instead, the Special Section provided an easy, though temporary, entrance for the commercial engineer to attain the highest professional rank. Its effectiveness relied upon the increasing looseness by which members regarded professional standards. It seemed, at first, that there would be no problems in getting the constitution adopted, even with the Special Section. The amendments went from the committee through the Board and then to the membership within six months. Indeed, 92 percent of the 44 percent who voted favored the amendments. Admittedly, they had been encouraged by President Dunn's unqualified recommendation of the amendments as "something that will add to the Institute's strength, prestige and usefulness." However, in August, when the secretary delivered the first list of Special Section applicants to the Board of Directors, protests came at once from among the directors.

This was the first meeting for the new president, Ralph Mershon. In the twenty years since he left Ohio State and entered Westinghouse with Scott and Lamme, he had become a wealthy electrical consultant with an international reputation and clientele. He had distinguished himself with his investigations of high-voltage phenomena and would join in the attempts to develop submarine detection devices with radio during the world war. But in 1912, Mershon had attained the presidency after a decade of service as director and vice president of the Institute. In that year also, he found himself at the head of the group of engineers who dissented from the Board of Directors majority interpretation of the special section. The first sign of conflict followed the secretary's presentation of the list of more than one hundred and fifty names seeking transfer to Fellow and Member grades. Immediately, "the question arose" about the propriety of approving applicants without having the "accompanying certifications" approved by the Examiners as "statements of fact." The constitutional provision had not explicitly excluded the Examiners, it simply specified that the secretary would certify the applications. For assistance in resolving the issue, the Board of Directors asked for an opinion from the Law Committee at the next meeting. Meanwhile, the Examiners had called a meeting to determine their position on the issue. This action froze the status of the first Special Section applications—including eighty-nine applicants for Fellow. Serious conflict had so far been avoided. No votes had been taken, and positions had not yet hardened since only a clarification had been requested. In October, however, both the Law Committee and the Examiners agreed, in their reports, that the Constitution required the Examiners to pass on the applicants. The Examiners had also adopted a resolution requesting the Board of Directors to refer pending applications to the Examiners "as required by Section 53 of the Constitution." But a minority report had also issued from the Law Committee that ruled out the need to go through the Examiners, a position concurred with by the Institute's legal counsel. The majority of the Board of Directors took these latter opinions as their guide. The original motion to allow the transfers in grade was adopted, with only the president and two directors voting against it. Mershon opposed the policy of relying solely on the secretary's judgment of a candidate's legitimacy: "So important a matter," he told them, "should [not] be left to any one man" [41].

In November, a compromise was attempted. After rejecting one examiner's motion to return the matter to a vote of the membership, a resolution was offered to alter the Special Section procedures to allow participation by the Examiners, but only for applications received "in the future." The first list would stand. New applications would be inspected by the Examiners, however, and, if one were found inconsistent with the constitutional requirements for transfer, Examiners would return it to the certifiers, "asking if they persist." If so, the matter went to the Examiners "for action." This "means of escape from a very serious situation," as Mershon described the resolution, was laid over until the next meeting with instructions to Dunn and another director to revise it for reconsideration.

So far, matters had been discussed in procedural terms; when Dunn returned in December with a distinctly different resolution, the issue had become substantive —revolving around issues of professionalism— and political. The question of professional standards, in short, would be decided on the basis of the power held by contending groups within the society. In Dunn's new version, the Examiners could inspect the applications, but with two restrictions. The first was the familiar one: only applications "hereafter received" would go to the Examiners. The second qualification, however, moved into distinctly new territory: allowing the Examiners to pass on all new applications yet forbidding them from considering "questions of professional standing." Dunn's resolution veered sharply not only from the earlier version, but from statements he had made as president in the spring. Even he was concerned. Omitting professional standards from consideration, Dunn told the Board of Directors, violated the position of his February letter to the membership, which had been published with the proposed amendments. Nor did it agree with presentations he had made to various section meetings. Thus, Dunn took the rare step of asking that the minutes indicate the reasons for his reversal. He had reversed himself, Dunn explained, because not to do so could cause "serious dissension and check . . . the Institute's progress and development" [42].

But Dunn had misjudged. Already from among those few who knew the circumstances rose men who disapproved. Just minutes later, when Dunn moved for approval of the initial list of applicants, five directors who had applied for transfer to Fellow under the Special Section asked that their names be removed from the list. In a manner that would become repetitive, the Board's action at one monthly meeting led to counteraction at a later meeting. The Examiners refused to give their assent to the December resolution that ordered the omission of questions of professional standing, asking, instead, that the directors resubmit the names without the "limitations imposed." Thus, rebuffed by the Examiners, the Board rescinded the resolutions that were passed at both the November and December meetings and set up a "special committee" of three members to receive the names of applicants processed by the secretary. But the Examiners were again specifically ordered to disregard questions of the professional standing of applicants. When Cary Hutchinson and George Gibbs resigned from the Examiners, Comfort Adams and W. S. Rugg, both members of the special committee, were named in their place. In January, the Board revised the minutes of November and December without dissent, so as to eliminate "matter not relating directly to motions voted upon." However, Mershon was absent, and so sought at the March meeting to have the original minutes restored. Nevertheless, a resolution to keep the revised minutes passed with Mershon's lone dissenting vote. When the special committee reported a new list of applicants for transfer under the Special Section, he voted no again, as he would on all lists of Special Section applicants until his last meeting as president in June [43].

Yet, as early as January, Mershon's lonely dissenting position had become considerably less isolated as complaints began to come in from members not on the Board. One group of members asked that an "eminent counsel" be engaged in view of the differences in the legal advice given to the Examiners and to the Board. This request came from Cary Hutchinson, still chairman of the Examiners, and from engineers like Frank Sprague, Louis Duncan,

Schuyler Wheeler, Francis Crocker, and Michael Pupin. Most had, the year before, sat on the Jury of Eminent Members, who passed on the Code of Ethics. But the Board regarded it as improper "to delegate in advance any of its powers" and rejected the notion. Letters to the Board, in February, added new names to those requesting impartial legal counsel. However, Elihu Thomson, Lewis Stillwell, and Charles Steinmetz now asked that the issue be submitted to the Appelate Division of the New York State Supreme Court for an interpretation. Again the Board refused [44].

There were other forms of dissent. Besides Mershon's "no" votes, some of the earliest protests came from members requesting that their names be removed from the lists of applicants for transfer under the Special Section. Harris Ryan was among the first. He had entered the profession and the society during the 1880's, with the rising tide of engineering science, and was now a professor at Stanford University in Palo Alto, California; he was also a leading researcher in high-power transmission. A. H. Babcock, of San Francisco, who was on the original list of applicants, asked that his transfer be nullified. Another San Franciscan, A. M. Hunt, apparently began this form of protest in February when he publicly criticized the first published list of Special Section applicants. He noticed that one name, Henry Stott, sat off from the names listed for transfer. Hunt thought that the decision of "a past president of the Institute, and a man of the highest standing," to be transferred through the "regular course of procedure," suggested that, although references might take the place of the Examiners, "they cannot be considered as equivalents." Hunt did not mention any of the other names, besides Stott, among the new Fellows which led him to question the procedure. Yet he could have learned much from the positions held by many of the new Fellows: from the vice president and general manager of the Niagara Falls Power Company to the assistant-to-the-vice president of Westinghouse. More simply, however, Hunt might have been reacting to the single name of Samuel Insull, who was listed as the president of the Commonwealth Edison Company of Chicago [45]. That Insull held the membership grade of Fellow in great esteem was demonstrated when he published his first book of speeches and papers. On the title page, he gave his rank in several organizations: including Member status in England's Institution of Electrical Engineers, the Franklin Institute, and past president of NELA and the Association of Edison Illuminating Companies. However, first on the list was "Fellow of the American Institute of Electrical Engineers."

The increasing number of members who shared Hunt's sense that something was wrong at headquarters did not have long to wait for a public record of what had been taking place. In the next month's *Proceedings,* Mershon supplied a full accounting of what had transpired since the previous August. His letter alone covered seven pages in the *Proceedings* and included the most significant of the Board's resolutions. He appended nine pages of documents, including letters from men who had served on the committee that, in 1912, prepared the proposed constitutional amendments. Mershon's chief criticism of the Board's actions was that by the judgments of even the Institute's lawyers, the Board retained "full power to examine into the eligibility of applicants . . . [as] if the Special Section did not exist." Since the directors "have committed themselves not to do this," Mershon believed it would "be possible for a man to be transferred to a higher grade . . . whether he has the least claim to such transfer or not . . ." [46].

The great irony of Mershon's position was that, as chairman, he was formally responsible for the Board's actions. Thus, as he arrived to chair the April meeting, he was served with a "summons and complaint in an action against the Institute brought by Louis Duncan, Francis Crocker, and Michael Pupin, relative to transfers under the special section of the Constitution" of the AIEE. As Board members arrived for the meeting, they, too, received summons. The case effectively ended in June, though an appeal by the plaintiffs carried it into September before the plaintiffs withdrew it. The State Supreme Court judge who heard the case denied the injunction, seeking to enjoin the AIEE Board of Directors from transferring members under the Special Section. He ruled that the Special Section was:

> a convenient, practical and temporary method whereby old members of the society who are eligible for transfer to the grades of Fellow and Member may quickly assume their new title and standing without being compelled to go through the routine provided for in the other sections of the constitution [47].

The judge's view was the one advanced by the Board, and it was a convincing one. Yet Lewis Stillwell, who helped write the Special Section, had recalled, in a letter to Mershon, that although the Special Section sought to lighten the work of the Examiners, the drafters had wanted to ensure "that no man should be transferred" who was not "eligible." John J. Carty had wanted to require that endorsers of applicants "certify to the facts." Besides, as *Electrical World* editor William Weaver wrote, not even the dissenters intended to deny the new grade to those "men who at one time were leaders in their profession" but who, now, might not qualify "under conditions which arose after they had made their careers." Even Michael Faraday would have "only an emeritus status in electrical circles were he to receive a reincarnation in the year 1913" [48].

Indeed, the problem was not the old members. There was another class of members under the old constitution that men like Mershon, Stillwell, Carty, and the petitioners wished to exclude from the new Fellow grade. To them, adding a new, higher grade provided an opportunity to undo the degrading of the membership criteria, which had been going on for some years. The dispute over the Special Section applications, Weaver explained, could "in almost every case be traced to the advancement . . . of members who, under a strict construction of past membership requirements, would not have received promotion to the former senior grade." Though this observation applied to a great number of the new Fellows, Insull, alone, illustrates the case. Joining as an Associate on December 7, 1886, he had remained so until 1912, when he skirted the Member grade to become a Fellow of the AIEE. The lawyers for the plaintiffs had presented arguments similar to Weaver's. But the judge accepted the argument of the Board's counsel: that the Special Section was simply an entry point for older members. At this point, the opposing lawyers could have, first, pointed to the list of applicants accepted into the Fellow grade and, then, to the member's *Handbook of 1912* for prior grades of the new transfers. Yet, they did not; so the Special Section transfers remained in the professional grades, and the judge's ruling stood. As it became clear that this would be so, Weaver advised that the dissenters now leave it to "the scythe of time" to weed out those members who "have taken advantage of an easy promotion they do not deserve" [49].

The problem was, of course, not with just one man. The AIEE's crisis of professionalism rested on economic and social changes that had deeply altered engineering culture in America and, thus, within the Institute during the decade after Charles Scott's presidency. The Special Section owed as much to these larger forces as to the direct influence of power company engineers and executives within the AIEE. It was this combination of cultural change and specific influence that moved the AIEE further from the values of engineering science and toward those of commercial engineering. Because the concentration of power engineering interests had narrowed the spectrum of technical fields within the society, the dissenting engineers could not hope for support from the exponents of the new engineering fields in their defense of engineering professionalism. Without these numbers, the constitutional defense of professional standards in 1912 and 1913 was lost. For it left a mere handful of engineers lined up against the great majority of the AIEE's more than 7000 members, a majority that had apparently grown indifferent to the engineering values of 1901.

The AIEE and the diversity of electrical engineering

Although the Special Section expired in May, 1913, the attack upon professional standards affected the future of the AIEE. In part, the Institute had always retained a touch of the trade association so prominent in the beginning. Even so, the Special Section was a temporary measure, and the executive engineer who rose to the top of the society would generally do so on the strength of engineering achievement, as well as of administrative accomplishment. During these years, in fact, the AIEE was marked more permanently—and more deeply—by the disciplinary narrowing that came from the dominant position of the power engineering field. The damage this caused was not to the integrity of individual members but to the diversity that had been the Institute's strength and, thus, to its early position as a society for the full spectrum of professional electrical engineers.

Such narrowing had not affected the field of power engineering itself. As Michael Pupin's toast to "Electrical Engineering" at the 1912 annual banquet in Boston suggested, the field was tremendously vital [50]. Pupin remembered the time when attempts were made to contain electrical engineering not just within the bounds of power engineering but also within the confines of direct current power alone! Only "an ignoramus like myself," Pupin recalled, fresh "from a German University with a doctor's degree in his pocket," would have delivered a paper before the AIEE in 1889 on the practical aspects of alternating current theory. He advocated that the a.c. system should be "immediately adopted" and was as quickly damned for "electrical heresy." Had not the president of the Institute, Elihu Thomson, approached him and told him, "That is a splendid paper that you read," then sat beside him, Pupin feared he would have been crushed by the response. Such an outcome seemed imminent when a member of the audience said, "Dr. Pupin, my alternator runs 1000 volts and the ammeter tells me that it delivers fifteen amperes, and yet the belt is slack, nothing doing, how do you explain that?"

Pupin was at a loss until Thomson leaned over and whispered, "'Big angle of lag.' I caught on, quick as lightning, and I said, 'Look here, Mr. Jones, you did not mention what your phase angle was.' He blushed with shame, and I blushed with pride. . . . That was the beginning of my career."

From such members of the AIEE, Pupin said—and from members like Tesla, Thomson, Steinmetz, Kennelly, and Stanley—had come the "great progress" made since he had innocently intruded into the battle of the systems. He was impressed by power transmissions of 150,000 volts. And "who would have expected, even fifteen years ago, that today we would obtain light and distribute power by means of vacuum tubes constructed very much like the old Crookes' tubes with mercury electrodes?" There were many things, which

> to produce . . . electrical engineers had to dive into the theory of the black body, into selective radiation and absorption, and into many other deeply scientific things, which fifteen or twenty years ago were considered to be entirely outside of the province of electrical engineering.

Yet this remained a problem for the Institute: deciding what lay within and what lay outside the province of electrical engineering. Pupin's toast made the point. Though gratified to be asked to speak on electrical engineering, it was "a matter of fact" that "I know very little about electrical engineering, not even enough to be a professor of electrical engineering. I teach mathematical physics." This disclaimer came from a man who, after early work on a.c. theory, went on to become professor of electromechanics at Columbia. In 1902, he sold a patent for electrical tuning circuits to the Marconi Company and provided a mathematical solution to the problem of electrical transmission over telephone wires. Out of this latter work came his most significant contribution. He devised induction coils to be placed at specific points along the lines to effect long-distance telephonic transmission. His disclaimer, as it turned out, made sense only at that time in the life of the society, for in 1925, he would become president of the AIEE.

Pupin's perception came thus not from the narrowing of a discipline but from the narrowing of an engineering society. The situation had been explicitly recognized in 1909 when the Board of Directors took steps to widen "the scope and influence of the Institute," and so to prevent the further splintering of the profession. Recently, the American Electrochemical Society had organized and had attracted to its leadership men like Carl Hering, Institute president in 1900, and Willis R. Whitney, who directed GE's Research Laboratory. And already in 1909, radio engineering societies had been initiated in Boston and in New York [51].

The Board turned to the technical committees as a way to keep the profession intact. In March, the Directors appointed a committee of five to study

> the question of the development of the Institute along the lines of special interests, as illustrated by the Transmission Committee, Industrial Power Committee, and others, so as to extend special activities to such an extent that the necessity for the formation of special associations outside of the Institute will be removed. . . .

The difficulties of the AIEE situation appeared in the composition of the committee: four and possibly five members were power engineers. Nonetheless, the committee's final report, at the end of the year, led the Board to make the Telegraphy and Telephony Committee independent of the Meetings and Papers Committee and to establish an Electro-Chemical Committee. The roles of other Institute committees were also seen differently in response to the Institute's loss of diversity. Thus in 1910, the Committee on Membership dropped the words "Increase of" in its title, since it was becoming a committee on "solidarity of membership." In August, the new president, Dugald Jackson, instructed the committee's members "to see what can be done for the assimilation of our large and somewhat loosely associated membership" [52]. But the technical committees, as the leadership had hoped, would serve the Institute most in containing the centrifugal forces fragmenting the profession. Within the next few years, committees were established on topics as wide ranging as Electricity in Mines, Marine Work, Electrically Propelled Vehicles, and Records and Appraisals of Properties.

In 1911, GE engineer David Rushmore joined with Pupin to try to establish a committee on electrical science. Rushmore urged Jackson to organize a committee to "handle the subjects of theoretical electricity . . . in which line but little is being done at present." He thought the subject of "the very greatest importance." Rapid progress in the field required "that the engineering profession be kept in close touch with these developments, which is not now the case." Rushmore's urgency came from Pupin's remarks at a recent meeting of the society. The Columbia professor had reminded his audience of the breakthroughs in electrical science that the Institute had regularly recognized in the past. When Hertz "made his great discovery of the electrical waves," Rowland spoke to the society on "the Hertzian discovery." William Thomson had lectured on high potential discharges, and Tesla spoke on high-frequency electrical discharges. "I think it is high time," Pupin said, "for the American Institute of Electrical Engineers to get some men to give one, two or three lectures on the electron theory — not the theory so much as the phenomena which forms the physical basis of the theory." Rushmore's request for a committee on electrical science went to Jackson and to Board members. Nothing came of his attempts at the time, however, and his motions were tabled without action [53].

Later that year, the Board once again considered organizing a technical committee on science when it acknowledged the "great practical importance of engineering experimental research in subjects relating to high-tension transmission, electrochemistry, the behavior of insulating materials, and electric illumination." Concluding that these areas drew significantly from the field of electrophysics, the Board established a Committee on Electrophysics, with Bureau of Standards physicist E. B. Rosa as chairman. The committee included some of the society's most distinguished engineering scientists: Pupin, Ryan, Steinmetz, and Samuel Sheldon of the Brooklyn Polytechnic Institute [54].

And yet in August, 1912, in the month Pupin declined to be counted as an electrical engineer, a resolution was moved to create a committee in the most critical area of all. Introduced by President Mershon, the resolution sought discussion on the possibility of establishing a committee on radio transmission, "with the end in view of bringing radio engineers into the Institute's membership ranks." A related resolution sought "a course of three or four lectures on radio activity." Mershon offered to raise the funds for this from sources outside the Institute. The Board favored the resolutions, passing them on to the Meetings and Papers Committee for recommendations. In November, the committee unanimously approved a Radio Transmission Committee and suggested that it have ten members with the chairman sitting on the Meetings and Papers Committee, as did the chairmen of other technical committees. But it was too little, too late. Mershon was unable to find a chairman for the committee, and the motion for a radio committee was tabled until such time that the president might find feasible. That time was not to come, both because of the absence of an appropriate chairman and because the battle over the Special Section had begun in earnest, leaving Mershon and the Board with neither the time nor the spirit to continue [55]. In any case, the question for the AIEE was already moot, since earlier in 1912, a group of radio engineers had met in New York and organized the Institute of Radio Engineers (IRE).

The birth of the Institute of Radio Engineers

In his presidential address in June, 1913, Mershon advised that the proliferation of engineering societies was not all bad. Multiple societies, he argued, were more economical since the duplication of dues helped pay for the duplication of services. Yet as Mershon had been in the fall, other AIEE spokesmen were concerned by what seemed the defection of the radio engineers. Cyprien Mailloux made the strongest public assertion of regret in his presidential address before the AIEE in 1914. Elected president after nearly thirty years of membership and active service in the AIEE—only illness prevented him from reading a paper at the first convention in 1884—Mailloux spoke from an intimate knowledge of the older society. However, he had become familiar also with radio science. In his studies under Pupin at Columbia, he had come to know some of the founders of the IRE. Unlike

Mershon's recent comments, he believed that the society need suffer "no loss of cohesion or of solidarity as the result of increased specialization." Mailloux wanted all electrical technologists in the AIEE, because "in union there is strength" [56]. He allowed that a separate organization for the electrochemists may have been warranted. But in a comment, apparently aimed at the radio engineers, he insisted that any group that contemplated forming a separate society should have first "tried to utilize other facilities, . . . especially, those which the Institute could offer them." In this respect, the technical committees represented "an honest and earnest effort . . . to secure diversity in technical work" and to give scope to diverse interests within the Institute.

In the middle of this call for diversity and consolidation, however, there appeared that prime symbol of the power industry: the central station. While arguing for technical committees to "foster diversity," Mailloux turned to the "fundamental principles of efficiency and economy," contained in the notion of "diversity" and "load" factors in central station operation. Thus, to talk about the idea of diversity and the Institute's dilemma, Mailloux could only draw upon the image of the central station itself:

> The Institute is . . . a kind of central station for the generation and distribution of a certain kind of electric "power" which is useful in the production of electrical "work" of greatly diversified character and of extreme importance, as a whole, to the vocation of the electrical engineer and to the standing and the advance of the electrical engineering profession.

In spite of the single-mindedness that invaded the AIEE after 1900, there had always been a place for the engineer interested in radio technology, of what would later be called electronics. During the 1890's alone, besides Pupin's provocative paper published in 1890 as "Practical Aspects of A-C Theory," there was: Thomson's paper on a.c. induction; Steinmetz's papers on hysteresis; Kennelly's papers on "inductance," "magnetic reluctance," and on "impedance"; as well as a number of other papers by electrical engineers "who showed no hesitation in treating electric power and communications transmission problems as two group manifestations subject to the same laws." From another source came the Institute's first paper by Edwin Houston on the "Edison effect," and others before 1905, that discussed electron theory, X-rays, and the flow of electrons in a vacuum. In 1907, Samuel Sheldon devoted his presidential address to "The Properties of Electrons" [57].

Other, more direct, contributions to the growing radio field appeared in these years. Though the application of Ambrose Fleming's diode tube of 1904 was not reflected in the AIEE papers or activities, the importance of the new field was recognized even before Fleming's work. Recognition came in topical discussions on wireless telegraphy, held in 1897 and 1899, and in a dinner given for Marconi in 1902, at which Pupin gave a stirring address. Five years later, an American founder of the radio field, Lee deForest, published in the *Transactions* his first paper on "The Audion—A New Receiver for Wireless Telegraphy." During the several years after 1907, deForest participated actively in Institute discussions on radio matters and, in 1909, served on the Telegraphy and Telephony Committee. The year after deForest's article on the Audion, the American radio engineer who was working most persistently to achieve continuous voice transmission, Reginald Fessenden, delivered a paper to the AIEE on "Wireless Telephony." Fessenden, an entrepreneur as well as an inventor in the field, suggested how far the technology had moved toward becoming an industry in the United States, when he ended his long paper on the history and present state of the radio field with an angry account of "how wireless telegraphy has been throttled by governmental action." And yet, none of this belied the truth of an IRE founder's answer to the question of why the radio engineers started a separate engineering society. John V. L. Hogan's answer was simple: "The radio men . . . were not satisfied with the idea of perhaps one or two radio papers per year, sandwiched in between meetings devoted to what the Germans call 'heavy-current' electrical engineering" [58].

This then was the reason for establishing an engineering society in a field that had entered its own takeoff decade. Like power engineering in 1884, by 1912, radio engineering had entered the decade in which both radio technology and the industry would be consolidated. To the radio engineers, the growth of the field led to the desire for a place in which their particular interests would receive adequate attention. For the men in wireless telegraphy, in fact, this time of congregation had begun to be felt even earlier. In 1907, John Stone Stone organized the Society of Wireless Telegraph Engineers in Boston. At first, membership was apparently restricted to employees of the Stone Wireless Telegraph Company, but soon included engineers from Fessenden's Boston-based National Electric Signaling Company. When deForest succeeded Stone as president in 1909, the Society looked even more like a formal engineering society where engineers could explore problems of mutual interest [59].

That same year, a society with a broader scope began in New York City. The Wireless Institute's principal founder, Robert H. Marriott, had distributed a circular in 1908 urging an organization on the basis that wireless "would be developed faster if those engaged in it would work together more." The recent move of Fessenden's firm from Boston to New York supported Marriott's efforts, as did the demise of Stone's company in 1910. Chiefly responsible for the success of the Wireless Institute, however, was the broad professional basis on which it existed. To achieve such maturity at birth, as explained in the circular letter, Marriott generously borrowed from the AIEE. The logic of the step appeared obvious to Marriott: The AIEE "has helped to make better Electrical Engineering, better Electrical Engineers, and better feeling between competitive firms." Thus he asked the radio engineers, "Why should not we form the Institute of Wireless Engineers and pattern it after the American Institute of Electrical Engineers" [60].

This he did. With sixty responses to his letter, in January, 1909, Marriott founded his radio engineering society on a solid basis. Taking on the same membership structure as existed in the AIEE, the Wireless Institute admitted to the Associate level all persons "interested in Wireless" and to full membership persons "having done valuable, original work." A Meetings and Papers Program was established, along with a full complement of administrative committees; a library and journal were also planned. Alfred Goldsmith, a young student of Pupin's at Columbia who had been a laboratory partner to Mailloux, assumed the editorship of the *Proceedings.* However, within a few years, it became clear that the Wireless Institute was doing no better than the Boston society. And so representatives of the two groups met in rooms at Columbia University on May 13, 1912, to found the Institute of Radio Engineers. Marriott was elected president with the other officers drawn from the two groups. Goldsmith was chosen to edit the radio society's *Proceedings.* Managers included Lloyd Espenshied, John H. Hammond, Jr., Hogan, and Stone. During the first year, a Standardization Committee was established, which published its first report in September, 1913, the year in which their first *Proceedings* appeared [61].

The IRE's attentiveness to the AIEE appeared especially during the early years. After adopting Associate and Member grades in 1912, two years later, the IRE followed the AIEE's example and added a Fellow grade. The new Institute even adopted a special mechanism to carry out transfers from Member to Fellow that followed the constitutional changes of 1915. Yet the Board chose an easier route of simply allowing members to automatically transfer without reapplying, naming a committee of David Sarnoff, of American Marconi Wireless Telegraph Company; Emil J. Simon; and Lloyd Espenschied, a Bell engineer, to compile a list of eligible members. More substantive similarities existed in the understanding that members brought to defining membership criteria and stating the purposes of membership. In 1914, secretary-treasurer of the Washington, D.C., section, C. T. Pannill, who worked for the Baltimore office of American Marconi, wrote the vice president of the company, E. J. Nally, to join the IRE. Though Nally's offices were in New York, Pannill advised him to join the Washington section because it was "composed largely of representatives of the government with whom we do business." The opportunity to discuss matters informally would "keep up the friendly feeling between the government and our company by a little closer contact with these representatives" [62].

In other cases, influential members worked to bring in new members without following constitutional procedures. Thus when Alfred Goldsmith, a professor at City College of New York, proposed Gano Dunn for direct entry to the Fellow grade, the board decided that "some information . . . would be required," as if Dunn himself had sought election. But Goldsmith argued that too much paperwork would be necessary, including duplicating abstracts of papers "which could be found in 'Science Abstracts.'" Though no action was taken at that meeting, the following month Dunn entered alongside Pupin into the IRE as a Fellow. The Institute president in 1915, Louis W. Austin, went even further than Goldsmith in laying aside procedures when he asked that the Board allow transfers from Associate to Fellow without application—though he thought Associates who were not "well known" should have to apply. He reasoned that, in the other cases, foregoing the need for an appli-

An IRE banquet at Luchow's Restaurant in New York, 1915; a banquet during the early years gathered a significant portion of the few hundred members of the society.

cation was "more dignified than that of expecting our men of higher standing to apply for the privilege" [63].

Such pressures were common to all professional societies, and the IRE was no exception. The radio society's uniqueness came rather in the positive emphasis put on technical values. This attitude appeared when Lee deForest was named to a committee on radio equipment on steam vessels. When it was realized that deForest would be the only Associate on the committee, a Member moved to transfer him to full membership. Goldsmith, who did not find deForest personally acceptable, would vote for deForest because of the definition of "good professional standing" recently advanced by President Austin. He would vote for deForest, in short, with the understanding that professional standing in the Institute referred to "a high scientific and technical standing" [64].

Clearly, the IRE had not simply spun off from the AIEE. It had risen from an engineering field that, though close intellectually and institutionally to power engineering, possessed a distinctive social and technical basis. As a result, the IRE contained many men, like Stone, who had never joined the AIEE, who had simply found in it too little to sustain and feed their technical interests and industrial concerns. However, similarities between the AIEE and the IRE still existed. Like the telegraph operators of the early electrical community, the IRE had members ranging from young radio operators trained in the field and possessing some night school education in electrical matters, like David Sarnoff, to trained engineering scientists, like deForest and Goldsmith. The two societies also shared some members, not only men like the recent AIEE participant, deForest, but stalwarts of the older society like Arthur Kennelly, who was a charter member of the IRE and served on its first Standards Committee. As suggested by the wide-ranging membership, the new Institute rose from conditions strikingly similar to the electrical world of the 1880's. Technically, the field was in a state of becoming, with continuous voice transmission and vacuum tube technology no more developed than were centralized electric power systems and alternating current technology in 1884. The radio industry similarly resembled the early lighting and power field with its many small firms working with a raw and embryonic technology and struggling to survive in a competitive atmosphere.

Had the field risen in the 1890's, alongside the power industry, it is conceivable that a structure would have been worked out in the AIEE to contain the technical interests of men like Charles Scott and John Stone Stone or Charles Steinmetz and Reginald Fessenden. But the situation at the end of the first decade of the new century presented a radically different order in power engineering and in the AIEE. The period of startling innovations that followed the electric power field into the twentieth century was, as Arthur Kennelly's periodization of standards work indicated, if not over by 1912, nearly so. Kennelly had seen the period of "applied science, engineering, and technology" replace the concern with units and measurement by the end of the nineteenth century. This in turn would, by the end of the 1910's, be pushed aside by work relating to "production and manufacture." Yet this latter work came to the fore in the power field before the world war. The radio industry, on the other hand, was, at this time, fully involved in questions of units and measurement — in questions, that is, of wave length — and with scientific and technical questions that were prior to matters of extensive application. Out of this environment came the young organization's commitment to honor, above all, the supreme engineering value of "valuable, original work."

It was this emphasis, then, that distinguished the Institute of Radio Engineers. Except for the problems over the constitutional changes of 1915 and in cases where basic policy still needed to be clarified, the early organizational history of the IRE appeared mild besides the controversies that had stirred the AIEE. On the contrary, the energies of the radio engineers would seldom stray from the technical path. More than to constitutional matters, their early activities were tied to fundamental changes within the radio industry and to changes spurred by war and national concerns. Even when the Institute strayed into the political economic arena, as it did during the war years, its interests with rare exception related to the engineering aspects of technical standardization or, increasingly during the years between the world wars, to the technical complexities of governmental regulation and the building of a national and international radio broadcasting system.

5/THE RADIO ERA: THE EXPANDING CONTEXT OF ELECTRICAL ENGINEERING

Time was when radio was simply a form of telegraphy and telephony without wires. In the past few years, however, its by-product developments have mushroomed out to an extent that those pioneers who first worked in this field of electrical engineering can now hardly believe their recollections of the simplicity of its beginnings.

Frederick E. Terman, 1941 [1]

The engineer and the radio phenomenon

Writing in 1933, in an early volume of *Electronics,* radio pioneer John V. L. Hogan labeled the ten years before 1913 a "Lost Decade" in the radio art. During that period, he explained, the "experts" had ignored Reginald Fessenden's work in continuous-wave transmission and reception. Instead, Hogan believed, Fessenden's historic 1906 Christmas Eve transmission of a program of music and speech should have left him inundated with monetary and professional support. However, there was logic in the oversight. Fessenden worked with a transmitting device — the alternator — and a receiving instrument — the "electrolytic detector" — whose days were numbered. Though the alternator was a standard piece of apparatus until well into the twenties, attention rightly came to radio only when similar experiments were undertaken with that progenitor of electronics, the vacuum tube.

The radio era began, therefore, after 1913, when the tube experiments of a man who was as young as the technology he utilized began to come to fruition. By 1930, Edwin Howard Armstrong's work had helped both to revolutionize radio technology and, as historian Hugh G. J. Aitken writes, to establish "an industry fundamentally different in structure and function from the industry of 1900–1914" [2]. Though Armstrong was only one of the inventive engineers who helped bring radio into existence during these years, the career of this outstanding individual contained the basic ingredients whose mixture resulted in modern radio technology. During the half-dozen years after 1913, when Armstrong's investigations of the medium won increasing recognition, he not only remained on the inventive edge of the field but also became involved with the organizations that gathered around the embryonic technology — business, government, military, and professional society. Of course, other scientists and engineers also helped advance radio, and they, too, worked amid these varied interests. So, to a large extent, it can be said that Armstrong's experience was theirs as well.

Though the idea of a formative decade is an abstract one, Armstrong did not experience it as such. After he had spent three years investigating the properties of deForest's three-element vacuum tube, in December, 1913, agents of the American Marconi Wireless Telegraph Company arrived at Michael Pupin's laboratory at Columbia University to examine his achievements. A recent graduate of Columbia, Armstrong had that same month filed for a second patent on his feedback circuit, having filed for an initial one in October. Among the Marconi group drawn to the "Radio Laboratory" was David Sarnoff. Like Armstrong, still in his early twenties, the young radio operator understood that the inventor's regenerative circuit promised to be a fundamental contribution in the making of a continuous-wave transmitting system. This rested on what was increasingly called radio and, as Sarnoff well knew, on the opening of new markets for the company. Though the new circuit would come to have value as a transmitter, Armstrong used it in this instance to receive and amplify radio signals.

In a report to the company's chief engineer, Sarnoff clarified the importance of Armstrong's "receiving outfit" and its pertinence to the company's technical goals. After describing the experimental apparatus, he assured his boss that "there is no question" that the stations being received were of "the continuous wave type." The remainder of his remarks demonstrated how close Armstrong and the industry were to a full-blown radio system: "The detector used is of the vacuum type. The other receiving apparatus consisted of ordinary loading coils, variometers and air condensers." Armstrong had explained that "the unique feature of his arrangement" was the action of "the received energy . . . on an electric valve, which in turn produced considerable amplification of the signals." Sarnoff was further impressed that, "when receiving continuous waves, the static could be eliminated to quite a considerable extent" [3]. What Armstrong had invented, Sarnoff in effect reported, was an effective feedback circuit made up of several radio tubes arranged to amplify the received signal.

Of crucial significance to Armstrong's success was his persistence. From his first days at Columbia, Armstrong resisted the power-station orientation of the department, studying, instead, the far weaker currents used in wireless telegraphy. Several years before entering college, he had mastered the art of the radio operator and had read the relevant texts in the wireless field. When he arrived at Columbia, his teacher in the wireless course quickly realized that his young student knew the field better than he. As Armstrong matured professionally, he utilized the IRE's papers program to further his concerns. In the fall of 1913, Armstrong attended one of these programs, which was held at Columbia, in order to hear a paper by the inventor of the triode, Lee deForest ("the two men," Armstrong's biographer asserts, "disliked each other at sight"), and to demonstrate his receiver. With the device concealed within a closed box, deForest recounted, Armstrong "led two wires to my amplifier input to demonstrate the squeals and whistles and signals he was receiving from some radiotelegraph transmitter down the Bay" [4]. Early in 1915, Armstrong followed the publication of an article on the radio tube in the *Electrical World* with a paper before the Institute of Radio Engineers on the same subject.

That Armstrong's patents were original and potentially remunerative inventions intensified his relations to the shapers of the radio context, introducing him early on to the search for capital to finance his work. Though Sarnoff's 1913 visit led to no Marconi purchases, Armstrong had a growing business with the Germans as they sought a means to secure communications between the United States and Germany. An American subsidiary of the German Telefunken Company acquired a license to utilize the regenerative receiver circuit. When the British severed the Germans' cable connections to America in 1914, Armstrong's circuit enabled the Germans to reestablish communications with their home country through their station on Long Island. But Armstrong's financial position was shaky. Until 1916, the $100 the Germans' license brought in monthly was the sole earnings of Armstrong's invention. Besides, deForest and the owner of his patents, AT&T, were threatening litigation over Armstrong's claims of originality for his feedback circuit. However, at that time, American Marconi bought licenses under several of Armstrong's patents. By 1917, therefore, the inventor received a steady flow of royalties, which, with an assistantship in Pupin's laboratory, supported his investigations. Though this arrangement ended when America

Edwin Howard Armstrong, on the right, helped the Army improve its radio communications system during the Great War.

entered the war that spring, never again would Armstrong worry over income for simple sustenance.

Armstrong found himself engaged, rather, in a battle for survival being waged on a much greater scale. After a brief stint in an officer's training course, the Army shipped him off to Europe as a captain to help the Signal Corps solve the problems plaguing its radio communications systems [5]. Yet the war affected more than the fortunes of a single inventor. For during these years, President Woodrow Wilson and his Secretary of the Navy not only moved closer to governmental ownership of radio in America but also, in 1917, when America joined the world struggle, Wilson granted the Navy authority to consolidate all the scattered components of radio into a comprehensive radio system. The private companies that owned these components had wanted also to consolidate the parts into a state-of-the-art system. But the war brought new arrangements: Not only did military-governmental agencies like the Navy, Army, and Post Office want possession of this promising means of communications, but with the war, the power of decision had shifted to them. And, thus, the military and the government became, for a while, a vital part of the world of the electrical engineer.

What stood out in the profession's role in World War I, however, was the central position of the nation's engineering societies in assisting the military in preparing for the war. The AIEE had previously been involved with interests of the military. Navy Captain O. E. Michaelis' role during the 1880's in encouraging standardization activities within the AIEE was an instance of that long relationship. The society's direct concern with the military, however, began in 1903 when the AIEE Board of Directors formed a committee "to consider the advisability" of creating "an electrical engineering reserve." A Board resolution notified the Secretary of War of the Institute's support for a reserve "to be available for immediate service in time of war or threatened hostilities" [6]. A dozen years later, after war broke out in Europe, Major General Leonard Wood of the U.S. Army revived the idea of an engineering reserve within the AIEE. General Wood wanted "an organized reserve corps of engineers" to be available if necessary "to cooperate immediately with the permanent military establishment." Institute directors promised "enthusiastic support and cooperation in any plan . . . on the part of the War Department" to establish a reserve [7].

The engineering community had not actually moved into strange territory. Maintaining a close relationship with large-scale public and private interests had been a persistent part of the history of the profession. Yet something new was afoot during the two years before America entered the Great War. For the nation's political and military leaders called on electrical engineers to help organize research directed at specific military problems arising during the struggle. In the process, sometimes working through the professional societies, sometimes through individual engineers, the war initiated the making of yet another world for engineering workers. More precisely, it intimated an era of military research and development, an activity which, after another world war twenty-five years later, became a strong presence in both university and industrial research. Though instances of military research had existed since before the Civil War, most of those activities had been short-lived. In the years just prior to the world war, however, governmental involvement in science and technology shifted radically as "exploration" gave way, in the words of historian, A. Hunter Dupree, to "the application of science to the weapons of war" [8].

For a period of less than two years, this shift in national concerns pulled a great number of electrical engineers into the heady and exhilarating waters of wartime scientific and technical research. A generation would pass before these activities made up a major segment of the profession's base. However, this early effort to induce innovations in technology by a nationally coordinated program in basic and applied research and development contained the form of later and larger ventures.

Electrical engineers and the Great War: Intimations of a new order

The relative absence of governmental research on the eve of the world war could be inferred from John J. Carty's 1916 presidential address to the AIEE. Carty discussed the "Relation of Pure Science to Industrial Research"—a favorite topic of his during the next decade—out of his observations of the war then going on overseas. The "startling agencies of destruction" being used in Europe appeared to Carty to be "the product of both science and the industries," a fact that revealed "the deplorable unpreparedness of our own country." However, Carty believed that the American people's awareness of the new weapons of war plus their knowledge of the "brilliant achievements" of electrical engineers had awakened in them a sense of "the vital importance . . . of science in the national defense" [9].

Though Carty distinguished between industrial research, as "applied science," and "pure scientific research," whose "natural home . . . is to be found in the university," he failed to mention government-sponsored research or the work of national laboratories. But what could be omitted in 1916 would be impossible five years later. Carty correctly implied in his address, the war would soon draw electrical engineers into the nation's new departures in research and development. When Carty gave his address, Congress had already, in 1915, established the Naval Consulting Board (NCB), an engineering body, and was considering, in 1916, a proposal by the National Academy of Science for a National Research Council (NRC), which would be composed chiefly of scientists. Finally, in August 1916, Congress established the Council of National Defense (CND). But though founded last among the three main agencies, the CND emerged as the directing agency, with the NRC and the NCB as its research and development branches. Composed exclusively of members of President Wilson's Cabinet, with the Secretary of War as chairman, the CND became the President's War Cabinet, its resolutions having authority second only to presidential orders [10].

Electrical engineers played leading roles in all these activities. The Council of National Defense came in part from the efforts of Hollis Godfrey, an

MIT-trained electrical engineer who had made a career in municipal engineering in Philadelphia. After earning a reputation as an administrator and writer, in 1914, Godfrey assumed the presidency of the Drexel Institute of Technology in that city. Already, he had won national attention in 1908 with his futuristic novel, *The Man Who Stopped War,* in which he imagined new weapons for use in future wars. With the United States drawing closer to the war in Europe, a trio of influential, national figures — ex-secretary of state Elihu Root, Major General Wood, and mechanical engineer Howard E. Coffin, who had brought standardization to the automobile industry — won Godfrey's help in getting Congress to establish a defense council for America. Godfrey readily joined, lending his pen to the campaign for the CND. Not since George Washington agonized over the effects of national disarray on the state of the military in 1780, Godfrey wrote, had there been more of a "necessity for some governmental means" of coordinating "industries and resources for the national security and welfare." He pointed out that national defense councils had historically acted as "the expert advisor of the government in regard to technical matters." Godfrey served as a member of the CND's working group, the Advisory Council, which included men like banker Bernard Baruch and Sears president Julius Rosenberg. Godfrey's special responsibility became that of coordinating technical education for the war [11].

Another engineer who stood at the center of the research efforts was General George O. Squier, an active AIEE member who headed the Army's Signal Corps. The Corps held as strong an interest in radio as it had in telegraphy a generation earlier. A West Point graduate with a Johns Hopkins Ph.D. in electrical engineering, Squier even hoped for the appointment of a Secretary of Sciences as the newest Cabinet officer. Through his efforts, not only was the National Research Council named to function as the Signal Corps' Department of Science and Research, but Squier also managed to get most members of the Physics Committee of the NCR into uniform. Even closer to the new research agencies were several prominent AIEE members who served on the National Research Council: John J. Carty, Gano Dunn, and Michael Pupin. Both Carty and Pupin lent important aid to George Ellery Hale, the astronomer who led the campaign to establish the NRC. Hale argued that true preparedness required the support of "every form of investigation, whether for military or industrial application, or for the advancement of knowledge." It was electrical engineers, however, who helped acquire the funds for this work. Gano Dunn, as head of the Engineering Foundation in 1916, and Pupin, as his successor, arranged for the Foundation to provide the major funding for the National Research Council [12].

Beyond the involvement of individual engineers, however, the wartime efforts at research and industrial mobilization drew most heavily on the engineering societies themselves. In planning an inventory of the nation's industrial resources, the Naval Consulting Board's Industrial Preparedness Committee called on the presidents of the five leading engineering societies. In this work, the AIEE's influence was particularly prominent. To head a state-by-state inventory, President Carty arranged for the appointment of

Bell's chief statistician to manage the survey. Another instance of society involvement occurred when, acting on their own in June, 1917, the four founder societies established a War Committee of Technical Societies. At first made up only of members from the Engineering Council — the American Institute for Mining Engineers, the American Institute of Civil Engineers, the American Society of Mechanical Engineers, and the American Institute of Electrical Engineers — the War Committee came to include representatives from eleven societies. And although, like the NRC, supported mostly by the Engineering Foundation, a formal relationship emerged between the committee and the Inventions Section of the Army's General Staff. Later, to establish another link, its chairman became a member at large of the NCB [13].

That engineering should be central to the government's efforts to mobilize the industrial and research capacities of the nation was natural. As one of the most challenging new technologies of the era, it was also natural that the radio was the vehicle that carried electrical engineers into the wartime-research arena to work alongside the physicists. But although radio technology was the focus of wartime research, it soon became clear that the electrical engineers' advanced position was derived from the high status won by the achievements in power engineering. In fact, Secretary of the Navy Josephus Daniels got his idea for a research agency from reading an interview with Thomas Edison in the *New York Times* in May, 1915, giving the inventor's "Plan for Preparedness." "Modern warfare," Edison explained, "is more a matter of machines than of men." Since to Edison's mind machines were "simple matters," the bigger challenge lay in merging technical and military roles. He would, thus, have military officers supplement their academy educations with periodic spells of industrial work to train them to "big knowledge." However, as the submarine war between England and Germany threatened American neutrality, more immediate steps were needed. National preparedness required both an inventory of the country's industrial resources and an elaborate laboratory to conceive new weapons, develop prototypes, and plan for mass production [14].

Edison's bold plans led Secretary Daniels, in July, to ask the inventor to head the Navy's efforts to carry out a similar scheme. What Daniels wanted most was an independent board to assess the numerous inventive suggestions received by the Navy from private citizens. A similar canvass had been carried out during the Civil War, and Daniels wanted to repeat it. But to "meet the new conditions of war," Daniels also wanted a "department of invention and development." Like Edison, he wanted a facility for "experimental and investigation work." Daniels cited, as an example of "one of the big things" the Navy confronted, the "new and terrible engine of warfare . . . the submarine." A research agency "composed of the keenest and most inventive minds," he believed, could help "meet this new danger with new devices." Working to found a vast laboratory to research new weapons for war dominated much of the Navy's wartime research and development efforts. But after examining thousands of useless inventive notions sent in by private citizens with only a vague conception of military needs, the NCB's engineers focused their attention on the challenge Daniels had first singled out: that of devising a suitable means of detecting German submarines [15]. It was this work that drew the engineering societies into their most significant participation in the war, as Daniels turned to them to staff the Naval Consulting Board. This was Edison's doing. At the inventor's instigation, Secretary Daniels asked eleven leading engineering societies to appoint two representatives each.

Although Daniels' respect for Edison enabled the organizing in Washington to proceed smoothly, the task of appointing AIEE representatives to the Naval Consulting Board raised old disputes over the role of the electric utilities. The trouble began when, after receiving Secretary Daniel's letter in July, 1915, the AIEE's Board of Directors sought nominations from directors, past presidents, and section chairmen [16].

As if anticipating pressures from the power industry members of the AIEE, John C. Parker, the section chairman from Rochester, New York, sent a circular to the nominators asking for guidelines. Parker argued that this unique opportunity for "the great national scientific and professional societies" required "the most sober and careful choice." He suggested that researchers in "pure or applied science" be appointed by the chemical and physical societies and that the AIEE choose "designers" while avoiding consultants, "Central Station men," or "men engaged primarily in executive work." In addition, the AIEE's nominees should come from "the broad phases of design work in large manufacturing organizations." Corporate engineers would bring with them the experience of their organizations, thus ensuring the cooperation of "such manufacturing facilities as might prove necessary." Parker's suggestions brought a quick response from Dugald Jackson, who wanted men like Elihu Thomson and John Lieb, "the designing being left to . . . the Navy." Parker responded that he "had in mind not so much the men who have one eye glued to the microscope as the men capable of handling design in the bigger way." Some of the names he had in mind were "Sprague, Rice and Lamme." And indeed, the AIEE did name Benjamin Lamme and Frank Sprague as its representatives [17].

By the summer of 1915, the disputes over the Naval Consulting Board nominations were finally settled, and by October of that year, the NCB had formed fifteen technical committees, a number that grew to twenty by the end of the war. However, now that the NCB had gotten started, another problem arose: that of funding. With all the agencies that had been founded and charged to initiate research on weapons, neither the government nor the military provided the level of funding that the industrial corporations had learned was necessary for such work. Ample and continuous funding was particularly required for research aimed at both basic and applied problems. For the investigators, the absence of funds meant continuous changes in the research activities, with projects shifting between the committees and from agency to agency as responsibilities were distributed. In the shaping of technical policy, therefore, the lack of public funding led to a triadic relationship between industry, the government, and the military. For the Naval Consulting Board, this arrangement meant a persistent narrowing of functions and tasks. The trail left by this devolution, however, revealed the essential story of technical research during the war.

Rather than coordinating engineering research, as the Secretary of the Navy hoped he would, Thomas Edison spent most of his time on his own inventions for the war.

Because Edison declined to assume administrative leadership, devoting his time to making inventions for the war, the committees served as the locus of the Naval Consulting Board's activities. Among the committees were the Committee on Chemistry and Physics (later separated), the Committee on Electricity, and the Committee on Wireless and Communications. The Industrial Inventory Project was the first to leave the NCB, being abruptly moved, in late 1916, to the National Defense Council. This project, which had begun a year earlier as the Committee on Production, Organization, Manufacture, and Standardization, signified, by the title, its broad reach as clearly as the committee's new title, Industrial Preparedness, expressed its larger purposes [18].

Another major departure from the NCB was the research begun in aeronautics by Elmer Sperry, who represented the American Society of Aeronautical Engineers on the Naval Consulting Board. He initiated several projects for the NCB, but his attempt to devise an "aerial torpedo" attracted the most attention, gaining $200,000 support from two naval bureaus. The aeronautical work, however, was soon taken up by the National Advisory Committee on Aeronautics, an agency formed in 1915 to "supervise and direct scientific study of the problems of flight" [19].

The narrowing of the NCB's activities did not leave it without purpose. It retained the most advanced technical research of the war: the search for dependable means of detecting submarines. Though many of the ideas tried were mechanical — steel nets, for example — the use of underwater signals was quickly perceived as the most promising research area. The nature of this work was explained in two pamphlets dealing with the problems of submarine warfare, which were published by the Naval Consulting Board in 1917. Though written to instruct would-be inventors, their content indicated the nature of the research work being undertaken by the NCB's engineers. One pamphlet suggested that a listening device might be constructed to pick up sounds "by a telephone receiver." It could then amplify the sounds "by means of relays, such, for instance as the audion and other similar apparatus." The second pamphlet recommended that citizen-inventors avoid attempts at using electromagnets, since they were too weak to deflect torpedos. It also stated that inventors should reject any notions of using electricity for "charging the sea" to repel submarines. "On the other hand," the pamphlet stated, "applications of the transmission of electrical energy by means of alternating or pulsating currents — as used in wireless systems, for example — belong to a different class of electrical development" [20].

Because the importance of this research was not immediately realized, work on submarine-detection devices had sprung up on several of the working committees. Such uncoordinated activities were tolerated during the months of preparation, but the lack of centralized control became less tenable in February, 1917, when Wilson ordered the German ambassador out of the country. Only days later, the Naval Consulting Board organized the Special Problems Committee to consolidate the work on submarine detection. The Special Problems Committee was further divided into a half-dozen areas, with the most advanced work going to chemist Willis R. Whitney, head of the GE Research Laboratory, and to Westinghouse engineer Benjamin Lamme. Whitney pulled together the work on "detection by sound," and Lamme assumed the responsibility for efforts in the area of "detection by magnetic, electromagnetic, and electrical means" [21].

Of the electrical engineers who participated in wartime research, then, Lamme wound up at the heart of the most advanced work, even assuming the chairmanship of the Special Problems Committee when the first chairman left. In March, the committee sponsored a Submarine Defense Conference, to which forty of the nation's leading scientists and engineers were invited. Thirty-five names had been submitted to Secretary Daniels, who added the names of five Navy officers. The list demonstrated that the work in submarine detection was not exclusively the NCB's but, rather, represented the special preserve of electrical engineers and physicists. Among those present, for example, were physicists Percy W. Bridgman and G. W. Pierce from Harvard and Albert A. Michelson and Robert A. Millikan from the University of Chicago. There were also a number of engineers from the power and radio engineering fields: Reginald A. Fessenden of the Submarine Signal Company of Boston, Irving Langmuir and Elihu Thomson from General Electric, R. B. Williamson of Allis-Chalmers, Charles Scott of Yale, Harold B. Smith of the Worcester Institute of Technology, and consulting engineer Ralph Mershon. From the makeup of the men attending the conference, it became clear that the research in submarine detection was the main point of contact between the war's two chief research agencies in physics and electrical engineering: the National Research Council and the Naval Consulting Board [22].

The conference quickly established that there was no reasonable way to stop a torpedo and turned the search for a solution to the "submarine menace" toward the idea of detection. The chief form of detection would be by microphones and oscillators with some means of filtering out sounds from the ship carrying the apparatus. A key question asked was: "In case submarines become noiseless what are the limitations upon sending out sound waves and listening for the echo?" The participants in the conference concluded that because the Germans were making great efforts to reduce the sound coming from their submarines, "the problem seemed to be to get a faint sound detector of exceeding delicacy" plus a method of determining direction and distance [23].

Once the Special Problems Committee decided that the goal was to develop listening devices, the experiments could begin. This happened when the Navy followed the suggestion of Fessenden's company that a station be built at Nahant, Massachusetts. The station was completed in April, 1917, soon after a Navy conference with representatives from industry and the National Research Council had led to the creation of a Special Board on Anti-Submarine Devices. Following a series of NRC conferences in June, which included European scientists and representatives from GE, Western Electric, and the Submarine Signal Company, a new station was set up at New London, Connecticut, to be staffed largely by physicists under the National Research Council. The Nahant and New London stations became the Special Board's research centers. With the Naval Consulting Board and the National Research Council organized under the Special Board, a single advisory committee was formed—containing Willis Whitney from GE and the NCB, joined by Robert Millikan from the NRC and representatives from the Submarine Signal Company and Western Electric.

Thus, the work of the Naval Consulting Board and the National Research Council had become part of a much larger, organized research force, with the engineers and the physicists working as partners in research on detection devices for the remainder of the war. The most productive research came out of the work with variations of Broca devices that were provided with microphones to be towed from ships. This began with an NCB team of GE engineering scientists headed by Whitney and including Irving Langmuir. Working with the Broca tube, the Nahant researchers were able to develop a C tube, as the Navy called it, that, with the MB listener (the M standing for multiple tubes), devised by University of Wisconsin physicist Max Mason who was working with Millikan's physicists at New London, provided the Navy with a number of workable detectors. Several variations were developed for installation on different types of vessels [24].

Another phase of the work that was assigned to the Special Board was that being done by Lamme and the Special Problems Committee. Lamme's group was charged with finding means of detection other than with the Broca-tube variations. That work took on an interesting complexion after Army Major Ralph Mershon was attached to the Lamme project, since Special Problems came more and more to resemble an ad hoc committee of the AIEE. Already Charles Scott and a group of assistants at Yale were working on magnetic needles while Mershon and Lamme experimented with setting up a

Working from his Pittsburgh office, Westinghouse engineer Benjamin Lamme coordinated the Special Problems Committee's attempts to devise a submarine detection system.

magnetic field with an alternating current magnet so as to detect changes in the field in the presence of submarines. Soon after beginning to work with high frequencies — the experimental work began at 500 cycles — Scott's group discovered a number of disadvantages and turned to investigations with low frequencies, using methods suggested by Mershon. But upon hearing that Vannevar Bush, who in 1916 had finished his doctoral studies in electrical engineering under Kennelly at MIT, was working more successfully with a high-frequency detector, the NCB sent Mershon to Boston to interview him. After that, the NCB committee concentrated on low-frequency research while giving support to Bush's work, which was being financed by the J. P. Morgan company, American Radio & Research Corporation, or AMRAD. Bush had devised "a magnetic bridge affair" by putting "an alternating current magnet on a wooden subchaser, with a pickup coil, and a rig for balancing out the voltage normally developed." In this way, he explained, "one simply creates a field and so balances a receptor that, when the field is disturbed in any way, there is a signal, usually a tone heard in earphones" [25].

Though Bush's work was thought to be especially promising, Lamme's group undertook a number of experiments which utilized a low-frequency alternating magnetic field. Besides fundamental testing carried out in the Electrical Engineering Laboratory at Johns Hopkins University in Baltimore, engineering researchers performed practical experiments at the Naval Academy at Annapolis, Maryland, and on board a submarine chaser. As designed by Mershon, the experiments utilized a device similar to Bush's. A large coil was placed at a distance and a small coil close by a magnet located on the searching vessel. After adjusting the smaller coil to be in phase with the larger, detecting coil, when a submarine entered the area, a delicate instrument could detect a disturbance in the magnetic field formed by the coils. Before the war ended, practical tests confirmed predictions made by the experimenters in Baltimore [26].

Although by war's end, the tube detectors developed at Nahant and New London received credit, along with the convoy system, for winning the battle against German U-boats, the research with high and low frequencies carried out by Lamme's committee contributed little to the prosecution of the war. As Bush later recalled of his research, "I detected many tame submarines." But he also remembered that after developing his detector to the experimental stage, the Navy told him that it would have to work on iron ships as well as wooden ones. Then, after six months of trying to adapt his device, he learned that the Navy wanted to use it on wooden subchasers after all. The lack of experience with directed research on such an ambitious scale, plus the short time period and inadequate funding, made such confusion highly likely. Yet there were also many examples of coordination, as shown in the events leading to the Special Board, which aimed at "the closest possible cooperation" between all Naval Bureaus and boards, including the NCB and the NRC [27].

But if problems in funding, coordination, and late starts plagued the actual research undertaken during the war, much was accomplished in the realms of policy. The NRC had succeeded, as Pupin concluded after the war, in cultivating "an intimate relationship between abstract science and the industries." Pupin thought this was "as it should be: national defense in its broadest sense demands such cooperation." However, Vanevar Bush disagreed from the vantage point of additional experience acquired during the Second World War. He complained that, during World War I, there was a "complete lack of proper liaison between the military and the civilian in the development of weapons." There had been no "centralizing group." But Bush wrote more from pique and poor memory than from a reasoned assessment. He had long chafed from perceived personal slights during World War I meetings with NRC members Pupin and Millikan. In recalling that Mershon "helped me more than the entire Navy," he had also forgotten that Mershon acted as a member of the NCB's Special Problems Committee. Between June, 1917, when, at Lamme's request, Mershon first saw Bush at Tufts College in Boston, and July, 1918, Mershon wrote a dozen times to Bush. He also wrote frequently to Naval officers to arrange for tests and reported continuously to Lamme on Bush's work [28].

Bush's actual situation during the first war was far from isolated. Instead, he worked within a network composed of military, governmental, industrial, and university representatives. Not only did Lamme's Special Problems Committee work within the CND-NCB-NCR complex, Special Problems itself was a multifaceted operation. With Mershon as his deputy, Lamme had overseen the work of five research branches, including—besides the projects in Boston—Nahant, Yale, Baltimore, and Lamme's own project at Westinghouse in Pittsburgh. Indeed, Lloyd Scott, a Naval officer who worked with the Naval Consulting Board and wrote its official history at the end of the war, best described what took place. At first, there was the disappointment of realizing that "the board" was not going to "evolve some invention that would conquer the Central Powers with one fell swoop." Then Scott realized

> that the best results come from team work on inventions, such as that utilized by the Special Problems Committee of the board in its work of developing listening devices at Nahant, especially so when such teams are made up of highly trained technical men, each contributing something to the solution of the problem [29].

The consolidation of radio technology

It seems highly ironic, given the importance of radio to the wartime research in submarine detection, that Secretary of Navy Josephus Daniels did not invite the Institute of Radio Engineers to join the Naval Consulting Board. However, when the preparedness movement got underway in 1915, the IRE was a small organization representing only a few hundred engineers; it was headquartered in New York in a single room that was provided by a wealthy member. That was apparently Daniels' perception when he received an IRE Board member's letter requesting a place on the newly created Naval Consulting Board. Daniels advised that he "could not at this time add the name of the I.R.E." But only at first glance does the irony appear. For in spite of its emulation of the AIEE in organizational matters, the IRE confronted a different sort of challenge during the war years. The IRE was pressed, instead, to deal with the problems of an infant engineering society. Externally, moreover, Institute directors chose to concentrate on the challenges that the Navy and President Wilson presented to an embattled radio industry [30].

In any case, it was far from certain that the IRE wanted a position on the Naval Consulting Board. As it turned out, the IRE Board had not approved the letter, which was written by a former secretary of the IRE Board, only recently replaced by David Sarnoff. Thus, during a meeting, the directors of the IRE simply instructed Sarnoff to thank the Secretary of the Navy for the courtesy of responding. At the same meeting, Alfred Goldsmith presented an elaborate set of amendments to the by-laws to ensure that such an invasion of the Institute's authority would not occur again. However, radio engineers did participate in the war. Lee deForest, whose dealings with the Navy went back ten years, produced radio tubes for both the Navy and the Signal Corps. Edwin Armstrong was one among many radio professionals and amateurs who staffed military laboratories and field stations during the war. And on the home front, men like Alfred Goldsmith ran radio schools for the military at their colleges [31]. These were the acts of individuals at a time when the IRE's directors worked chiefly to strengthen the organization. They worried over such matters as low membership numbers and the need for increased income from the sale of advertisements in the *Proceedings.* Until the mid-twenties, an IRE officer later recalled, the Institute of Radio Engineers existed mainly to hold periodic meetings in New York and to publish "radio technical papers" [32].

Although technical interests were central to the youthful IRE, with the presence of men like David Sarnoff, whose concerns tended more to the commercial, nontechnical values were necessarily present. They dominated Sarnoff's drive for the private control of radio in America, for which he drew the radio engineers into the public-policy arena for a brief, if critical, moment. Given the dynamic character of the industrial and technical scenes in radio during the first two decades of the century, it was not surprising that the two mixed so readily. Both spheres, the industry and the technology, had been a number of years in the making. Indeed, the consolidation of radio during the first year of the war rested on twenty years of research, invention, and development. Wireless began during the 1890's when Guglielmo Marconi took basic discoveries from the physical sciences and built a crude, but workable, system of wireless telegraphy. Though he did not achieve what was later called *radio,* Marconi had moved wireless technology into new territory. By introducing his antenna and coherer, the Italian inventor revealed areas that required new understanding and promoted theoretical work on antenna design, propagation, and transmission lines. By the early 1900's, Hugh Aitken has shown, "in each of these fields, Marconi's work was already generating new data and new problems." In 1907, Marconi developed his "disc discharger," which carried spark-generated radio transmission to its ultimate development before the achievement of continuous-wave transmission. His arrangement of three rapidly rotating disks created a continuous oscillating spark, approximating a system of continuous-wave transmission. With its installation that year at Marconi's Clifden station on the west coast of Ireland, the discharger represented "the 'state of the art' in spark radio transmission" as well as the furthermost point in the development of spark technology [33].

Other engineers, specifically in America, were already taking radio further. Reginald Fessenden began his experiments in continuous-wave transmission around 1900, though Nikola Tesla had earlier attempted this with high-frequency alternators. In 1900, University of Pittsburgh professor Fessenden left his faculty post, joined the U.S. Weather Bureau for a short term, then convinced two Pittsburgh capitalists to finance the National Electric Signaling Company. He sought to move beyond spark apparatus that transmitted only with gaps to the transmission of speech. He arranged for GE to build an alternator to help him do this. At first, Steinmetz worked on Fessenden's problem, building two high-frequency transmitters for him around 1903. Then Ernst Alexanderson was assigned to the task, and in 1906, he completed an 80,000 hertz alternator. Using this apparatus, Fessenden broadcast his Christmas progam of voice and music that year, which was picked up by ships in the North Atlantic. But its successes came too late for him to benefit by it. Fessenden continued to work on a continuous-wave receiver, but the full realization of voice transmission and reception was to come with deForest's

The radio era emerged from the pioneer work of inventors like Italian Guglielmo Marconi (at left, above) and American Reginald Fessenden, who was the founder of the National Electric Signaling Company, Brant Rock, MA, shown here (below).

and Armstrong's work with the Audion, or radio tube, as it was called by the end of the war [34].

Like all great inventions, the vacuum tube was an achievement of many. However, Lee deForest added the element that made it not only the key component of a modern radio system, but also the progenitor of the electronics age. Early in 1899, shortly after receiving his doctorate from Yale with a dissertation on wireless—"The Reflection of Short Hertzian Waves from the Ends of Parallel Wires"—deForest took a job with Western Electric in Chicago and began to seek, as he wrote, "a new form of detector for wireless signals." An erroneous interpretation of the flickering of a gaslight in his lab while he was operating a spark transmitter set deForest on the path to the Audion. Thinking that the flame detected the spark rather than (as later realized) the sound waves from the spark dimmed the flame, he began experimenting with gas-filled bulbs composed of two electrodes. Shortly afterward, he added another electrode—a grid—between the cathode and anode and found that he could control the flow of current between them more effectively than when he simply heated the cathode in the diode, invented by Ambrose Fleming in 1902. deForest's work on the triode came to a head in 1906 and 1907, first in his paper to the AIEE—in which he "made only veiled reference" to his still unpatented triode—and then with his patent. After the patent, he quit work on the Audion. In 1912, however, after the collapse of another of his enterprises—nearly sending him to jail for mail fraud—he took a position with a California telegraph company and turned once again to his device, this time to explore its potential for radio. The next few years would be his last work on radio; though, as he was to make clear during his 1930 IRE presidency, it was not the end of his interest [35].

While the development of an adequate receiver rested on deForest's three-element radio tube, its possibilities were chiefly worked out by others. The number of workers in the area went far beyond either deForest or Armstrong. The Marconi team's visit to Pupin's laboratory, in 1913, and to later demonstrations at the Belmar station, in 1914, indicated how important vacuum-tube development was to the nation's largest radio firm. Marconi himself reacted to the reports of Armstrong's receiver with an enthusiasm born out of the extensive work on the tube going on in Europe. He reported that his company had made "tests which have given extraordinarily good results in connection with modifications to the circuits of the so-called Audion and other vacuum detectors." Investigating the device of the young engineering instructor at Columbia was part of the wide-spread efforts to exploit the promise of the tube. For even with the experimental work going on in England, Marconi thought it was "necessary for us to know exactly what the device of Mr. Armstrong is," and, he added, "whether it has been patented, and if so, when" [36].

But the conjoint work going on was even more extensive. Besides the developmental activities underway in New York, at Federal Telegraph on the West Coast, and at Marconi's station in England, a German by the name of Alexander Meissner was working on a regenerative circuit with a gas-relay tube, and in America, AT&T energetically pursued the vacuum tube's potential. AT&T began its quest for a continuous-wave system with the purchase of deForest's Audion patent in 1912. In 1914, AT&T purchased additional rights on deForest's patents and, three years later, acquired all of them, whether existing or pending, including two in contention with Armstrong's key patents. Furthermore, AT&T's researchers — particularly physicist Harold D. Arnold, a former protégé of Robert Millikan's in Chicago — were making vast improvements on deForest's designs. After making a number of innovations, by the summer of 1913, AT&T had achieved a tube with a life of a thousand hours compared to the fifty-hour life of deForest's triodes [37].

But GE had the strongest position. So advanced was its work that on the day Armstrong filed, Irving Langmuir filed for a patent on a regenerative circuit utilizing the vacuum tube. With GE's possession of Ernst Alexanderson's alternator, plus his 1912 electronic amplifier, by 1915, the company almost possessed an effective continuous-wave system. Two years later, when GE's engineers had perfected the alternator, the Marconi company moved to obtain exclusive rights to it. With Alexanderson's alternator, Marconi could go after the one prime commercial market still available in the second decade of the century. The wireless could not compete successfully on land with either the telegraph or the telephone for commercial and governmental traffic and the ship-to-shore market was already being exploited by commercial and military interests.

However, transoceanic communication, though already tapped by the cable, extended beyond the reach of the telephone and presented a ready market for wireless telephony. Yet it was for transoceanic telegraphy that GE's alternator had at last provided a powerful enough transmitter so that old commercial desires could be satisfied. As a Marconi annual report explained, the company's "principal aim and purpose" had been "the establishment and maintenance of transoceanic communication." Moreover, this had "been always considered . . . the greater and more profitable business" than ship-to-shore communications or the sale of apparatus. The British radio firm's goal was to achieve the successful transoceanic transmission of voice and, thus, to assure its dominant position in commercial radio in America [38].

Yet the growth of the British-owned company had been closely observed by other powerful interests in the country besides the corporations, especially the Navy, though the Post Office also had a natural interest. The Navy had actually begun to seek control of radio in 1912 when it supported amendments to the Wireless Ship Act of 1910. The 1912 act had sought domestic control of wireless when the Marconi company had begun to grow rapidly following its reorganization in 1910. Seeking more than American dominance at this point, the government also wanted favored positions on the electromagnetic spectrum for the Navy and, most potent for the future, authorization for the President to assume control of radio in a national emergency [39].

This obvious political act had no less obvious technical implications. During the 1910's, the crystal set, a wireless telegraph system, had been supplanted by the radio, which had the potential to be a wireless telephone system. Yet on the eve of America's entry into the war, radio technology remained fragmented. As Armstrong described the situation, "it was absolutely impossible to manufacture any kind of workable apparatus without using practically all the inventions which were then known." However, as a Navy investigation found: "Not a single company among those making radio sets for the Navy . . . possessed basic patents sufficient to enable them to supply, without infringement, . . . a complete transmitter or receiver." Thus, when Congress declared war in April, 1917, among Wilson's first acts was to order radio amateurs to cease operation and to instruct Daniels to take over all radio apparatus as needed for military communications. It was a momentous event for radio as well as for the nation, for the fruit of twenty years of experimental efforts to devise a practical system of radio had ripened [40].

Sarnoff, the IRE, and the consolidation of the industry

The drive to consolidate radio technology possessed a twin in the issue of control. This political and economic issue was similarly a leading radio question of the day as men in both industry and government asked: Who will control radio at the end of the war? But they did more than raise the question. While the Navy had consolidated the technology of radio, the political economic status of the medium had been frozen by the war. A thaw would require the efforts of a David Sarnoff who could focus the issues and prod legislative bodies to act. It was for this that Sarnoff sought an endorsement from the radio engineering community. Sarnoff had been active during the IRE's first half-decade. On the list of founding members prepared at the end of 1912, the twenty-one-year-old Sarnoff's name appeared as number 112, just before that of Arthur E. Kennelly's. Sarnoff rose rapidly. Soon after gaining a position on the board, in 1915, he assumed the responsibilities of secretary until Alfred Goldsmith took the job two years later. In 1917 also, the twenty-six-year-old Sarnoff became a Fellow of the Institute of Radio Engineers. In spite of this high rank, Sarnoff's technical credentials approximated those of the key operators who worked the wires for Western Union in the middle of the nineteenth century. Like most of them, he educated himself, a process completed in 1911 and 1912 with night classes in electrical subjects at the Pratt Institute in Brooklyn [41].

However, joining the IRE had not reflected technical concerns so much as business interests. That orientation appeared starkly in Sarnoff's sole IRE paper on "Radio Traffic." Delivered in 1914, he spoke not of a new application of the radio tube or of a more rapid printer but of the need for operators to speedily handle messages. During a discussion the following year, Sarnoff again sought to impress upon his audience the ideal qualities of the "proficient radio telegraphist." "Brevity," he pointed out, was "all-important" since a profitable return rested in part on the ability to transmit "as many messages as possible within the shortest period of time." Sarnoff defended the Marconi Company's recent purchase of its largest rival on the grounds that the industry was strengthened by placing "all operators under the control of a single organization." As Samuel Insull had done to power engineers so often, Sarnoff lectured radio engineers on the commercial needs of radio. Though some criticism of operators by engineers was warranted, he allowed, "on the other hand, something may be said about the radio engineer who, when designing radio equipment at the laboratory, failed to appreciate the operator's difficulty on shipboard." Nor had "designing engineers" paid sufficient attention to the need for detectors to be both sensitive and stable [42].

Sarnoff's active involvement in IRE affairs became intensified during his two years as secretary from 1915 to 1917. This was a time when advocates of public ownership and defenders of private possession contended for control of the medium. Sarnoff thus guided the IRE Board in its efforts to keep radio in the hands of private capital. In 1916, he was instrumental in drawing up a formal Board resolution that advised the government on the proper context for the development of radio technology. The directors acted in response to a bill pending in Congress "to Regulate Radio Communication." Before final consideration, several governmental agencies established an interdepartmental committee on radio legislation, which in turn scheduled a conference to hear comments from interested parties. Besides the radio engineers, testimony came from various departments of the government and "commercial interests . . . represented by their executive heads." Letters from the society's president, Arthur Kennelly, and from Alfred Goldsmith, who, besides his position on the engineering faculty at the City College of New York, was a consulting engineer for GE, amply expressed the Institute's position. But principally, there was the Board resolution, which had been prepared following a "discussion of the general subject of Government ownership or competition" with private radio interests. Composed just two weeks before the Washington meeting, its author was apparently Sarnoff, who had already appeared before the committee and prepared now to present the Institute's position to the Washington group [43].

The resolution contained a series of clauses, alternating statements of fact with assertions of belief. Its premise was that the London Radio Telegraphic Convention of 1912, to which the United States had been a signatory, contained "unwise provisions" that allowed governmental interference. That interference had retarded "the engineering development of the radio art" and threatened to continue to do so just when the radio field presented a number of technical challenges. These included especially the need to develop long-range radio telephony, reliable systems of selectivity, and call systems as well as to eliminate static caused by atmospheric strays. Solving these problems required "the highest engineering and inventive talent and research." If lawmakers wanted to promote creative engineering, the IRE Board asserted, they should know that foreign governments had been most successful when assisting rather than opposing "individual initiative and private enterprise in developing the radio communication systems of their countries."

Mixed with these statements was a political-economic credo, which the IRE Board confidently stated as a tenet of the Institute's. Its gist was this: that in the "new arts of electric communication," governmental interference "always" impedes technological creativity, and "free individual initiative" always fosters development. The Board's assertions left no room for exceptions. Inventive achievement "can only exist" within a social order where private enterprises dominate and "healthy competition" prevails. "Government competition, or confiscation," on the other hand, "would effectively stifle inventive effort." Military control of radio in peacetime constituted an "inquisition into private correspondence, an undemocratic and dangerous institution"; on the other hand, private enterprise, the Institute resolution argued, would keep "the inventive and engineering resources of the nation" at its "highest pitch and . . . broadest scope." Private control, in short, would ensure a dependable system of radio communication should a "time of sudden national peril" arise.

The IRE Board's resolution, as Sarnoff told the Washington meeting, expressed "the spirit . . . of the Institute." It also captured the heart of Sarnoff's position. The 1916 appearance in Washington was one of several he made in the campaign to ensure private control of American radio. Presenting the engineers' statement was important to Sarnoff since he believed the support of radio technologists essential to the campaign to place radio in private hands. As he later asserted, testimony by the actual inventors of the medium supported the commercial interests to which he was much closer [44].

Sarnoff's arguments before the Washington committees were given urgency by the Navy's assumption of absolute control of radio. Except for private ownership of manufacturing and a few radio installations on merchant ships, the Navy had taken command of all existing radio technology. However, when the time came to reestablish radio as a private undertaking, the various patents consolidated by the Navy remained the property of the original corporate owners. How consolidation would take place did not become clear until the end of the war when the Marconi Company reopened negotiations with GE for Alexanderson's alternator and drew the President once again into the fray. Wilson convinced GE to retain its patents and help establish an American-owned company. This GE did and, in 1919, founded the Radio Corporation of America (RCA). RCA bought out American Marconi and placed its patents with the Alexanderson alternator and other GE-owned patents. This time, it became clear that the issue of American control was not the paramount issue. More compelling was the desire by engineers, managers, and most government bureaucrats and Navy officers for private ownership [45].

The fundamental issue of control did not end with the founding of RCA. The medium was still recognized as a public as well as a private means of communication. The issue of control came up again in 1920 during a Washington meeting of the London Radiotelegraph Convention. The conferees called for a Universal Electrical Communications Union with jurisdiction over radio, wire telegraphy, and cables. The idea of an internationally controlled radio still meant to American firms a form of public ownership. However, that possibility receded as the Radio Corporation's position strengthened with the acquisition of AT&T's patents. From AT&T came patents which would allow the transmitting of radio signals between stations in a national broadcasting system and the rights to deForest's radio tube. By the end of 1921, when other companies had also contributed their patents — including those of Westinghouse, the Wireless Specialty

Apparatus Company, the United Fruit Company, and the Tropical Radio Telegraph Company — RCA had gained the potential to become synonymous with radio in the United States [46].

Gaining this unique position still left Sarnoff and RCA with the need to negotiate with other powerful parties, especially with AT&T, whose interest in the medium persisted into the twenties. Nonetheless, RCA prospered from Sarnoff's aggressive search for new sources of revenue and from the rapid acceptance of the medium by the public. Thus, the company's earnings jumped from $427,000 in 1921 to nearly $3 million a year later. When RCA's first radio set, the Radiola I, went into production in 1922, instead of the estimated sales of $7.5 million, the company sold $11 million worth of sets. And by 1924, instead of an estimated $45 million, sales amounted to $50 million. And, yet, that barely touched potential sales. RCA and its manufacturing associates, GE and Westinghouse, were unprepared for the radio boom of the early twenties, being simply unable to produce enough radio sets to meet the demand. Even with the Radio Corporation's unexpected revenue, industry sales of radio sets came to $60 million in 1922 and $358 million two years later. The differences were made up by new companies entering the radio manufacturing business. Crosley Radio Corporation was founded in 1921, Zenith in July, 1923. Other companies begun at this time were Emerson, Atwater-Kent, and Philco. By the early thirties, when most of these manufacturing firms were weakened by the depression, and some like Atwater-Kent and Grigsby-Grunow failed, many of these smaller companies remained standard fixtures in the radio manufacturing segment of the industry [47].

This vast expansion also had meaning for the IRE. With the rapid growth of the industry came the need for standardization and the founding of new industry groups like the Radio Manufacturers' Association (RMA). The same movement involved the AIEE and older trade associations like the National Electric Light Association (NELA). The IRE too was drawn into this movement for standardization within the industry. Though it had established a committee and issued its first report on standardization soon after its founding, not until after 1925 did the Institute of Radio Engineers begin to develop formal relations with the manufacturers of radios and electrical apparatus or with the American Engineering Standards Committee (AESC). However, the Institute of Radio Engineers entered reluctantly into formal cooperation with the manufacturers and national standards bodies, a reluctance made manifest when the Institute's Board carefully defined its standards role relative to the other groups. It strove to commit its energies to engineering values in contrast to production and marketing values.

As IRE representatives viewed the world of standards in the first half of the 1920's, they saw an industry in flux. No single governmental agency had responsibility for the industry. Though the Department of Commerce acted as if it did, its coordinating activities were mostly extralegal, leading the IRE to respond first to private initiatives. Late in 1925, the IRE Board authorized the Standardization Committee to request an exchange of representatives with the Radio Manufacturers' Association and the Associated Manufacturers of Electrical Supplies. This was one of a number of steps taken over several years to define its standards responsibilities in relation to the manufacturers' groups [48].

Though representatives were appointed to meet with these groups, the IRE Board vacillated. During the fall of 1926, the IRE Board appointed a Committee on Broadcast Engineering to act as a "public relations" committee in seeking stronger links in the area of popular broadcasting. However, after receiving authority to act, when the Radio Manufacturers' Association did request a joint committee two months later, the Board reversed itself, deciding that "for the present no such contact will be made." The issue of coordinating standards work with the manufacturers moved closer to a resolution in 1927, when the IRE Board prepared a policy statement contrasting the Institute's standards work to that of the trade associations. It became an important guide during the next few years as the IRE slowly moved into formal relationships with the nation's standards groups. The Institute's primary concern, the IRE Board explained, was to establish terms, definitions, symbols, and methods of testing materials and apparatus. Manufacturers, on the other hand, sought to standardize the size and physical characteristics of apparatus to accommodate the interchangeability of mechanical or electrical parts. Manufacturers also were interested in standard ratings for properties and performance. The IRE's policy had been captured in a simple statement, but one which served well to delineate the society's basic engineering orientation [49].

During his IRE presidency, in 1928, Alfred Goldsmith reiterated Institute policy before the annual meeting of the radio division of the National Electrical Manufacturers' Association. The Institute's special contribution to the industry, he pointed out, had been to create a "language of radio." Goldsmith used the opportunity to prescribe for the manufacturers broader responsibilities to the radio engineering community. He asked the association to urge its member companies and their employees to encourage and support the IRE not only by using the technical information and ideas transmitted at meetings and in publications but also by contributing to that flow of information. In short, the manufacturers must remember "that what they sell . . . is nothing more than a mixture of raw materials and engineers' brains" [50].

In limiting its standards work to engineering matters, the Institute of Radio Engineers readily fit the mold of the American Engineering Standards Committee. The IRE Board, nonetheless, in early 1928, declined an AESC invitation to become a member-body "on the grounds that the work of the standardizing bodies did not seem to be sufficiently stabilized." The AESC had that year reorganized by expanding and changing its name to the American Standards Association (ASA). Even before the change of name, the ASA had moved beyond being simply a collectivity of engineering societies, a fact the changes recognized. Its central duty had become that of assuring that its sectional committees contained, besides engineers, "a suitable balance of power among the producers, distributors, consumers, and general interests." For the ASA's Sectional Committee on Radio, for example, this meant "that practically every important organization in the radio industry" was represented. In 1929, the year before the Institute of Radio Engineers formally affiliated, these important organizations included, besides the IRE and the AIEE, such groups as Bell and RCA (represented by IRE members Lloyd Espenchied and Alfred Goldsmith), the National Electrical Manufacturers' Association, the Radio Manufacturers' Association, the National Electric Light Association, the National Association of Broadcasters, and the Commerce, Navy, and War Departments [51].

There was little drama to the IRE's hesitant approach to formal affiliation with the national standards community. Its complaints in part were legalistic since, long before defining Institute duties in regard to industrial standardization or becoming a member body of the ASA, the IRE had coordinated its standards work with other agencies. Representatives had been named to joint standards-setting bodies and to cooperate with the AIEE as cosponsors of the ASA's Sectional Committee on Radio. Its reluctance was understandable. The IRE's tendency to define membership and aims in pure technical terms placed standards concerns that went beyond nomenclature and methods for testing somewhat distant from its central purposes. During these formative years, then, IRE leaders made careful attempts to establish a standards policy that was separate from commercial aims. However, its purposes were different from the manufacturers' aims, not antagonistic to them. To the contrary, as a 1929 IRE statement on "Radio Standardization" made clear: besides the cost savings to industry and the "ultimate consumers," uniform standards encouraged "industries . . . to grow — and continue to grow, and the savings made become available for new research and for improvements in the methods and processes" [52].

As one of the final acts of consolidation, then, standards work had only partly involved the Institute of Radio Engineers. As engineers whose chief employer was industry, individual engineers were involved from the beginning in bringing standards to apparatus and performance. The IRE, moreover, while concentrating on definitions, symbols, terms, and methods of testing, which primarily served research and technical communication, had, nonetheless, cooperated with manufacturers and the national standards community. While this work was going on, however, another act in the consolidation of radio was taking place under the auspices of the federal government, only this time, the task was to consolidate, not the technical components of the apparatus, but those of a national system of broadcasting. Again, David Sarnoff was a major player. In time, this new act would require from the IRE, an additional part in standardization, but first, there were pressing questions regarding the extent of public responsibility toward a technical medium with such obvious social implications. As it happened, this work too involved, at the center, an engineer. During this critical decade in the making of modern radio, the mining engineer and public administrator, Herbert Hoover, wielded great power over national policy — and radio policy was a special concern of his.

Supersystems and the context of radio engineering

Radio was just one of many technologies that had historically concerned the federal government, including for example, the railroad, telegraph, and airplane. As historian and political scientist Charles A. Beard explained in his 1930 study of "the republic in the machine age," such innovations were special in that they overrode "historic political boundaries." And in the ten years following the world war, the "technological revolution" had intensified, so that more than in any previous age, technology had "thrust itself into all the institutions and practices of government." Increasingly, it seemed to Beard, technical change brought "new perils in its train," leading to problems ranging from "the pollution of streams" to "technological unemployment." He concluded, therefore, that

> Few indeed are the duties of government in this age which can be discharged with a mere equipment of historic morals and commonsense. Whenever, with respect to any significant matter, Congress legislates, the Courts interpret, and the President executes, they must have something more than good intentions; they must command technical competence [53].

Beard used radio to illustrate his contention that inventions sometimes introduced "unexpected conflicts and confusions into society." Radio, he explained, especially needed the guidance of precise, technical rules so that "chaos would [not] prevail 'on the air.'" By 1930, when Beard wrote, such a prescription rested on an entrenched tradition. Since the late nineteenth century, regulation by existing governmental agencies or by independent regulatory commissions had become widely accepted as a proper response to such problems as radio presented. In the early 1920's, moreover, radio represented a special case. The railroad, in contrast to radio, had been brought under the Interstate Commerce Commission, the nation's first regulatory commission, in 1887, after forty years of development. The rapid spread of radio following World War I, however, required extensive controls before the technology and the industry had matured. The building of a national broadcasting system, in short, had need of regulation before the technical knowledge existed to regulate or build it [54].

But this complex new technology had special characteristics that demanded more from engineers than technical competence. On the one hand, radio lent itself readily to the idea of a national, interconnected system of communications. On the other hand, radio broadcasting constituted a communications medium of great potential social power. And in each of these respects — complexity, expansiveness, and application — the technology and the medium involved the radio engineering profession. Besides participating in its invention and development, the radio engineers of the twenties made it technically feasible for corporate leaders to achieve vast organizational and physical systems. Thus did they help shape the bureaucratic context in which they worked and, in part, the social uses of the technology they had created.

Plans for a national radio system began soon after technical and organizational consolidation had been achieved in the 1910's. No one better captured the physical character of that system than radio's organizing genius, David Sarnoff, nor did more to achieve it. Sarnoff was a remarkable man. At one time a poor immigrant boy, as he liked to relate, in 1930, he attained the presidency of the Radio Corporation of America just before his fortieth

Always courting the support of scientists and engineers in his radio ventures, here at RCA's transoceanic station in New Jersey in 1921, David Sarnoff (leaning into the center of the picture) gathered Albert Einstein, Charles Steinmetz, and other luminaries under the antennae.

birthday. Sarnoff first gained national attention in 1912 as a radio operator when he remained at his post at Wanamaker's store in New York for seventy-two hours during the sinking of the S.S. Titanic. On that acclaim, he rose through the ranks of American Marconi: first to an inspector's position and an instructorship in the company's radio operators' school, then to chief radio inspector and assistant chief engineer, and, in 1914, to the position of contract manager for the nation's largest wireless company. While a Marconi employee, he worked intensely to serve the company, mastering the art of the corporate memorandum and using it to the utmost both to call attention to himself and to promote and clarify the interests of his company. Later, when the core staff of the Marconi company transferred to the newly created RCA, Sarnoff drew on his corporate skills to make the new company and himself supreme in the radio field [55].

Sarnoff was not remarkable, then, as his chroniclers have asserted, because he could talk the language of both engineers and managers. As he told a congressional committee in 1918, when considering the regulation of radio, the legislators should not simply heed "commercial men" like himself, who were biased. The opinions of the "technical man and inventor" must also be sought. Around 1915, as he described it, Sarnoff's work for Marconi moved into the "business producing end of the company" and to the "creation of new fields." Sarnoff was notable, rather, because of his ability to acquire power and because of the singleminded, effective manner in which he wielded it. His was not the entire story of the commercialization of radio technology; yet in intertwining his aims and purposes with the radio community, Sarnoff helped shape its character in the years between the two world wars [56].

With his ready sense of promotion, Sarnoff quickly incorporated the idea of *super* systems from the electric power industry. The phrase appeared shortly after the end of the war, when plans for linking power systems into a national network gained the label of *superpower* in the electric power industry. An early exponent and active promoter was electrical engineer William S. Murray. A consulting engineer, Murray had directed the electrification of the New Haven Railroad before being named, in 1920, as chairman of the engineering staff responsible for carrying out a national superpower survey for the Interior Department. This had come after years of Progressive agitation for natural resource conservation by entities as diverse as crusading individuals, the Department of the Interior, and the engineering societies. The AIEE had a long involvement with the movement, drawn by its close links with the power industry to seek protection for the nation's water-power sources. Beginning in 1906, the American Institute of Electrical Engineers established a forest preservation committee, which evolved into the Natural Resource Committee, which, in turn, became the Public Policy Committee in 1911 [57].

The ideal of economy so critical to the conservation movement stood at the center of Murray's superpower program. America had been a "profligate and wasteful nation in the extreme," he asserted in a 1919 talk to the Connecticut Chamber of Commerce. "Drunk with the wealth of our natural resources, we have eschewed their conservation . . . . " Murray repeated the themes of the Connecticut speech several months later at the midwinter convention of the AIEE. Impressed with the concept, the AIEE directors appointed a Committee on Super-Power Systems and named Murray chairman, along with an advisory group to include electrical engineer Magnus Alexander of the National Industrial Conference Board, governmental administrator Herbert Hoover, and chemical engineer Arthur D. Little, who ran an influential management and engineering consultant firm [59].

A simple logic was attached to the idea of vast, centralized technical systems in the electric power industry. Following the patterns of thirty years of growth, *superpower* expressed what was already happening as the dreams of men, like those of Samuel Insull, came ever closer to reality. In the embryonic radio industry, however, the idea rested less on established patterns than on the plans and ambitions of men and corporations. Given the nature of the industry and the person, this at first meant RCA and David Sarnoff but later came to include others like William Paley, who founded the Columbia Broadcasting System (CBS) in 1928. The idea, Sarnoff told a congressional group in 1926, pointed to "superpower transmission in radio broadcasting." This implied a system of "superpower stations, joined by wire or eventually connected by radio relay" to form "a nation-wide service." Three years earlier, he had spoken of "one gigantic broadcasting station" for the country, predicting fewer rather than more stations. Twelve would do in 1923, Sarnoff informed a House committee, yet before long, "a single station will do what these twelve can do" [60].

Sarnoff perceived the superpower movement in the electric power field and his vision of a national radio system as being similar. After hearing Westinghouse board chairman Guy E. Tripp advocate superpower systems for the nation's electric power companies, Sarnoff persuaded him to promote the idea of "super-broadcasting" too. To assist Tripp, he even prepared a statement applying the idea to radio. Sarnoff was amply satisfied when Tripp's "Plan of Super-Broadcasting for the Nation" received prime coverage in the *New York Times.* The Westinghouse executive reiterated his belief that national transmission systems would "usher in a new and greater era of industrial development" in electric power. "As great in its potentialities," moreover, was the "distribution of a constant flow of power through the air to carry education, entertainment and news by radio to millions of homes" in America. "Call it super-broadcasting as distinguished from super-power," he said [61].

Tripp's statement, Sarnoff observed, came "at the psychologically correct moment" and had "publicly strengthened my position." For a national system emanating from one giant station, while technically fanciful, could be achieved if taken as an organizational metaphor. Its translation into reality was called *chain broadcasting,* the first expression of which appeared in 1926 as the RCA-spawned National Broadcasting System. Before the decade had ended, moreover, both the Columbia Broadcasting System and the American Broadcasting Company had been established.

Achieving a centralized national system would not have been possible without engineers to devise ever more powerful transmitters to serve both Sarnoff's corporate dreams and the vexing problem of interference. Whereas the original stations had transmitters of between 100 and 1000 watts, by the mid-twenties, high-powered transmitters of 5000 to 50,000 watts were in place. In 1928, forty high-power stations covered the entire country. The use of greater power helped broadcasting deliver "high quality radio service." The consumers of that service additionally required adequate receivers. This too, research engineers satisfactorily achieved during the twenties as they gave increased attention to problems in receiver design and manufacture. Besides reducing the number of tuning dials from as many as five to one, engineers also dispensed with the need for batteries as they accommodated the home receiver to standard electrical outlets [62].

The tremendous growth of the industry fueled the growth of the system, as it did both Sarnoff's vision and engineering developments. Total revenues grew from $2 million in 1920 to a half-billion dollars by 1927. Public listenership rose from zero to an estimated 40 million persons by 1928. Similar expansion came to the electric power industry. Though it had begun to grow in the 1880's and 1890's, great increases in electric power generation came about during the twenties, accompanied by the increased manufacture of electrical machinery and appliances. Other parts of the economy also grew rapidly at this time: aviation, automobile production, and the motion picture industry, to name three other areas of new growth. But no area, taken in terms of both quantitative expansion and cultural impact, surpassed the growth of radio [63].

"Popular broadcasting" and the defining of radio

The rapid expansion of radio was accompanied by qualitative changes which were equally momentous. The medium was being increasingly used to advertise consumer goods for the domestic market and, by the middle of the decade, its primary activity had become the advertising of goods and services. Sarnoff had earlier looked to the day when radio would be "a household utility." To that function, he had given the name *radio broadcasting.* It was "not an invention," he explained; "it is the application of an invention to public use." The nature of that application was implied in Robert Marriott's phrase of 1929, "popular broadcasting" [64]. This shift meant, as social theorist Raymond Williams has explained, that radio had moved from its primary role of serving "the needs of an established and developing military and commercial system" to the status of being a "a new technology of social communication" [65].

The social power of radio, reinforced by its physical reach, made it an obvious candidate for regulation. During the twenties, moreover, the business community was in a position to be able to achieve whatever regulatory scheme it deemed appropriate for this new technology. Financier Andrew Mellon's position as Secretary of the Treasury from 1921 to 1932 illustrated its strength in the government. Mellon was more than a symbol, for while he served three Presidents, he led a series of Cabinets whose members were said to be worth $600 million. (Mellon was reputed to be the second wealthiest man in the nation.) The second of the three Republican presidents of the period, Calvin Coolidge had, thus, confidently proclaimed that "the business of the American people is business." And Mellon's definition of the federal bureaucracy simply turned Coolidge's definition on the government itself: "The Government is just a business and can and should be run on business principles" [66].

At the heart of the government Mellon referred to was the Department of Commerce. The Department's importance appeared when its budget was the only one not cut during this era of conservative Republicanism. In part, this stemmed from the superior administration of the Department by its head, Herbert Hoover. Hoover spent eight years in the post before becoming President in 1929, and was consistently praised for his administrative abilities. Kansas journalist William Allen White summed up Hoover's chief talents: He was "a great administrator, a splendid desk man" [67].

Hoover occupied an ideal office for this work. The Commerce Department's duties were far ranging. In the case of radio, except for the war years, the department had been, since 1912, its sole regulator. During his first year in the Commerce Department, then, it seemed that Secretary Hoover would soon be able to establish a national structure to regulate the fast-growing industry. Within a few months of taking office, he had arranged discussions in Washington between the interested parties, including corporate groups — AT&T, GE, Westinghouse, and RCA — and technical bodies — the IRE, the American Engineering Standards Committee, and the Radio Division of the Commerce Department. Maintaining private ownership in the radio industry in America was a basic purpose of the meetings, which, specifically, meant opposing the policies expressed in the document issued the year before by the international radiotelegraph meeting held in Washington. IRE representative Alfred Goldsmith had objected then to the "tendency toward government control which is felt throughout the draft." Thus, Hoover's meetings of 1921 sought to counter that tendency by promulgating two basic positions: that the Commerce Department would preserve radio in private hands and that broadcasting would be nurtured as the dominant use of the medium [68].

Hoover's method of gathering all interested parties to establish common positions before seeking legislation seemed eminently workable. Even on the matter of determining the content of commercial broadcasting, there appeared to be no dispute. Even Sarnoff had defined broadcasting to exclude advertising. In 1922, he compared radio with libraries, and suggested that a "public benefactor" might be found who would establish an endowment and remove "the broadcasting company . . . from the atmosphere of being a commercial institution." Entertainment would entail bringing existing cultural offerings in music and thought to a vast, popular audience. At the first of the Commerce Department's series of formal meetings, the Washington Radio Conferences, engineers, commercial broadcasters, and governmental representatives met to both define radio and establish an appropriate regulatory order. At this initial meeting, then, Hoover moved specifically to exclude advertising. The technological revolution which had transformed radio during the past six years, he said, had made the leading issue

> primarily a question of broadcasting, and it becomes of primary public interest to say who is to do the broadcasting, under what circumstances, and with what type of material. It is inconceivable that we should allow so great a possibility for service, for news, for entertainment, for education, and for vital commercial purposes, to be drowned in advertising chatter, or [be used] for commercial purposes that can be quite well served by our other means of communication [69].

But who was to make decisions regarding issues such as advertising? In January, 1923, when the House of Representatives passed a bill to regulate radio — the first new legislation since 1912 — it appeared that the Commerce Department might win the authority to bring order to radio and to preside over its evolution. However, the corporate owners of radio had other purposes, the chief one being, as RCA board chairman Owen Young stated, that there be no "regulation in advance of profits." So while the House acted, RCA representatives were urging members of the Senate to proceed more slowly. Though competing for control of commercial radio, AT&T joined RCA's campaign, protesting particularly the stipulations that forbade monopoly corporations from owning stations [70].

Whatever was to be decided had to be decided soon. At the second Radio Conference held in 1923, Hoover explained that whereas a year earlier "there were 60 broadcasting stations . . . ; today there are 588." Then "there were between 600,000 and 1,000,000 receiving sets, today it is believed there are between 1,500,000 and 2,500,000 persons listening." At the third conference, held the next year, the corporate lobby cooperated even more closely. Though AT&T planned a fifty-station national network to compete with RCA, each company agreed that the high costs of managing stations prohibited the proliferation of stations. Even the struggles between public and private entities were put aside when a broadcasting band of frequencies was arrived at that removed amateur and military stations from places on the spectrum desired by commercial interests. By 1924, then, the basic tenets of broadcasting had been decided: it would be a national operation; advertising would provide the chief source of revenue; transmitters would become more powerful to meet the problems of interference and to increase station audiences; and, in the spirit of super radio, the number of stations would become fewer. Secretary Hoover, moreover, now accepted both advertising and concentrated ownership as acceptable tenets of a national policy [71].

Sarnoff's antiadvertising position had, in any case, served more as a "tactical maneuver" in the battle to win control of radio. By 1925, an RCA committee submitted a report prepared by its chairman, Alfred Goldsmith, then chief broadcast engineer of RCA, which strongly suggested that advertising would become the company's chief source of revenue. Written in a question-and-answer format, the report asked, "Is there any way" RCA could support its operation "without going into the advertising business?" Goldsmith's committee answered that "there is no way in which the Radio Corporation could secure such financial returns outside of the advertising business." So unpopular was advertising with listeners and the radio magazine editors, however, Goldsmith concluded that RCA's radio sales, dependent on "the good will of the broadcast listener . . . might be jeopardized" by advertising [72].

Despite these listener preferences, by the time Congress established the Federal Radio Commission (FRC) in 1927, advertising had become the dominant mode of financing radio. When, the year before, the courts struck down the Commerce Department's right to continue its extralegal regulation of the industry, "a mad scramble to get on the air ensued and a broadcast of bedlam resulted," threatening "the very existence of the industry." The final result was the creation of the FRC. Granted both administrative and quasi-judicial powers, the bipartisan Commission was empowered to classify and regulate the operation of stations as to frequency and type of apparatus, to act to prevent interference, and to assign areas of service. Additionally, it could summon witnesses when investigations required them. But the settlement was not permanent. After one year, all the powers of the Commission would return to the Department of Commerce, and the FRC would become an appellate body for hearing disputes or conflicts. However, new radio acts, in 1928 and 1929, ensured that the FRC's authority would never go to the

Commerce Department. Then in 1930, two more acts achieved what President Hoover urged: "the reorganization of the Radio Commission into a permanent body." The powers given the FRC had in effect gained permanency, so that when the Federal Communications Commission (FCC) was established under President Franklin D. Roosevelt in June, 1934, it received all the powers of the FRC, as well as those held by the Secretary of Commerce with respect to radio and those held by the Interstate Commerce Commission over the telegraph and telephone [73].

Even during the Federal Radio Commission's first two years, in spite of vacancies and uncertain tenure, the Commission made decided progress in bringing order to the ether. The Institute of Radio Engineers played a critical role at this time. Since the Commission hired no engineers until August of 1928 and did not establish an Engineering Division until the end of 1929, the Commission asked the IRE to study and report on a proposed system to assign broadcasting frequencies. The IRE Board assigned the task to the Institute's Standardization Committee, which submitted its report in May, 1927. When additional requests came from the Federal Radio Commission, the Institute of Radio Engineers' Board established a Special Committee to work with the Commission and, in November, formed a permanent Committee on Broadcasting. Its assignment was to deal with the "problems brought before the Institute by the Federal Radio Commission on broadcast." With Lewis M. Hull, a young Harvard-trained physicist, as chairman, the Committee continued into the thirties to respond to requests for studies of radio problems. So close was the IRE to the FRC during these early years when technical problems predominated that, in 1929, two of the five commissioners served also as IRE Board members [74].

For radio engineers and their professional society, the simultaneous establishment of a national business and governmental order in radio had required both cooperation and accommodation. The process of adjustment, which had begun when the nation's radio interests gathered under the aegis of the Commerce Department at the beginning of the twenties, had effectively ended in 1934. The radio order embodied in the Federal Radio Commission was absorbed without change by the Federal Communications Commission. Only then, when the FCC moved between obeisance to the needs of RCA and the social questions raised by the Great Depression, did the leading members of the IRE turn to other concerns, especially to the rising field of electronics. But first, some of the leading radio engineers of the Institute found it necessary to declare their independence from the industry.

Radio broadcasting: The engineering response

In an address before the National Association of Broadcasters in 1931, President Hoover characterized the work of the Washington Radio Conferences as being of "unending importance." He celebrated their work for establishing radio channels as "public property" to be "controlled by the government." Proudly recalling that the conferees had rejected the European taxing device of licensing receiving sets, placing channels, instead, "under private enterprise where there would be no restraint upon programs." Thus, a greater variety of programs had been made available "without cost to the listener." Hoover apparently forgot the bold challenges he presented at the 1922 conference, when he warned against filling the waves with advertising chatter and generally misapplying radio technology. However, despite Hoover's forgetfulness, radio engineers took up these issues at the end of the twenties, publicly questioning a radio settlement that had left radio almost entirely a mass entertainment medium [75].

Criticisms of this settlement exploded upon the IRE in 1930 when Lee deForest made the state of radio the central theme of his presidency. Although, by then, he had entered fully into his experiments to add sound to motion pictures, his early innovations on the vacuum tube placed him forever among the founders of the medium. deForest had finally abandoned radio in 1923 when he sold the deForest Telephone and Telegraph Company to a group of "Detroit capitalists." From that time, he decided, "radio progress was to be none of my concern." He meant its technical progress, however, for deForest was ever after concerned with the quality of popular broadcasting.

Lee deForest, on the eve of his presidency of the Institute of Radio Engineers, attacked the domination of radio by business values.

He used every opportunity offered by his presidency to lambaste the "degraded standards" of "commercial broadcast": first in his acceptance speech in New York in January; then in a welcoming speech at the annual conference held in Toronto, Canada, in August; and finally in a valedictory address in early 1931 [76].

Even in 1923, he had believed radio "was already aboard a raft, 'sold down the river' and traveling fast." By 1930, it seemed to deForest that radio had arrived at its destination, though not at the port he would have chosen. The pilots of radio's disastrous journey, the inventor charged, were the "broadcast chains" as well as the individual stations that had given in to "the greed of direct advertising." As a result, they had destroyed the "usefulness" of this new means of communication, "which we engineers have so laboriously toiled to upbuild and to perfect." And their chief weapon, deForest insisted, was "this reptile of etheric advertising." Though admitting the impact of a "general great depression," deForest attributed the decline in receiver sales—which was hurting manufacturers especially hard that year—to listener dissatisfaction with both "distasteful advertising" and the poor quality of many of the programs. This, he predicted, would lead to the loss of large numbers of listeners and, eventually, to a decline in advertising revenue as well [77].

Criticisms of advertising appeared commonly in the editorials of radio magazines and were expressed by other leading radio engineers who had been around at the birth of broadcasting. In a paper to the IRE the next year, in which he offered a broad critique of the state of public broadcasting, Admiral Stanford Hooper listed advertising as one of the fundamental problems of the medium. Similarly, the chief of the Federal Radio Commission confessed that he too feared the loss of radio listeners from the excessive advertising. Some engineers, including deForest, went so far as to devise "advertising eliminators." Of several announced in 1934, one operated manually to short the antenna-ground and "reduce the talk to mere audibility." deForest's device was electronic; he placed a photocell so that he could control the on-off switch from a distance with a pocket flashlight. At a meeting of the American Association for the Advancement of Science in 1933, a professor from Tufts College demonstrated an "announcer killer." His device used a detector amplifier similar to an automatic volume-control system, which, on detecting a silence of one-quarter second, turned the radio off for ten seconds [78].

But the response to the state of the radio industry went beyond critiques of advertising. deForest did not restrict his comments to complaints about the quality of radio programs. Rather, he entered the realm of political protest, urging engineers to seek control of the industry by means of the federal government. deForest's concern rested on his belief that radio was "the most . . . potent means for entertainment, culture, and education which mankind had ever devised." Compared to electric lighting and power, radio was "something finer, more powerful," exhibiting a "peculiar . . . influence upon the human mind, the intimate daily home life, nay, the very spiritual life of man" [79].

In the face of radio's capacity to influence, deForest urged radio engineers to use the Institute of Radio Engineers to change the social applications of radio. He deplored the attitudes of engineers who thought the Institute should stick to the "engineering aspects" and allow the manufacturers and "outside businessmen"—only "recently come into the picture"—to control radio broadcasting. "We members of this Institute," he declared, "should be jealous" of the radio engineer's achievements and insist on a "wise supervision" of broadcasting. Because the Institute had too long failed to protest, it was necessary for the members to "take active steps (in Washington if necessary) to rid ourselves of this . . . killing avarice" in the radio industry [80]. In deForest's final message to the IRE, he moved beyond tactics to begin to sketch an alternative scheme for organizing and financing the radio industry. It seemed "hopeless" to him that the Congress would levy a tax on receivers and radio tubes to support "fine programs" or "authorize any censorship of radio programs." In their stead, he recommended an economic structure that retained private ownership, yet put broadcasting stations into the hands of manufacturing companies whose profits came from radio sales. This, of course, was Sarnoff's source of profits in the early twenties, as it was Westinghouse's when they opened the first commercial broadcasting station in 1921. Alternatively, program sponsorship could be retained with advertisements reduced to "the barest announcement" naming the sponsor [81].

deForest did not singly call for reform in the radio industry. After 1930, a growing number of engineers protested both the state of radio programming and the engineer's lack of power amid the degradation of their achievements. Like deForest's criticisms, these protests came in part from the growing possibility that the Federal Radio Commission might soon be fundamentally revised. But the engineer's complaints ranged far beyond a legislative act. The industrial consolidation of radio during the twenties by private interests had changed the character of discourse. IRE members, like Stanford Hooper and Lewis Hull, assumed, as had deForest, that while private interests dominated the radio order, the solution necessarily involved government.

Because the problems with the radio industry were specific, however, IRE spokesmen advanced specific reforms. At the IRE's 1931 convention, Admiral Hooper discussed at length the radio engineer's responsibility in the face of the misappropriation of the radio art. Hooper spoke both as the director of the U.S. Naval Communications and as a distinguished member of the Institute. In an address entitled "The Spokesman for the Radio Engineer," Hooper reiterated deForest's basic critique: "too much advertising," "too much competition," and the "depression in radio manufacturing" (which he attributed to overproduction). The confusion in the profession on these issues stemmed from the absence of a "standard bearer to speak to and for the public . . . before the government." His initial assumptions were that "engineers are in need, greed is ruling, and the interests with the best legal talent most often obtain the desired channels." Such a situation, he believed, was "an indictment of the radio engineer." The fundamental question before the profession, therefore, was that of the control of radio technology:

> Engineers as a rule have vision and are practical men. They evolve an idea for something worth while and then invent the apparatus, but why

does their original idea for the proper application of the system have so much difficulty in realization [82].

Part of the engineer's dilemma lay with the narrow education they received, Hooper believed. Engineers lacked "the practical knowledge and confidence" needed to promote their interests in economic and political arenas. Hooper's assessment was a familiar response to the problem of the engineer's lack of control over his technical achievements. Yet he did not think that understanding alone would achieve the engineer's goals. For that, engineers needed the power to influence the character of the engineering context. Besides a broader education, Hooper called for political clout through an appointed spokesman backed by an organized body that instructed its agent "in the policies of the engineer" [83].

To form a policy that would make the IRE "a vital factor" in the life of the nation, Hooper outlined a scheme for opening up the organization. He wanted to devote one-third of the IRE's *Proceedings* to political and economic matters. Beyond that, the journal could be used to poll members on the pertinent issues. Thereby, the radio engineer would empower himself and his inventive accomplishments by making known his views on the vital issues: advertising; the "question of ownership and operation of radio"; qualification for radio commissioners; the question of a "tax on radio sets"; whether there should be a "radio patent pool"; the efficiency of the government's radio stations; and the "cooperative regulation of production." Yet the basic question was the one of control: If the radio engineer "continued as at present," Hooper told the engineers, "the fortunes of fame and wealth resulting from your achievements, and the great wealth of your imaginative genius will pass to manufacturers, wholesalers, jobbers, politicians, and lawyers; you will only be the slaves" [84].

Hooper had raised the big question, the one most liable to marshal the forces of corporate enterprise, when he opened up questions of ownership and management to the engineering community. Not even deForest had left those questions open. At the IRE meeting in Toronto deForest had declared that unless business interests "voluntarily cured" the evils of advertising or unless "engineers organized protests to force a cure, then we are headed straight for government regulation" and "control." It was a path he did not want but was willing to recognize as a possible solution. Hooper, to the contrary, who had not closed down any possibility, had not judged public ownership and control as inadmissable. Though he had long defended the right of private interests to develop radio, Hooper now raised the question of responsible ownership and judged it ripe for democratic balloting within the Institute of Radio Engineers [85]. The admiral's stature within the profession added an importance to his recommendations, which assured attention from the defenders of corporate control. And so Sarnoff, having only recently become president of RCA, arranged several talks with Hooper. As a result, Hooper, although he had grown increasingly critical of FRC policies in recent years, turned staunch defender of the Commission and emerged as a spokesman for the status quo in commercial broadcasting. Testifying before several

Washington groups, Hooper argued that both radio itself and the stability of the corporations that controlled broadcasting were essential to national security.

In 1933, Hooper still criticized radio performance. He chided the thirty stations in the New York area for the virtual sameness of their programming and for their habit of filling the air with popular entertainment and advertising. Yet his defense of the political and economic organization of the radio led President Roosevelt to appoint him—along with ex-IRE president J. Howard Dellinger, a senior physicist with the Bureau of Standards—to the President's communications policy committee. That group became the mainstay in the battle to keep radio in private hands as Congress considered the bill that led to the Federal Communications Commission of 1934 [86].

However, the engineer's dilemma existed outside the regulatory reach of the FCC. deForest's and Hooper's addresses had dealt with questions of the application of radio, of its ownership and control as a public medium. Each had concluded with a call for engineers to develop a political voice. In the one-page statement published in the *Proceedings* in 1933, IRE president Lewis Hull issued a similar challenge. In a brief critique of the National Recovery Administration's (NRA) recently issued "Code for Fair Competition for the Electrical Industry," Hull, thus, delineated another area in which engineering values were slighted. Only recently approved by Roosevelt, the electrical industry code was one of ten that had been hammered out over a three-month period by General Hugh Johnson and the NRA. Hull questioned an electrical industry code which omitted "formal recognition of that select class of professional laborers to whose creative effort the principal commodities of every radio manufacturer owe their origin." He attacked the pricing wars in the industry and the manufacturers' predominant use of engineering skills to design less expensive rather than better apparatus. No officer of the IRE or AIEE had spoken so pointedly to the issue of the engineering worker. Edwin Armstrong had earlier raised the issue in a personal way. Angered by the tendency to let profit-seeking dictate design changes, Armstrong complained to Sarnoff, in 1930, that RCA had reduced the number of circuits in the superheterodyne. Such practices, Armstrong's biographer concluded, had begun, by the time of the depression, to diminish "engineering integrity." Increasingly, competition meant "cutting all possible corners, making minor 'gadget' improvements" at the expense of "basic engineering changes." In a similar vein, Hull had raised to a level of official concern not just the use of the radio medium but the use of engineering creativity by the industry as well [87].

Hull's belief that price-lowering tactics degraded the product was widely shared within the industry. In 1931, an editorial in *Electronics* worried over the effects of "price whittling." The writer feared that the "market is being flooded with receivers which 'just get by' " and that "poor radio reproduction" was becoming "the standard." Later, an *Electronics* writer advised that it was time "to go before the public . . . and let radio buyers really know the dangers of buying skimped sets." Thus when Roosevelt established the National Recovery Administration in 1933, the industry immediately sought "agreements on price-fixing." They aimed to end price lowering by large companies, who were able to compete by selling under cost, and to remove inexpensive radio sets from the market. A 1932 editorial in *Electronics* advised letting "the public know." Customers should be instructed to purchase receivers for their "fidelity of tone or for other technical or artistic reasons — not because one particular set is half-a-dollar cheaper than another" [88].

Although the sentiments of the Radio Manufacturers' Association (RMA, later the Electronics Industries Association) in part supported Hull's aims, the issue of creative engineering was not among the RMA's objectives as it sought a specialized code for the radio industry. The RMA appointed a Committee on Industrial Recovery, and, by the end of June, 1933, had submitted a code to the National Recovery Administration for the radio manufacturers. It described a Radio Emergency National Committee composed of six divisions, including radio receiving and television sets; radio tubes; and radio parts, cabinets, and assessories. Yet the Radio Manufacturers' Association's code veered sharply from the goals of the National Industry Recovery Act (NIRA), whose principal aims were to reduce unemployment and raise wage levels. The RMA, rather, wished exemption from the antitrust act in order to centralize and stablize the industry through controlling prices and bringing standards to the industry. Standardization was to be accomplished in such ways as presenting, in the case of television and radio sets, a "detailed classification of products by lines, and by numbers of tubes." The Radio Manufacturers' Association would forbid introducing "new radio receiving tubes except for experimental purposes" without approval from the Radio Emergency National Committee [89].

To achieve these and other goals, the RMA persisted in seeking its own code, separate from the electrical industry and the National Electrical Manufacturers' Association's approved code. However, when the RMA had almost won its argument, the codes began to fade in importance. In 1933, as the industry "bathed in the warm sunshine of prosperity, and customer orders," the issue lost its edge. Then in 1934, in spite of studies by a presidential commission and the Brookings Institution which showed that the codes fostered monopoly and hurt small businesses, Roosevelt installed an administrator who turned most authority over to the industrial groups. As the RMA's near victory was anticlimatic, so was the act of the Supreme Court the next year, declaring the National Industry Recovery Act unconstitutional [90].

Even though the Radio Manufacturers' Association's code had shared Hull's disdain for price lowering (*Electronics,* a voice for the RMA during this episode, reprinted the radio engineer's statement in full), its campaign for recognition had not touched Hull's broader criticism. Hull wanted to force recognition of the radio engineer's achievements and of the engineer's right to a voice in directing the course of the industry. "Unless the teachings of industrial history are mockery," Hull declared, the future of the radio industry is essentially dependent "upon inventive thought, both technical and artistic." The radio engineers and their employers were faced with a clear choice: "Whether to revive conditions favorable to inventive engineering effort or to continue with price-lowering as the main objective of engineering thought." Hull's choices were not the RMA's. And like deForest and Hooper, Hull desired no end to private ownership of the industry. He insisted, instead, that "competition in ideas, rather than competition in prices, is still a sane and profitable activity." Hull concluded by calling for a rehabilitation of the industry" founded on a "revival of creative engineering."

The programmatic ideas advanced by engineers like deForest, Hooper, and Hull faded as quickly as did the National Recovery Administration and its codes. Engineering creativity was not to be revived by assaulting the big companies. It would come — when it did—with new conditions and a new industry. Within the engineering profession, moreover, as suggested in a 1935 statement on the work and objectives of RMA's Engineering Division, another element figured in this engineering equation: there were contending engineering values within the IRE. The contrast between Lewis Hull's critique, for example, and the ideas of the director of the RMA's Engineering Division, Walter R. G. Baker (who, after World War II, also gained the IRE presidency), reveals the spectrum of engineering styles within the Institute. As Baker explained, the engineer's responsibility to the radio manufacturing industry was to provide a safer and better product, to work toward incorporating new "arts and services" into the standards framework, and to coordinate "the technical effort of the industry" in order to bring "better service to the consumer." Baker drew the contrast himself, asserting that while "the work of the Engineering Division must not be at variance with technical societies, such as the Institute of Radio Engineers," it "must satisfy the practical merchandizing and business requirements of the organization of which it is a part." The Engineering Division did this mostly through standards works, which dominated the Division's activities; this was the "measure of [its] effectiveness." It was in the area of television standardization, in fact, that Baker made his mark. In the mid-1930's, television was "in such a state of flux" and "so far from commercialization," that any serious attempt to standardize would have been futile, Baker believed. Yet insofar as the Division assisted the industry, it did so by developing standards and, thus, providing a common "language." For this reason, despite the unformed state of the medium, RMA's Engineering Division undertook to establish standards for television. This work, as it turned out, provided a measure not only of the Division, but also of Baker. His television standards work not only brought order to the art but also carried Baker to the leadership of the National Television Standards Committee in 1939 [91].

A growing reputation also brought Baker, by 1939, to the position of manager of GE's radio and television work at both the Schenectady and Bridgeport plants. But Baker had been moving toward this position since entering radio work at the end of the First World War. After earning a bachelor's degree in 1916 and a master's degree in electrical engineering at Tufts College two years later, Baker began at GE, testing transmitting and receiving apparatus for the government. By 1926, he was in charge of radio development, design, and production, being responsible, as well, for a lengthy research program in short-wave propagation. After RCA Victor was organized in 1929, he spent six years there as head of radio engineering, then of production, and, finally, as vice president and general manager in charge of research and engineering. Because Baker outspokenly favored frequency modulation (FM), Sarnoff at first planned to put Baker in charge of a task force to push the development of FM and television. However, Sarnoff became fearful that Edwin Armstrong's innovation might cause a revolution in radio broadcasting by supplanting the already established amplitude modulation (AM) system. Thus, in 1935, Sarnoff arranged for Baker to return to the General Electric Company. There Baker joined the work in radio, which GE was just reentering [92].

Although Baker had long been associated with the Radio Manufacturers' Association, in 1935, he expanded his participation as the association became interested in the development of television. Following an RCA demonstration of an electronic television system at an RMA meeting that year, the RMA board requested that Baker's Engineering Division take up the issue of television standards and establish a technical committee on television. The next year, when the FCC asked that the "various branches of the industry" come to an agreement on television standards, the RMA set up an additional committee on allocations. Though the technology was in flux, in 1936, the RMA established standards very close to the 525 lines, 60 fields per second, and 4-Mhz video bandwidth that would be adopted in 1941. This latter decision, in fact, issued from Baker's efforts, since, when the FCC formed the National Television Standards Committee in 1940, he led the exhaustive study that followed. However, Baker's career at GE took him into an area far broader than the work of determining standards for monochromatic television, when, in 1941, he took charge of GE's new Department of Electronics [93].

The changes then going on in the field of radio engineering were reflected by the founding of GE's new department. Yet on entering this world, Baker converged with an individual who revealed more dramatically than Lewis Hull the complex character of radio engineering on the eve of World War II. This came during a struggle within the IRE in 1940 over who would be president. The contest was between Baker and Frederick Emmons Terman, an energetic California engineering professor. The victor, Terman, was a petition candidate recruited by Haradan Pratt, an industry engineer who had been president just a few years before, and other members, especially a group from the Midwest. They believed the nominating committee had been filled to ensure Baker's nomination. However, as Terman later explained, the dispute was not an academic-industry split, for his recruiters also came from industry [94].

To Terman, rather, Baker represented the "Boston-New York-Washington axis." But, more specific to the dispute, Baker's interests centered in engineering management and standards. During the 1930's, for example, while Terman fed papers to the IRE, "several per year, sometimes as many as half a dozen," Baker was publishing articles on "Building an Engineering Organization" and "Stimulating the Engineer" in industry magazines. During these years, Baker rose as an engineering manager through the ranks of RCA and GE, emerging also as a leader of the standards movement for the radio and television manufacturing industry. In this area, too, Terman veered from

Baker. Terman was not, he said, "of the 'Standards Type.'" Because he wrote books, Terman once complained, "people seem to feel that . . . I am interested in the details of definition, nomenclature, etc., whereas, actually, these things to me are means to an end." Terman made significant contributions to the knowledge base of his field, and believed, moreover, as both the IRE and AIEE Constitutions stipulated, that the highest order of engineering values involved the accomplishment of original technical work [95].

Yet it was Terman's recent work in creating a strong West Coast branch of the IRE and his eminence in the emerging field of electronics that led Pratt and the others to seek him out. Their choice — and the outcome — held great significance for the profession as well. When Baker later won the IRE presidency in 1948, therefore, he did so only after Pratt and Terman gave their assent — which they did only after deciding that Baker's efforts in a building fund campaign had been creditable and that "the way the Institute is organized now" ensured that Baker would be unable "to force any radical changes" [96]. It was not just that Baker dwelled on the commercial end of the engineering spectrum. Rather, he was a traditional broadcast engineer, whereas Terman's reputation rested on achievements in the rising electronics field. Terman, therefore, dwelled on the technological end of the engineering spectrum. And it was in this direction that the IRE traveled, carried by engineers like Terman. Early in the twenties, he traveled to the East Coast for an advanced education (much as nineteenth-century American students studied in Europe) which fired his interest in the vacuum tube. Terman then returned to the West to transmit his knowledge to his students. By 1940, he had written widely-used textbooks and handbooks. Terman, in brief, was helping redefine the field of radio engineering and, with it, the constituency of the IRE. The technical crest that Terman rode, moreover, would reorient the profession no less than a world war and the changes which came in its wake.

6/THE NEW WORLD OF ELECTRONICS ENGINEERING

There was the Steam Age, then the Electrical Age, and later came the Radio Age. Now I think we are on the eve of a new age—the Electronic Age. Just as electricity electrified industries and life in general, so these developments that you gentlemen are producing with tubes and circuits will electronize industries and create the Electronic Age.

David Sarnoff, 1941 [1]

The tree of electronics

Signs of an embryonic electronics engineering field appeared widely in the early 1930's, being most visible among the activities of the maturing radio-manufacturing industry. One such sign was a trade show held in 1930 by the Radio Manufacturers' Association in Atlantic City, New Jersey. As usual in periods of growth, it was larger than any held before. Nearly 200 manufacturers exhibited radio sets, tubes, and parts and accessories in a massive auditorium that, the exhibit organizers pointed out, had cost $15 million to build. On the second day of the trade show, the Institute of Radio Engineers sponsored a special session, a session not unlike a number of others that would be held by the RMA's Engineering Division throughout the week. Besides the engineering sessions, meetings of the Radio Wholesalers and the National Federation of Radio Associations represented the commercial side of the Radio Manufacturers' Association [2].

Yet there were far more spectacular signs of the electronics age during this period, most of which the newly founded *Electronics* magazine dutifully reported. In the year of the RMA's trade show, *Electronics*' editors bragged that three city blocks were to be leveled in New York and a Radio City raised as a "temple of electronics." In 1933, the magazine described the World's Fair in Chicago as "an all-electronic exposition." A "quantum of light energy from the star Arcturus caught by a photocell opened the fair," and 15 miles of "gaseous-tube lighting" outlined building facades, illuminating and coloring sprays of water and steam. One exhibit explained the Edison effect; in another, a large vacuum tube contained an "emitting filament and two fluorescent plates" to indicate the flow of current. A foot-high radio tube showed "how plate current is controlled by grid potential." Numerous exhibits celebrated the tube's wide usefulness with "demonstrations of amplifying processes, of oscillation generation and control, of trigger type tubes—thyratron and grid-glow—of capacitive effects, of continuous wave modulation . . . and of many other electronic phenomena." These, the editors promised, would "cause the visitor to stop and look, and, shortly, to understand" [3].

That the vacuum tube stood at the center of the exposition made the fair an accurate picture of the rise of electronics. For, by design, the exhibits displayed the tube's uses in ways beyond broadcasting, extending to the "many non-radio applications of the electronics tube and associated circuits." The founder and editor of *Electronics,* Orestes H. Caldwell, stressed the tube's breadth in a 1930 address to the IRE, titled "Radio's Contribution to Modern Civilization." Speaking in the year of the magazine's founding, he predicted that within a short while most of what "the average man hears, sees, or buys" will have been touched in some way by the electronic tube. In a manner that became common in the pages of his magazine, Caldwell ended his paper by listing almost 200 areas of application. Similar lists appeared regularly in *Electronics,* frequently as a slim column of words below the masthead and beside the lead editorial: "radio, sound pictures, . . . carrier systems, beam transmission, . . . therapeutics, traffic control, . . . machine control, television, . . . crime detection, geophysics" [4].

Several times during the 1930's the lists were augmented graphically by a growing "Family Tree of the Thermionic Tube." The tree of electronics sprouted from the seminal work of Edison, Fleming, and deForest. Its roots reached into "previous research," and the trunk rose out of the Edison effect of 1883. Ambrose "Fleming's valve — a two-element rectifier" of 1905 — and "deForest's audion — introduction of grid or electrostatic control" of 1906 — sat at the top of the short section of solid trunk before the tree divided into three main branches. From Fleming's achievement came one branch of "Two-Element Tubes" consisting chiefly of rectifying tubes and power magnetrons. After deForest's invention, the remaining two of the three main branches reached up from the trunk. One led to "'Soft' tubes" — chiefly thyratrons — and the other to numerous types of "Three-Element Tubes" — including power oscillators, loud-speaker power tubes, alternating current tubes, and the power and screen-grid pentodes.

The bushing out of the tube tree outlined the growth of electronics during the three decades after Fleming's and deForest's breakthroughs. Single items continued to appear on the tree, such as Langmuir's 1912 "pure tungsten emission." In the mid-twenties, the tree's shape turned abruptly out, depicting the great multiplication of electronic tubes from the late 1920's. The caption below the 1935 tree reinforced the picture of rapid growth, pointing to the addition of "tubes unthought of in 1930."

The magazine and its tree depicted the years around 1930 as pivotal, especially in suggesting the onset of a period of rapid increase in the uses of the tube. The explosive growth rested on a web of events and circumstances. Electronics — as an industry and a technology — had evolved from the work of Fleming and deForest early in the century and would continue to advance well beyond mid-century through innovations in electronic devices and the expanding potential and applicability of the electronic computer. Nonetheless, electronics — implying not only the physical devices but also the fact of wide applicability — moved decisively beyond the radio tube during the twenty years between the world wars. The increasing uses of the vacuum tube appeared in the school curricula — first as communications and radio courses, then electronics; in entrepreneurial efforts across the continent; and in institutional departures ranging from new technical magazines to efforts by an engineering society to expand its technical territory.

But however advanced this movement by 1940, World War II brought unparalleled growth, deeply transforming both discipline and profession. Swollen wartime expenditures in research and development concentrated work in the field of electronics at the same time that it inaugurated a technology of nuclear fission. And although private industry would continue after the war to employ most electrical engineers, the unprecedented continuance of the wartime military organization and of military funding in peacetime deeply altered the world of electrical engineering.

Electronics magazine and the decade of the tube

Beyond the reportorial aims of *Electronics,* this self-described "engineering journal" had another critical mission. It promised to nurture a "scientific vision to look above and beyond the present." Anticipating "a thousand new uses" and "great and greater achievements" for the tube, the statement of purpose in the first issue concluded with the speculations of a handful of leading engineering scientists on "the future service of electronics to mankind": from Thomas Edison and the two men who revolutionized the early world of electronics with their innovations, Ambrose Fleming and Lee deForest; from two of the captains in the movement to bring national science into partnership with military and governmental agencies, Robert Millikan and Henry Davis; and finally, from the chiefs of research for General Electric and Bell, Willis Whitney and Frank Jewett [5].

Edison, introduced as "the discoverer of the first-known evidence of electronic action in a vacuum," foresaw an increase in "powers and capacities in future vacuum-tube design and operation." He believed further investigations of the three kinds of tubes would "open a field for research in physics, chemistry, electricity, heat and light, beyond imagination." Edison foresaw improved rectifying tubes that would simplify long-distance power transmission. The potential of direct-current high-tension transmission similarly impressed deForest. But deForest contrasted this with a vision of "oscillator tubes of minute dimensions," enabling "physicists to generate undamped wave-trains having frequencies approaching the infra-red." Besides extended uses in radio and television, deForest saw applications of the tube also in the fields of medicine, agriculture, and aviation.

Millikan, the physicist, contrasted the tube to the German-American Ottmar Mergenthaler's typesetting machine. Mergenthaler's machine rested on the accumulated inventions of thousands of years, whereas the electron tube, though only a "physical appliance," represented a "new physical principle." Unlike Mergenthaler's achievement of the late nineteenth century, then, the tube marked the beginning of a technical event. From his vice president's position at Westinghouse, Henry Davis viewed the promise of the tube from the perspective of the electrical industry. Although its benefits already proliferated, Davis looked ahead to its "great future . . . in other fields than radio," specifically in the power industry. Already, the power ratings of electronic tubes increased at a "rapid pace," presenting seemingly "unlimited possibilities . . . as rectifiers, converters, transformers, arresters, etc., on power lines and in industrial applications."

Whitney's and Jewett's views from the nation's great industrial research laboratories came last. They too focused on applications, especially those that made the tube more versatile to the researcher and the manufacturer. Since so much had been achieved with small vacuum tubes, Whitney thought the new metal tubes would take engineering researchers far. Jewett was struck by the transformation of "erratic and inefficient tubes," which handled small amounts of energy, into "rugged relatively efficient devices" moving "many kilowatts of energy." Though many uses of the tube remained obscure, Whitney predicted that, as in the case of the X-ray tube, "having seen our bones, we ought now to see what more we can see."

More than any of the commentators, Whitney caught the quality of the vacuum tube that made it the device upon which modern electronics could rise:

> Heretofore, when electric power was shoved about, rotated, reversed, switched or modified, it was necessary to move large masses of metal, but electronics seems to separate the mass or weight of apparatus from its electrical properties, so that in a sense we may leave the masses fixed, and just move or direct, put brakes on, or stop, the electricity itself.

In short, the vacuum tube made it possible to control electrical energy by electrical rather than mechanical means, and to do so rapidly. It would be replaced within two decades by the solid-state transistor, already glimpsed by scientists and engineers during this decade of the tube [6]. Yet the transistor, a slightly altered chip of germanium — later silicon — did not function differently so much as it advanced the technological areas opened up by vacuum-tube development. Transistors, rather, extended the qualities Whitney and Jewett had seen as issuing from the improvements made in vacuum tubes during the twenties: those of efficiency, sturdiness, durability, the ability to carry high and low currents, and reliability.

By the end of the decade, then, it was clear that the early signs of the electronics age only hinted at what was taking place. The fuller story appeared in a more permanent form than represented by the trade fairs in Atlantic City and Chicago and less abstractly than in Radio City. It appeared in the pages of *Electronics* itself, as the magazine closely followed industrial and engineering events in the expanding world of electrical engineering. They indicated, moreover, as editor Keith Henney declared in 1940, that there had been "no depression in science or engineering!" For in spite of the depression, marked growth had taken place both in the radio-tube industry and in radio tubes. Though in 1940 Henney was still looking to the future for the technological revolution envisioned in the magazine's first issue in 1930, his optimism arose from a perception that steady technical advance, not revolutionary change, had characterized electronics during the 1930's.

Electronics confirmed, in short, that what it had promised in the first issue had indeed occurred. A central goal of the new magazine had been to promote and report on the growing uses of the tube in "factories, mills, machine shops, packing plants, canneries, laundries, printing plants, garages, repair shops, restaurants, office buildings, stores, and hotels" [7]. The pages of the magazine further documented the broadening of electronics into industrial application and engineering research. Between 1930 and 1935, the terminology alone illustrated the expanding view, moving from "vacuum tubes" to "electron tube" and finally to "electronics," in the designations used in the annual indexes of *Electronics*.

In focusing on the collective efforts of researching, developing, and producing the vacuum tube, *Electronics* had amply covered the work going on in the established firms in the East. In its first year the magazine reported the appearance of GE's FP-54 tube as a "sensitive electrometer" to be used in such places as research laboratories and astronomical observatories. There was news also of Westinghouse's ignitron, whose high power gave "resistance welding . . . a new and powerful tube," and of independent inventor Stuart Ballantine's variable-mu tube, which he developed at the Boonton Research Corporation laboratories in New Jersey. These reports were only a few of the many that documented the electronics work done in companies like RCA, Bell, the electrical manufacturing companies, and the independent laboratories. Reports of the use of vacuum tubes in receivers for automobiles and for controls in industry accompanied notices in the magazine of continuing advances in tube design. Cathode-ray tubes and metal tubes were among the scores of developments mentioned [8].

For all their careful reporting on industrial and engineering developments, the editors of *Electronics* focused on the activities of the large East Coast companies and failed to see the signs of a West Coast industry in the making. The history of the electronics industry in the West was similar to that in the East with the Marconi Company dominating the California wireless industry before World War I. Two small companies, United Wireless and Kilbourne & Clark, competed in the sale of spark sets until Marconi bought out United Wireless.

However, the competition picked up when a recent Stanford graduate returned from the East, in 1909, with rights to the electric arc-transmitting system of Vladimir Poulsen. With Poulsen's apparatus, the Federal Telegraph Company began as a manufacturing arm of the Poulsen Company and, for twenty years, played a large role in the West Coast industry. Lee deForest found work and a laboratory at Federal to develop the Audion when his own company became embroiled in charges of fraud. With the Poulsen arc system plus Navy contracts, the Federal Telegraph Company prospered only to shrink in the twenties as Poulsen's arc system declined in use, reducing Federal Telegraph to solely an operations company. Then in 1928, the year after the International Telephone and Telegraph-controlled McKay Radio Company of New York purchased the firm, Federal Telegraph turned back to developmental work. This time, it concentrated on the vacuum tube and quickly became the largest employer of electronics engineers on the West Coast.

Bell Telephone was just one of the big East Coast companies whose researchers were transforming the tube from a radio device to an electronics device.

Thus, it was a serious loss for West Coast electronics engineers when, during the depression, IT&T moved Federal Telegraph's facilities to New Jersey [9].

Other small companies remained, however, and these too were required to define themselves against a background dominated by big eastern companies, specifically RCA. The case of Heintz and Kaufman makes the point. In 1919, Stanford graduate, Ralph M. Heintz, started a small company that made and repaired scientific instruments. Joining with Jack Kaufman a decade later, the firm of Heintz and Kaufman began with a large contract from a San Francisco steamship company to build radio stations and equip a shipping fleet of thirty-five ships. To develop a unique system on which the company could build a profitable business, Heintz devised a tube he called the Gammatron. A principal challenge was to get around RCA patents, which he did, though at the cost of some efficiency. Developing the tube formed the chief activity of most West Coast companies. And, as with Heintz and Kaufman, success normally required consideration of RCA's position [10].

This was clear to William W. Eitel, who headed Heintz and Kaufman's tube laboratory. He started his own firm with Jack A. McCullough in 1934. Rather than skirting RCA-controlled patents, the small company took another familiar path that was accessible to new West Coast companies: designing tubes for a market — in this case amateur radio — that the Radio Corporation was not interested in. Frederick Terman, who had returned from graduate school at MIT in 1925 to begin teaching in Stanford's electrical engineering department, succinctly described the patent situation as Westerners perceived it: "RCA dominated the patents and you couldn't leave RCA out, and if RCA was brought in it wanted to boss everything."

The corps of talented engineers that grew up around Stanford and the area to the south had nonetheless succeeded in creating a regional electronics industry. When the Federal Telegraph Company moved to the East, its leading engineers stayed in California. Two who remained were Leonard Fuller and Charles V. Litton, a recent Stanford graduate. They had designed innovative tubes for Federal Telegraph as the basis for electronic systems that both improved and circumvented RCA patents. Litton even designed machine tools and specialized apparatus to manufacture tubes. When IT&T closed down the West Coast operation, Fuller moved to the University of California at Berkeley to head the electrical engineering department and Litton founded the Litton Engineering Laboratories, beginning what Terman described as a "one-man operation." Litton, "had to do most everything," Terman recalled; "he didn't have the ingredients for expansion." As a result, Litton Engineering Laboratories grew into a corporate leviathan only when it was purchased in 1953 and combined with seventeen additional companies. At that time, Charles Litton still operated "a small vacuum tube company" [11].

As Terman also remembered, Litton "had lots of good ideas." Terman sensed a general "creativeness" among the community of engineers gathered in the San Francisco Bay area, which explained the growth of "an indigenous tube industry." Litton's involvement in David Packard's research at Stanford captured the spirit of this community. Packard had been a student of Terman's before going East to work for General Electric. But when Terman found money that would support Packard and his research at Stanford, Packard returned to Palo Alto. Litton got the money from the Sperry Company, whose growing interest in radar led them to support Packard's vacuum-tube research. Litton's skills and ideas were, thus, combined with Packard's knowledge of GE's vacuum-tube work [12].

In 1939, Packard joined William R. Hewlett to form a company. The dominant position of the large eastern companies spurred their creativity, too, leading the two California engineers to build electronic instruments and to introduce an audio oscillator as their first product. Hewlett and Packard represented one of two kinds of electrical companies which were emerging in the West. In southern California, the companies in competition with eastern companies like Philco, Zenith, and RCA mainly made home receivers and generally copied their designs from the products of those eastern companies. The small companies gathering around Stanford in the north, however, were moving into new areas. Hewlett and Packard, Terman observed, were "creating . . . new kinds of instruments . . . from scratch," which "weren't available elsewhere." Heintz and Kaufman, too, produced vacuum tubes that "nobody else was making in the country." As for Litton: "Charlie never copied anybody else. That's the last thing in the world that he had any interest in doing" [12].

One of Litton's innovations that was thought to be capable of generating high-frequency oscillations went into a group of Stanford patents relating to the klystron. The klystron was designed to bunch the electrons before being made to oscillate at high frequencies. In the middle of the research on the klystron, which had begun at Stanford in 1937, were Russell and Sigurd Varian and two members of the Stanford physics department, William W. Hansen and David L. Webster. Hansen was working on a device to accelerate electrons, and the Varian brothers sought an electronic system to detect enemy aircraft. Although they were successful, klystrons at that time lacked the power required for an effective microwave radar system. That was achieved instead by the English, who produced the multicavity magnetron, a tube that successfully operated as a high-power, high-frequency tube. The power and frequencies at which the magnetron could operate (3000 MHz or 10-cm wavelength at 20-kW peak power) surpassed existing tubes by four orders of magnitude. From such an "increase in capability," wrote Ivan A. Getting, who worked on radar during the war, "an explosive technological revolution is possible" [13].

The magnetron resulted from the same impetus that motivated the Varian brothers: defense against enemy aircraft in time of war. But it had come in the heat of war in 1939 to a group of British scientists and engineers working under the immediate need to prevent the night bombing of the country by German aircraft. By 1940, then, the revolution foreseen earlier by the editors of *Electronics* magazine was already taking place. So close was *Electronics* editor Keith Henney to the event as he scanned the accomplishments of the thirties that he failed to see the abrupt advances about to take place. As he ended his anniversary summary of the magazine's first decade, Henney observed that "electrons continue to go into industry slowly." "There is no revolution as was dreamed of 10 years ago." Not only did "broadcast stations" continue to transmit "drivel" alongside the Metropolitan Opera and Orson Welles but the technical field of "radio has settled down." Where was the revolution in "science and engineering" to come from? Henney wondered and then accurately answered: "Continued research and development into apparatus for the ultra-high frequencies point toward a vastly increased communications service in the microwaves." Expanded control of the electromagnetic spectrum—"approaching the infra-red," as deForest had written in the inaugural issue—would bring the revolution.

Henney was not off in his belief that microwave techniques would find their chief application in communications. He had hinted at the actual area in his magazine five years earlier in a photographic essay entitled "Microwaves to Detect Aircraft." Developed by Telefunken of Germany, the device was described as a "mystery ray" system that could locate aircraft in "fog, smoke, and clouds." A magnetron transmitted beams at fixed angles, which in turn were picked up by receivers placed at intervals to catch the reflected beam [14]. However, the revolution Henney looked to was taking place not only in England and Germany but in America as well. It was being prepared for, moreover, not only in industrial firms but within educational institutions and professional engineering societies.

Frederick Terman and the rise of electronics

Remarkable for this new field of electronics, one man's life and activities caught the outline of the larger story and gave substance to much of its content. Frederick Emmons Terman was not so much a representative figure as an encompassing one. Terman's importance lay in his singularity and in the complex interweaving of events and organizations that marked his passage from a college student following World War I to the status of a leading national engineer. By 1940, he had achieved the headship of Stanford's electrical engineering department, then becoming a leader in the world of electrical engineering education. In 1940, too, Terman had completed a half-dozen years of active service with the AIEE, served as vice president of the IRE, and had been elected to serve as president for 1941. Before his presidency ended, moreover, he had taken over the directorship of one of the wartime research laboratories created by Vannevar Bush and the Office of Scientific Research and Development.

Terman's professional career began on the eve of the growth of this major new field in electrical engineering. His engineering education began at Stanford, where his father, professor of psychology Lewis Terman, had created the Stanford-Binet intelligence test and had gained fame from his studies of the gifted. By 1922, Terman had earned a master's degree in an electrical engineering department headed by nationally recognized power engineer Harris J. Ryan. Working under Ryan, Terman did his thesis on a high-voltage topic.

Ryan's engineering roots reached back to the founding years of the profession. Having begun his studies at Cornell with the physicist, William Anthony, he joined in a brief consulting venture with fellow graduate student Dugald C. Jackson. However, Ryan's basic interests in research and teaching came to the fore, prompting him to move to Stanford in 1905. As Jackson had done at MIT, Ryan made his program at Stanford responsive to industry, leading him to investigate problems for West Coast power companies, often without charge. When Terman left for MIT in 1922, Ryan was planning the High Power Laboratory, which would be completed three years later, crowning, as he saw it, the electrical engineering department's already distinguished record of power research.

Thus, as Terman prepared to choose a graduate school, it was also natural that he looked for a school atypical of the low state of engineering education William E. Wickenden had described in his investigation. As suggested in Wickenden's findings, there was a basic decision for a young American engineering graduate to make. Terman met it in the advice given by one of his professors at Stanford, in whose class he was excelling. After congratulating Lewis Terman on his son's abilities one day, the engineering professor advised that "the thing for him to do" when he finished at Stanford was "to go out and get some experience." This, as Frederick Terman later recalled, was the "prevailing view." Not many students continued to the master's level and certainly not to the doctorate. However, with his father's encouragement, Frederick Terman decided to get a doctoral degree. Though a friend of his father's recommended Columbia, Terman followed Ryan's advice, who insisted that "MIT is by far the best" [15].

There could have been no disagreement on that point from Ryan's old associate, Dugald Jackson. Jackson had always perceived MIT as the nation's premier technical institute. He argued this unsuccessfully to the Chicago power executive, Samuel Insull, in 1910, when trying to gain a contribution for the department. Still, Jackson labeled MIT a national institution. Certainly, it had one of the leading programs in electrical engineering when Terman entered in 1922. Jackson reported that Course VI (Electrical Engineering in the MIT catalogue) annually graduated eighty to a hundred students with bachelor's degrees and forty to sixty with master's and doctor of science degrees. Though the latter degree was rare, Jackson accurately described the department:

> Our advanced instruction and research has grown rapidly but at the same time soundly and the Department probably has the most notable group of students in advanced studies of electrical engineering that the world has ever gathered together. The staff of the Department is correspondingly large and strong [16].

Terman not only came to a strong department but arrived at a propitious time. The curriculum was changing to recognize new fields, the commitment to research was growing, and the faculty, old and new, contributed enthusiastically and capably. During Terman's tenure, the electrical engineering department developed a Communications Option as part of Course VI. This start of an electronics program at MIT rose chiefly from the department's strong, research-oriented faculty, which was represented chiefly by Vannevar Bush and Arthur Kennelly during the decade after World War I. Bush moved from Tufts College to MIT in 1919 as professor of electric power transmission. Though he continued as a consultant to the American Radio and Research Corporation (AMRAD) and to other companies, Bush quickly established himself as an active teacher and researcher. Kennelly had direct responsibility for communications courses, however, and prior to his retirement in 1925, taught the courses in "Advanced Alternating Currents" and "Radio." He took responsibility also for "superintendence of research as necessary." Terman described one of the courses he took with Kennelly as dealing with "communications circuits, properties of long lines, theory of the telephone receiver, a little bit on filter theory in the very early days of filters—Communication type topics" [17].

Though the Communications Option was Kennelly's concern, it originated out of graduate student Edward L. Bowles' need to construct his own vacuum tube in order to pursue his graduate thesis research. From that necessity came an enthusiastic and consuming interest in the vacuum tube. His education compared with what Terman later experienced. Bowles came to MIT in 1920 after taking a bachelor's degree at Washington University in St. Louis, intending to study for only a year before entering the electric power industry. He studied under Bush and Kennelly and served as a laboratory assistant for Kennelly in his course on "Electrical Communication of Intelligence." While preparing a master's thesis under Bush, Bowles gained access to AMRAD's facilities so he could build the vacuum tube he required but could not purchase. More than any course taught at MIT at the time, this experience introduced him to the tube. After finishing his thesis late in 1921, Bowles became deeply involved in radio research, even writing a radio column for a Boston newspaper. With Kennelly in Europe, he devised the first curriculum for the Communications Option scheduled to begin in the fall of 1922 [18].

Bowles' activities satisfied a general concern. When outside committees conducted personal inspections of the department in 1922, considering especially plans for the future, the Visiting Committee urged attention to "advanced courses in physics such . . . as underlie devices like the vacuum tube, suddenly come to importance in electrical engineering." Bowles had requested a laboratory, and the committee recommended "space . . . for laboratory work in electrical communications." Continuing the innovations encouraged by the committee, he added a course in vector analysis in 1923, then a special course, that fall, in electromagnetic theory and wave propagation. The option included a course in wire communications as well as Bowles' radio course. With Kennelly's retirement in 1925, Bowles took full charge of the option. Around this time, Jackson and others made a brief attempt to establish a radio research laboratory at MIT, to involve "particularly researches in radio phenomena, electrophysics and electrical engineering." A yearly budget for the laboratory was contemplated at $1.5 million [18].

As important as Bowles' innovations were, the department still concentrated on power transmission and machinery. During the years when the electronics courses were being introduced, the department's largest project involved the building of the network analyzer. Bush began this project out of his interests in computer-assisted analysis. More immediately, he sought to simulate large power networks to save calculation time. In 1929, when the model was completed, GE, numerous private power companies, and the publicly-owned Tennessee Valley Authority used the analyzer. This orientation to power engineering in the department led no one to discourage Terman when he wanted to continue his Stanford research at MIT. Thus, he wrote his doctoral thesis on superpower systems and the "theory of long power lines." Yet Terman perceived an exciting difference between the questions asked by MIT engineering professors and those posed by Stanford's faculty. Ryan's main interest centered on high-voltage insulation, for example; however, at MIT, Terman found the professors working on "systems problem[s]" [19].

An incident in the mid-1920's involving several members of the MIT faculty demonstrated the advanced state of the work in the East. It arose around problems that faced a large eastern Canadian power company when attempting to transmit power over 500-mile distances (little more than 200-mile lines had been built on the West Coast). Jackson's consulting firm was involved, and he brought in Bush. General Electric and Westinghouse engineers were already involved, but it was Bush who went to the heart of the problem. The great distance led him to fear a loss of "synchronism" should one of the three planned lines go out, shifting its power to the other two. This had happened in 1921 in Chicago, the nation's largest centralized urban-power network, and Steinmetz had been called in. Drawing on this precedent, Bush solved the Canadian company's problem by including transients in his analysis, whereas GE and Westinghouse engineers had used steady-state analyses.

Terman used a similar systems approach in his work on problems encountered in "the long distance transmission of power." He was specifically concerned "that the maximum power limit of a line diminishes with increasing length.This introduces problems for "operating . . . 'Super-power' systems" which do not arise with "the ordinary line." The increasing distances in transmission led to new questions. What previously had been asked, "Will it pay to transmit power this distance?" had become, "Can an appreciable amount of power be transmitted the necessary distance with the conventional type of line?"

Though power concerns dominated Terman's early engineering research, the pioneering spirit behind the Communications Option finally triumphed in him—helped by a protracted illness. As he prepared to begin an instructorship at MIT, an attack of tuberculosis forced him to return home. With the enticement of a radio laboratory established in the department in 1924, Ryan persuaded him to teach half-time at Stanford while recuperating. Terman resumed the old radio hobby that had first led him to vacuum tubes. Kennelly's course had strongly influenced him in this direction, as had a book by H. J. Van der Bijl, a Bell Laboratories engineer. Van der Bijl's book discussed vacuum tubes as associated circuits and as amplifiers and oscillators. Terman found most valuable Van der Bijl's theory about how vacuum tubes worked and the information on the latest work coming out of Bell and GE. Chiefly, however, it was Kennelly and Bush who prepared Terman to move into the electronics field. Their teachings coalesced as he built a radio receiver during his enforced break.

> I discovered that the circuit theory that I learned from Kennelly, telephone things and so on, could be tied with what I knew about vacuum tubes. Bush had taught me circuit theory too and all this tied together. I could put the vacuum tube circuits and the non-vacuum tube circuit theory that I'd learned there at MIT all together for a nice understanding of amplifiers and tuned amplifiers, and things like that [20].

In some ways, new technical understanding came easily to Terman compared to his attempts to build an electronics program at Stanford. The institutional problems he faced were far from uncommon. Stanford began a graduate program in electrical engineering between the wars, as did many colleges. But doctoral degrees were few even at the leading graduate programs in the country: Bush, in 1916, had earned the fifth MIT doctorate awarded in the field; Terman, in 1924, won the eighth. Unlike MIT, however, the distinction Stanford won in the 1920's with its high-voltage work did not extend to communications. Yet this was changing. As at Berkeley, Terman observed, most schools were "just beginning to fumble along" in the task

As a student of Harris J. Ryan (left) at Stanford University and of Arthur Kennelly at MIT (center), Frederick E. Terman (right), whose own students are active today, linked the lives of electrical engineers over the entire century.

of establishing a communication option. The AIEE's Communications Committee had reported in the early twenties, moreover, that American universities and technical schools were giving more attention to "instruction and research work in communication engineering." Besides the course at MIT, the committee cited courses at the Sheffield Scientific School at Yale and one in underwater cable at Columbia. In addition, CCNY had a "thoroughly equipped and organized communications laboratory." When Terman began his radio course at Stanford in 1926, other young engineering professors at schools like Ohio State and Wisconsin were also beginning to teach electronics courses and "to edge over into graduate work" [21].

As Terman viewed the situation, however, "no program had very much substance or volume or character." Attempts to develop strong graduate programs were even more difficult in a situation in which "the general idea among engineers was that you went to college and got a bachelor's degree in engineering and then the important thing was to go out and get some practical experience." The problems in engineering education, then, were plural: On the one hand, few faculties provided the stimulating and innovative air of MIT or Stanford, though engineering teachers like William L. Everitt at Ohio State were building strong programs; on the other hand, the American students drawn to engineering too rarely proceeded to advanced study. The problems of ill-prepared faculties were being recognized at that time by the investigation Wickenden was making for the Society for the Promotion of Engineering Education (SPEE). Another report, completed a decade later and published in 1940, gave full recognition to the problems of student ambitions. And yet, in its "Report on the Aims and Scope of Engineering Curricula," the SPEE committee acquiesced before the engineering student's near single-minded pursuit of the four-year college degree. Admitting that the bachelor's level conformed to "the interest and career requirements" of most students, the committee recommended that colleges do their best by pruning the standard curriculum to the essentials of a sound engineering course aimed at "strictly professional practice" [22].

The AIEE and the electronics engineer

In such an educational context—of curricula narrowed to practice and supported by a general acquiescence—engineers like Terman worked to build their programs. But educational institutions were not alone in struggling to respond to the blossoming of electronics and, indeed, to cultivate it actively. The closeness by which university engineering programs hewed to prevailing industrial needs left most of them ill-prepared to support revolutions in the making. This placed a special burden on the professional society. For as the application of the vacuum tube expanded beyond the traditional communications areas, the number of engineers in the field increased as well. Even accepting the IRE as the natural professional home for the new engineers, by the 1930's, the radio industry's dominant position in the profession had narrowed the IRE in much the same way that the power interests had done to the AIEE after 1900.

The narrowing of the IRE had become so acute by 1938 that the Board declined an application for transfer to the Fellow grade because the applicant worked for a motion picture studio rather than for a radio firm. In a letter to the IRE secretary, Frederick Terman urged a "reasonable interpretation" of the constitutional definition of a "radio engineer by profession." He knew the engineer in question to be "very active among the radio group in the Los Angeles territory." Just because "a man is applying his knowledge of audio frequency circuits and techniques" to the problems of motion picture sound reproduction, Terman noted, he is no less "a radio engineer" than an engineer "involved in radio receivers or transmitters." Terman wanted to "integrate the term . . . to cover all of those activities" that, though typically associated with radio, "have applications in other fields" [23].

With such attitudes dominant within the IRE, the new electronics field could not be safely relegated to the radio engineering society. Besides, at this time, the AIEE Board still considered its organization to be the "parent body" of all electrical engineering fields. Thus, in 1934, the members of the AIEE's Communications Committee decided that it was time to do something about the "Subject of Electronics" [24].

As electronics had emerged from communications programs in the schools, the AIEE's Communications Committee provided an apt home for nurturing the infant electronics field. The committee's beginnings reached back to the early century, its chairmen and members always including the electronics and radio elements in the society. The earliest version of a Communications Committee was the Meetings and Papers Committee's Subcommittee on Telegraphy and Telephony, which Kennelly chaired. Soon after his Audion paper, deForest was appointed to a newly-independent Telegraphy and Telephony Committee. The close relationship of this AIEE committee to IRE interests was signified by these men's later service as presidents of the IRE. This was true also of Donald McNicol, who began his career as a radio operator. He chaired the AIEE committee from 1918 to 1922 and, three years later, served as IRE president.

But though its lineage was long, the distance to be traveled was great, as suggested in McNicols' narrow definition of the field in 1921 as including "the arts of telegraphy, telephony and radio signaling." Yet the change of the committee's name from Telegraphy and Telephony to Communications in 1924 signaled the broadening of the field. This became explicit when the committee moved, a decade later, to respond to the electronics field. One of its first moves was to broaden the committee's membership, leading to Terman's appointment in late 1934. In accepting, Terman expressed his desire to see "the communication side of the Institute strengthened and broadened out." In addition to seeking new members, the committee wondered whether electronics should be organized as a separate technical committee or as a subcommittee of the Communications Committee. Instead, a joint subcommittee was established with representatives from areas "most directly involved" in electronics, that is, from the committees on Electrophysics, Electrical Machinery, and Communications. However, the burden of responding remained with the Communications Committee, which, besides bringing in new members, sought to attract electronics engineers to the AIEE through sponsoring conferences on vacuum tubes and by expanding its annual review of research to include electronics [25].

Besides Terman, the members of the committee included IRE's president for 1934, C. M. Jansky, Jr., a former professor of radio engineering at the University of Minnesota who worked as a Washington-based consultant. Other members were C. B. Jolliffe of the FCC and Bell Laboratories engineer Mervin J. Kelly, who had won his physics doctorate at the University of Chicago in 1918 under Robert Millikan. The committee's course had been mapped out by a temporary subcommittee appointed in the spring. It had considered the idea of a separate electronics technical committee but, instead, focused on the more substantive need for additional electronics papers at AIEE meetings and for more articles in the *Transactions.* Specifically to launch this new campaign, the members unanimously agreed to plan "a group of papers relating to vacuum tubes" for the convention in 1935. The papers they thought appropriate for AIEE members were ones that covered "the whole scope of electronics applications," treating only "general material of a broadly educative nature and not necessarily new material." The committee wanted the membership to recognize that communications was "only one of the fields" using electronics devices. If a separate committee was to represent the "obvious" electronics interest in the society, its members should be "electronics specialists in the fields of communication, both wire and radio, electrical machinery and equipment, electrical measurements, X-ray and ionization types of illumination." In spelling out these considerations for the AIEE, not only did the subcommittee contribute to a definition of the field itself, but it also reflected the society's sense of the limitations of its own membership in this expansive new area of electrical engineering [26].

This question of a place in the AIEE for electronics was linked to the issue of publishing technical papers with no immediate commercial concern. The problem had first arisen in the early 1910's, when the leaders of the American

Institute of Electrical Engineers confronted the loss of the electrochemists and radio engineers. Some members of the Board of Directors wanted to distinguish more clearly between the organizational news reported in the AIEE's general membership magazine and the original technical matter that they thought should be in the *Transactions.* In his attempts to reform the Institute in 1912 President Ralph Mershon also asked that the society be more discriminating about "the quality of the papers accepted" for conventions [27].

Reviving the issue in 1929, GE engineer Philip L. Alger urged changes in the Institute's publications policy so as to admit "high grade technical articles" like those found in some foreign periodicals. Alger asked the publications committee to consider separate sections for the major fields of communications, electrical machinery, and power generation and distribution. One member agreed that "present policy" sought "to publish only articles of immediate commercial importance," accepting "articles of fundamental scientific and engineering importance . . . only under great protest." Dugald Jackson wanted an "additional highly scientific journal," for which he offered his department at MIT as a home. But the problem persisted. A decade later, AIEE president and Consolidated Edison engineer, John C. Parker, worried about the poor reception of "highly scientific" technical papers in the *Transactions.* Those resembling a "running discussion" and those of a "scientific and mathematical character," Parker observed, had developed contending constituencies. But all agreed that probably less than 5 percent of the members could appreciate highly scientific material. The arguments of engineers like Parker rested, rather, on the long-held belief of some that the *Transactions* should constitute a "permanent . . . record for the highest type of electrical engineering material" [28].

Consequently, it was with a strong sense of this persistent dilemma within the AIEE that the Communications Committee devised its response to the new field. Accordingly, a 1934 list of thirteen proposed papers for a "Symposium in Electronics" was designed to introduce the field to members as well as to provide a forum for the work of the new electronics engineers. For the first objective, the committee sought a paper from MIT professor Edward Bowles on the "history of technological advances in electronics." The other papers also hewed close to the double strategy. There were papers on specialized tubes, but another dealt with theory, and still others treated special categories of tubes, including cathode-ray tubes, cathode-ray oscillographs, and the ignitron. Papers on ratings and measurements were also proposed, along with one on "the natural limits to amplification." Five remaining papers covered application, two of which dealt with the general topic of the industrial applications of electron tubes. The committee's attempts to reach into all corners of the field led them not only to innovate with conferences and publications but also to give attention, finally, to the West Coast [29].

This became Terman's work when, in 1936, the committee directed him to organize a West Coast subcommittee of the Communications Committee. This was only one of his efforts for the AIEE campaign, since, as Terman's reputation grew, requests from the AIEE increased. For the Summer Convention to be held that year in Pasadena, California, Terman agreed to preside at a Technical Conference on the Use of Electronics Tubes in Industry. In his report on the conference, Terman confirmed the AIEE's shaky position among those working with electron tubes. Although the sixty in attendance rivaled the number present at any of the other technical conferences, they were "primarily . . . interested in the practical application of electronic devices. No one had any extended experience in the design of such equipment." They talked mostly of welders, fader equipment, and photoelectric devices of various sorts, and gave some attention to the ignitron. Considerable time was spent on "the economic justification" for such devices when several representatives of large companies complained that "the possibilities of tubes had been overemphasized" at the expense of simple and more reliable mechanical controls. Overall, Terman added, active participants at the sessions on tubes were probably younger than attendees at most sessions at the convention, with several not being AIEE members [30].

Terman's characterization of the AIEE's Pasadena conference captured the problems of building a communications interest. The divergent styles had appeared when the Communications Committee first approached Terman to join. He was asked to evaluate the AIEE's San Francisco section. Because of its good reputation, this section had been seen as a possible model for other sections. However, Terman was unfamiliar with their meetings, knowing better the IRE's San Francisco meetings, which he described instead and compared them to what he knew of the AIEE's. Not only were the IRE meetings "less formal" and more regular, but the AIEE's engineers selected papers of chief interest to "practicing engineers," whereas the radio engineers generally reviewed current research literature [31].

A semblance of a solution to the committee's dilemma was found in the idea of joint AIEE-IRE meetings. Already common at the local level, the formal arrangements required for joint convention sessions were not broached until the thirties. They came largely in response to the broad reach of the electronics field—as power, communication, and manufacturing sectors increasingly utilized the tube. The Board of Directors first entertained the notion in 1935 when considering a joint meeting with the IRE, at that time approving the idea of a joint convention with an "engineering group" that had broken away on the basis that it would be brought "closer to the parent body" [32].

More concretely, however, formal AIEE-IRE meetings at the national level came about as the AIEE's Communications Committee sought additional means to draw electronics engineers into the society. Terman's suggestion for a joint session at the 1936 Summer Convention at Los Angeles received impetus a year later from the success of an AIEE-IRE session held during the Pacific Coast Convention in Spokane, Washington. IRE president H. H. Beverage had presided over three papers by Bell engineers, one paper on the Radiotron by an RCA engineer, and two papers by West Coast engineers. Ninety-five engineers from the IRE were registered, though "not over 10 IRE members" lived within "150 miles of Spokane." That one-third to one-half of those ninety-five engineers were also AIEE registrants suggested to Terman "a very healthy mutual interest." The "preponderance of . . . the telephone transmission type" of paper at the AIEE convention sessions in the past had not built up a good following, and the IRE "never before had anything other than a local section meeting on the Pacific Coast." Hence, the Spokane meeting suggested a new possibility: There would be no rivalry between AIEE and IRE, Terman assured Communications Committee Chairman William Everitt, because "the basic interests out here on the Pacific Coast are mutual and not competitive" [33].

For whatever reason, the Spokane meeting had been "a great success," Everitt wrote. A radio engineering professor at Ohio State —from which he received his Ph.D.— and now chairman of the Communications Committee, Everitt wanted the joint meetings to continue. He expressed to Terman his hope that the suggestion to the AIEE Board for formal action on joint sessions "will possibly come from our committee." Thus, at the meeting in January, 1938, as members planned the next convention in Portland, Terman successfully moved that the practice be made a permanent feature of the Pacific Coast Conventions. In March, the AIEE's Executive Committee formally approved the recommendation. This was an early instance of necessity forcing the two electrical engineering societies into cooperation. Yet, predictably, at the same meeting that considered Terman's resolution, a member raised the question of whether radio topics were proper for AIEE meetings. The answer unequivocally defined "the committee's job" as informing the AIEE membership on all communications topics, making radio, therefore, "a proper and necessary subject." In fact, the two engineering societies would be "stimulated and benefited by a competitive situation of this kind." And, in any case, the bylaws supported cooperation with other committees and organizations while insisting that "no activity within the field of electrical engineering shall be relinquished . . . if such activity is desirable in order that the Institute may completely serve all of its members" [34].

Certainly, the AIEE had tried during the 1930's to serve all of its members. It had created the Pacific Coast Convention to complement its Summer and Winter Conventions, and, through the agency of its technical committee on communications, attempted to create a base for engineers connected with the proliferating applications of the vacuum tube. Where the AIEE had long foundered in the task of providing a place for engineers who brought a more vigorous scientific and mathematical approach to engineering problems, the Communications Committee had some success. The literature review was a traditional source for the research engineer seeking answers to problems in innovation and in development and the committee surveyed the field several times during the 1930's. Terman twice contributed to these efforts, reporting, in 1936, on the year's work in the general radio field and contributing, in 1940, to a ten-year review by heading a group of engineers surveying advances in "ultra high frequency technique." The differences in coverage alone suggested that a long step had been taken in comprehending the complexity and character of the field. Terman's contribution to the annual report in 1936 had covered radio broadcast transmitters and receivers, television, and the ultrahigh-frequency field, including Class B and C amplifiers (to which he had contributed a study), cathode-ray tubes, and electron focusing. Four years later, when the committee sought to cover developments since 1933, the time of the committee's last full review of the literature, Terman coordinated the work of six other engineers in the area of ultrahigh frequency alone. They discussed technique, wave guides, antennas, and new methods for generating and amplifying ultrahigh frequencies [35].

Terman's personal contribution was even more specialized. It also precisely pointed to a key aspect of the revolution in electronics that *Electronics* magazine had been seeking. Reporting only on the work going on at Stanford, he announced that "a new type of ultrahigh-frequency oscillator termed the klystron has been developed." After describing in a paragraph how "an oscillatory system" was obtained in the tube, he briefly listed its characteristics and assessed its functioning, following with a bibliography of a half-dozen articles. The klystron, Terman noted, generated oscillations of "considerable intensity," at 58 percent efficiency, and could be adjusted over a limited range of frequencies. "An approximate theory . . . has been worked out," he wrote and concluded with a note on some of the successes so far won in gaining control of its functions as an oscillator and amplifier. The short list of articles revealed as much as the text. All six had been published in the *Journal of Applied Physics* during 1938 and 1939, one by the Varian brothers and the remainder by the two Stanford physicists, David Webster and William Hansen.

Of the many contributions flowing into the AIEE's Communication Committee that spring, Terman's brief report constituted a literal scoop on the IRE since its *Proceedings* had earlier rejected Hansen's original rhumbatron paper. It was to correct this that Terman, while president-elect of the IRE in 1940, urged the IRE secretary to take advantage of Hansen's and the Varian brothers' presence on the East Coast to arrange a session on their work for the January, 1941, annual convention in Boston. Terman bolstered his case before the staff at IRE's New York headquarters when he argued the importance of the contributions of the Stanford researchers to microwave circuits, noting that "RCA has gone to a rhumbatron type of resonator . . . rather than staying with the resonant transmission lines."

By this time, Terman had shifted his professional loyalties fully to the IRE. The AIEE continued, however, to pursue its campaign for workers in the electronics field. The older society's persistence came to Terman's attention when the AIEE invited him to talk on "transmission line theory . . . as used in ultra-high frequency applications" —blending his early research with the new developments— at a special lecture series on Ultra-Short Waves planned for 1942 and 1943 in New York. The Basic Science Group of the New York section of the AIEE organized the series. Terman immediately sent the program and letter to RCA engineer and vice president Arthur F. Van Dyck for an East Coast perspective on the AIEE's undertaking. Van Dyck knew of the series and had "long wished that we were the ones giving that course instead of AIEE." Terman felt as if the AIEE "had taken the ball away from us again." When the IRE's New York headquarters failed to respond to his suggestion for a similar series to be sponsored by the New York section, Terman became even more upset —a reaction that intensified when he realized that three of the speakers at the AIEE series regularly contributed to IRE publications [36].

But though the ball had gone to the AIEE, most of the players remained with the IRE. In spite of the efforts of the Communications Committee during the thirties, a 1940 canvass of the occupations and technical contributions of AIEE members revealed that only a few of them came from the communications field. Moreover, its constituency filled a shrinking corner of the society. As had become apparent in the first decade of the century, the AIEE remained a predominantly power engineering society. Over 50 percent of the membership was concentrated in the industrial fields of lighting and power and electrical manufacturing, and another 15 percent worked as operating engineers with railroad companies or factories or as consultants to the power industry. Communications firms, on the other hand, contained under 10 percent of the AIEE membership. And among the papers published in AIEE publications, those in the communications field had declined between 1938 and 1940 from 9.2 to 6.2 percent [37].

A survey of reader preferences carried out in 1940 by the *Proceedings of the IRE* revealed a decidedly different place on the theory-to-practice spectrum for the radio engineers. Of nearly 1600 respondents, 71 percent preferred papers that dealt with research and development. Among the preferred subjects, 660 members, or 42 percent, rated "theory, circuit and general" first. Next in preference came four categories, each drawing 100 members, or 6 percent. Thirteen additional areas drew significantly less, ranging from sixty-two choices for "antennas" to six for "facsimile." The four categories that drew nearly a quarter of the IRE's membership contained the core elements of the society. Predictably, the radio industry contained two of the four categories: "receiving apparatus" and "television." A third category suggested a cluster of members around "measurements and standards" and a fourth around the field of "ultra-high-frequency and microwaves." Alone in the specialized and new microwave field, then, the IRE contained a percentage of members just shy of the portion of AIEE members declaring for the entire communications field. The AIEE's attempts to win over electronics engineers with conferences, special sessions, and literature reviews had served the electronics field well, but the society had done little to draw the new engineering group into its own ranks. Van Dyck confirmed the point in a "confidential" report to Terman in which he reported "encouraging" news for the IRE: "AIEE is much disturbed over the rapid growth of the IRE & their own ineffectiveness in attracting members interested in communications" [38].

But an even clearer picture of the emerging place of electronics within the IRE resided in Terman's reason for declining to speak at the Ultra-Short Waves lecture series: He was too busy with his wartime research work. Indeed, the role of engineers and ex-IRE presidents like Terman and Haraden Pratt in the government's Office of Scientific Research and Development (OSRD) demonstrated the high currency of the field of electronics on the eve of

America's entry into the Second World War. The work of engineers like Pratt and Terman during the war fully reversed the positions held by the two engineering societies during World War I. Just as the AIEE's Benjamin Lamme and the Special Problems Committee had collaborated with the Navy and thus helped shape the research program in that war, so would prominent IRE members take leading posts in the far vaster governmental research and development efforts of World War II, shaping both wartime and postwar policy.

The "technological war"

On the face of it, the roles of the engineers of the IRE in the Second World War closely paralleled the parts played by AIEE members in the earlier war. But although an electrical engineer led in conceiving and directing the wartime Office of Scientific Research and Development (OSRD), Vannevar Bush identified closely with neither engineering society. And regardless of how much Bush's position suggested that engineers might once again play the key role in a world war, the scenario was not to be repeated. For this time, the engineer who conceived the military R&D program differed decidedly from Thomas Edison, who was director of the World War I Naval Consulting Board in name only. Although both understood the extreme importance of technology to modern war, Bush acted with a far clearer sense of the scientific basis on which a modern, technological society functioned. Instead of turning to engineering societies, inventors, and engineering designers for help in conceiving new weapons, Bush exploited the two most compelling areas of physical research in the 1930's: microwaves and nuclear fission. For this, he and the members of the OSRD's R&D arm, the National Defense Research Council, would gather, not inventors or even engineers primarily, but physicists.

Certainly, electrical engineers held prominent positions in both the OSRD and the high counsels of the military. In the shaping of R&D policy at both the Cabinet and committee levels, engineers made critical contributions. However, in the research and development process itself, electrical engineers definitely played a supporting role. This general situation plus the particular experience of Frederick Terman at the Radio Research Laboratory—self-consciously created, in part, to make a place for engineers—fashioned part of the legacy with which the engineering community would live for many years after the war. For the technical activities taken up during the war gained exaggerated influence as the wartime R&D program powerfully launched electronics as the nation's dominant technology in the postwar era.

The foundation for this rise to dominance came into place during the years of preparedness, from 1939 until the Japanese attack on Pearl Harbor in December, 1941, when engineers were virtually absent from the laboratory. Rather they reigned in policy councils as exemplified in the role of Vannevar Bush, who had gained ample experience to acquaint him with the cutting edge of scientific and technical research. After 1932, his responsibilities as vice president and dean of engineering at MIT took him beyond the electrical engineering department into the larger world of engineering. And by the end of the decade, he had entered even broader areas of research, as his talents for administration took him into the highest levels of national research and development policy. After leaving MIT to become president of the Carnegie Institution of Washington, in 1938, Bush joined the National Advisory Committee on Aeronautics (NACA) and, the next year, became its chairman. On the eve of the war, then, he stood over the world of engineering rather than within it. That both the Carnegie Institution and the NACA tended to support large-scale projects, moreover, fit well the research directions Bush would pursue during the war.

From this solid base, Bush began to gather scientists and engineers into a concerted program of research and development to devise new weapons for the war. Convinced that civilian scientists and engineers, rather than military officers, should direct the nation's wartime research, Bush took on his greatest challenge. It was a double one, moreover, for, besides building research facilities and hiring scientists and engineers, Bush needed an R&D policy from the government that would give civilian scientists significant influence.

In seeking a form for a body that might centrally direct the research activities of the war, Bush drew on his knowledge of the NACA as a central committee with satellite technical committees, composed of nonmembers, but with representation from the main body [39]. Bush began with only a rough idea of the organization he would need. He first sought from President Roosevelt a high-level committee of scientists and engineers whose members possessed a degree of scientific and technical discernment equal to his own. For its members, therefore, Bush chose such men as physicist and MIT president Karl B. Compton, Harvard president and chemist James B. Conant, and Frank Jewett, now a vice president of Bell Telephone Laboratories. National Advisory Committee on Aeronautics' secretary, John Victory, suggested to Bush that the group take a name similar to the still-existing World War I agency, the Council on National Defense, under which the new body would be created. Thus the agency created by President Roosevelt in June, 1940, became the National Defense Research Committee (NDRC), with Bush as chairman. Bush immediately formed additional committees around major technical areas: armor and ordnance, chemistry and explosives, communications and transportation, instruments and controls, and patents and inventions. Even before Roosevelt created the NDRC the members had been assigned tasks. Jewett, for example, had the task of surveying available industrial research facilities while Compton investigated the military's present and future activities, and attempted to determine what it should be doing. Bush gave himself the job of establishing relations with other governmental agencies and evaluating "special projects needing immediate attention, especially uranium-fission" [40].

But the NDRC's mission proved too narrow; it omitted developmental problems from its scope and lacked the close liaison with the military branches required to link its research and development work to battlefield needs. Bush also confronted the problem of military leaders wanting the NDRC to march

As head of the Office of Scientific Research and Development during World War II, Vannevar Bush (fourth from left) was, as one journalist described him, a czar of science in America.

strictly to the commands of generals and admirals. Therefore, Bush sought a new agency whose responsibilities extended to development as well as to research and whose authority skirted the military chain of command. For this, Bush won from Congress, in June, 1941, the Office of Scientific Research and Development, of which the NDRC became a committee, along with a medical branch and several other operations. He became its chairman, and Compton took the chair of the NDRC. Compton and the NDRC would operate the laboratories while Bush smoothed the way at higher levels, acting as liaison to the military and to the President. But the organizational developments were only preliminary to the task of developing new weapons. By the end of the year, research had begun both in nuclear fission and on a microwave radar system; this latter research was done mainly at the Radiation Laboratory at MIT, where, by then, physicists had also begun to develop a RCM, or radar countermeasures, program of research. That Bush had accomplished so much with so little fanfare led one journalist to comment: "He has done a tremendous job, and made less fuss about it than [Secretary of Interior] Harold Ickes would make about brushing his teeth" [41].

Besides his own adroit political ways, Bush had gotten his way, in part, through the status held by the NDRC members in scientific and engineering circles. Men like Secretary Ickes or War Production Board member and former congressman Maury Maverick wanted their own organizations at the center of the growing research establishment. They lost out because the wartime program required physicists and electronics engineers, groups Bush and his associates already dominated. During the early months, physicists were especially in demand since their discipline lay at the heart of the two great research efforts of the war. The nuclear physicists had moved even earlier than Bush to gain governmental support for the work on nuclear fission. At the suggestion of Leo Szilard, Albert Einstein, and others, Roosevelt, in 1939, appointed an Advisory Committee on Uranium, which, when the NDRC was formed, came under the new agency [42].

The nuclear-fission project differed from the quest for radar since success in achieving control over nuclear fission lay so far in the future. The most optimistic time estimate for the production of nuclear bombs was 1944 or, with exceptionally good fortune, six months to a year earlier. Others saw the project as even more remote. "At least two members of NDRC," the official historian of the Office of Scientific Research and Development reported, insisted that the agency had been charged to research and develop "instrumentalities of war." Yet the new project, they complained, aimed more at advancing the scientific field of "nuclear physics or . . . atomic energy for peacetime use." Just the reverse was true of radar, of course, which by 1940 had reached an applicable stage of development. Engineers at the Naval Research Laboratory had been using existing vacuum tubes for several years to develop long-wave radar systems operating in the high-frequency range. Already their work proved useful in direction-finding systems for use by aircraft and in detection systems as well. To give notice to the maturity of the quest for an electronic warning system, Navy engineers coined the word "radar" in 1940. It was derived from the phrase, "radio detection and ranging." The acronym even demonstrated the process, being reversible, coming back upon itself as did radio waves in the actual warning system [43].

Though the Army's Signal Corps also developed a long-waves system, the Navy's researchers made the critical breakthroughs. Significantly, these accomplishments had come from engineers. Two of the chief contributors represented the full spectrum of engineering: Hoyt Taylor and Leo Young. Taylor had received an engineering degree in 1902 at Northwestern University in Evanston, Illinois. He taught physics for four years at the University of Wisconsin before earning a doctorate, in 1909, from the Institute of Applied Electricity at the University of Gottingen. Taylor researched vacuum tubes in Germany. Leo Young, a member of Taylor's research group at the NRL, had experience first as a telegraph operator and then, in World War I, as a radio operator. "I started as a ham back in 1905," Young later said, "and I am still a ham." Young's talents lay with the construction of apparatus and experimentation. Even Taylor, who was superbly trained in theory and mathematics, described himself as a "dyed in the wool experimentalist" [44].

From these men came the Navy's first radar system. In 1933, the Naval Research Laboratory's engineers conceived a pulsed system that greatly improved the resolution in the image received. After winning a special appropriation of $100,000 for developmental work in 1934, the NRL and RCA each built models that were installed on two Navy ships in 1938. Successful tests led to a contract for RCA to build six more. By 1940, these pulsed systems operated on six additional ships. It was a worthy system; until 1942, long-wave radar was the only system being used by the Americans. Pulsed system or not, however, the system lacked the clear resolution possible in the microwave end of the electromagnetic spectrum. Since the next step would be a microwave system, as the war approached, the Naval Research Laboratory understandably wanted primary responsibility for developing such a microwave system. But just as the NRL's engineers and scientists had been unable to get any of the Navy's bureaus to authorize this research, the large sums needed failed to come from a government not yet formally at war [45].

Given that Roosevelt had firmly established the National Defense Research Committee's authority over research, the military first requested that the committee undertake basic investigations of ultrahigh frequencies and pulse transmission. From these requests came the Microwave Committee, which in turn instigated the Radiation Laboratory at MIT late in 1940. The Rad Lab, as many called it, received the assignment of joining the British cavity magnetron with the pulsed, long-wave radar to produce a system operating in the ultrahigh-frequency range.

The contrasting makeup of the two entities well illustrated the division of roles in the wartime research establishment. Whereas the Radiation Laboratory included almost wholly physicists, the composition of its directing agency, the Microwave Committee, was nearly the opposite. Its members were chiefly engineers from industry and the university. Just as the Rad Lab and the "atom bomb project" of the Uranium Committee were social microcosms of the national physics community, the committees themselves reflected the world of industrial and university research administration. Industrial representatives predominated in these committees, with only two members from the university community — MIT electrical engineering professor Edward Bowles and University of California physicist Ernest O. Lawrence. The rest of the committee came from Bell, GE, RCA, Westinghouse, and Sperry [46].

Engineers appeared in many similar positions during the years before Pearl Harbor. Since his curricular innovations in electronics at MIT two decades earlier, Bowles had gained a wide reputation for his knowledge of the field. From the early service as secretary of the Microwave Committee, he moved to the task of serving as Expert Consultant to Secretary of War Henry Stimson. Engineers were highly concentrated in the NDRC's section of "Communications." RCA's chief engineer C. B. Jolliffe headed this section. Division 13, as the Communications Section was later called, differed radically from the radar and atomic bomb projects. The section established no engineering laboratories, because, as Compton explained, independent research on the problems of electrical communication was unnecessary since the field had already been "highly developed for peacetime purposes by the great commercial companies" [47].

Though Division 13 mounted no direct NDRC research undertakings, on the score of past and future IRE presidents, the unit demonstrated the high participation of radio engineers in the war. Haraden Pratt, who had served as IRE president in 1938, was actively engaged, replacing Jolliffe as head during the last critical year of the war. Other ex-IRE presidents in Division 13 were John H. Dellinger of the Bureau of Standards (1925), Harold H. Beverage of RCA (1937), and Professor William Everitt from the University of Illinois, who would be president in 1945. Among the consultants to Division 13 was a much younger Lloyd V. Berkner, a lieutenant commander in charge of the radar unit in the Navy's Bureau of Aeronautics who would be the IRE president in 1961. Moreover, IRE founder, John Hogan (1920) served as special assistant to Bush, and Ralph Bown (1927) represented Bell on the Microwave Committee.

That engineers sat so densely on administrative committees and on policy-formation boards did not come about accidentally. Bush clearly sensed the importance of certain engineering values to wartime work. In a letter to Roosevelt, he compared a report from a National Academy of Sciences committee to an assessment made by the British on the prospects of the nuclear bomb project. Bush preferred the conclusions of the Academy's committee because they were

> somewhat more conservative [since] the Committee included some hardheaded engineers in addition to very distinguished physicists. The present report estimates that the bombs will be somewhat less effective than the British computations showed, although still exceedingly powerful. It predicts a longer interval before production could be started. It also estimates total costs much higher than the British figures [48].

While admitting the need for realistic time and cost estimates, there remained the anomaly of engineers being well represented on the administrative committees but being largely absent from the research projects. Bush was uncomfortably aware of this, as he demonstrated in late 1941 when the need for radar countermeasures became urgent. After the Japanese attacked Pearl Harbor on December 7, Lloyd Berkner immediately urged the Navy's coordinator of research and development to call a meeting. Four days later, Navy officials joined representatives from the Office of Scientific Research and Development and the Radiation Laboratory to discuss the situation. At the formal conference that followed, the participants decided to undertake a "project for the development of radar countermeasures receivers and jamming equipment" [49].

Assigned to the Radiation Laboratory as a Navy project, the new research unit collaborated with the Naval Research Laboratory and the laboratories of the Signal Corps. Radar countermeasures (RCM) possessed a military urgency that went beyond even that of the microwave-radar research, since it had to do with defending against the enemy's changing radar capabilities. Already the British had devised techniques to confuse the radar signals sent out by the Germans along with the Luftwaffe. In the United States also, NRL researchers had worked on a wide-band crystal receiver to locate enemy transmissions and determine their frequency. As Karl Compton described the new OSRD program: "The R.C.M. project is of such character that even greater care should be taken to prevent every unnecessary intimation that work of this sort is in progress. Only by such care . . . can the risk be avoided of losing the benefits of our work." The Radiation Laboratory's work on RCM had begun under physicist Luis Alvarez. Alvarez's interest derived from the British, both from their successful use of countermeasures against the Luftwaffe and through the continuing contacts since receiving the advanced magnetron. The Rad Lab had already contracted with the General Radio Company to build an

After getting the magnetron from the British early in the war, engineers and physicists at MIT's Radiation Laboratory worked to perfect a microwave radar system.

intercept receiver within a specified frequency range. By the time of Berkner's meetings in December, 1941, then, the NDRC's leaders had even begun to plan a separate laboratory for the exclusive purpose of developing countermeasures to the enemy's radar systems, leaving the Radiation Laboratory staff with the more basic research on an ultrahigh-frequency radar system [50].

Another matter being settled at that time concerned the makeup of the new laboratory. By December of 1941, Bush had decided that the "Radiation Laboratory . . . [was] too greatly stuffed with Physicists" and that "a few engineers would be good for" it. As Bush thought about who could both direct the countermeasures laboratory and recruit engineers, he thought of his former student, Frederick Terman. Terman was an ideal man to do both jobs. He, too, had gone far since studying under Bush at MIT. Terman had endeavored to make Stanford a center of electronics (already, in 1941, an electronics laboratory had been discussed). He had also gained a reputation for helping research colleagues over tight spots in their work and, from his professional duties, had made numerous contacts within the electrical engineering profession. So in late December, 1941, as Terman prepared to attend an IRE meeting in New York, Lee DuBridge, a California Institute of Technology physicist and director of the Radiation Laboratory, called him to discuss the RCM project. DuBridge arranged to meet Terman in Cambridge before the IRE meeting. By the time he arrived in New York, Terman had become director of the new program [51].

Beginning his war work enthusiastically, Terman penciled long lists of possible appointments while on the train to New York and, later, resumed the lists when returning to California aboard the Southern Pacific. Like Bush, he wanted to include the members of his profession, believing, as he later told IRE members, that the "demands of national defense" offered the radio engineer an opportunity to contribute "an extraordinary versatility of service to mankind. The swifter deployment of mechanized forces," characteristic of modern war, made "far greater demands upon telegraphic and telephonic communications" than in any previous conflict [52].

However, the work to be done in radar countermeasures went beyond conventional communications. So as Terman searched his memory for names of candidates for the new research program, he evaluated the personal qualities and technical abilities of each candidate. Not all radio men fitted the need. "Wrong experience" and "poor pickings," he noted of broadcast engineers. "Bdcst consultants" were "not really research men," he concluded, deciding the same about engineers in "receiver mfg." Nor could he at first think of appropriate engineers from the television field. Though Bush requested that he bring in engineers, Terman looked also for physicists, wishing, especially, for a "theoretical physicist." However, he found that all had "gone to RL." Except for William Everitt and a sprinkling of others, he could think of only a few appropriate or available engineers in the colleges. He did not believe those in industry were available, though he listed "Bell men, RCA Science lab, [Donald] Sinclair, Litton, [Simon] Ramo," and others. Graduate students in engineering, as in physics, were "almost gone." And yet the problem of recruiting engineers went beyond availability, since, as he noted, the "engr requires more experience to be a good Engr than physicist does to be a good physicist." Clearly frustrated, he finally wondered "who can be raided" in industry who were not already in "Defense jobs" [53].

Relying on all the methods he conjured, however, he staffed his program. For an administrative assistant, he talked CBS into releasing Howard Chinn, and when he heard that CBS was phasing out its color television project, he brought in the "whole Peter Goldmark group" that had been working on color television. Engineers were also brought in from such places as RCA, Bell, and GE. But most of all, Terman brought in engineers from California, over a hundred finally, twenty of them with Stanford degrees. Early in 1942, Compton had told Terman that the project would employ fifty to seventy-five individuals by the next fall; yet by then, project personnel numbered 205, seventy-eight of whom were research staff. By July, 1943, those figures climbed to 475, including 165 research personnel, and, by early 1944, to 744 and 214—the highest numbers reached. (In contrast, the Rad Lab's employment reached 4000.) When the RCM program soon outgrew the Radiation Laboratory, the Radio Research Laboratory was moved, first to another building on the MIT campus, then to Harvard [54].

By the fall of 1942, the Office of Scientific Research and Development had created a new division to coordinate the work on countermeasures. C. Guy Suits, who was assistant director of the GE Research Laboratory, became chief of the new Division 15. So important were British activities to the laboratory's work, that twice that year visits by Terman and Suits to England were followed by large increases in staff and responsibility. With the organization of Division 15, not all the OSRD's radar countermeasures work would be done in Terman's laboratory. Division 14, in contrast, under which the Radiation Laboratory was organized, centered its work in the Rad Lab, issuing only subcontracts for specified tasks. Suits' Division 15, however, issued prime contracts to industrial firms to undertake key parts of the countermeasures program. Bell Laboratories investigated radio-jamming techniques and sought means for overcoming the enemy's jamming as well. Beverage led a group at RCA that worked on similar projects; he also took charge of the work on antenna problems [55].

The work of the Radio Research Laboratory got underway when Terman returned from England in the spring, sure of the tasks that needed doing. Its earliest significant work in the electronics field involved the development of chaff, or window. These were strips of aluminum foil that, when dropped from aircraft, effectively confused German radar during Allied air attacks. The RRL contributed not only to the design of chaff but also to the development of a high-speed cutter to allow for mass production. Its importance appeared in the amount of chaff used in numerous bombing missions, amounting to 10 million pounds of aluminum foil dropped by the 8th Air Force in Europe. But Terman's group also worked in "communications CM and guided missile CM fields," as John Hogan reported to Bush toward the end of the war. Electronics work had in fact come much earlier, for in its first year when chaff had been devised, the RRL created one of the war's most successful electronic jamming devices. Named Carpet, the airborne jammer defied the small German radar units that controlled the flak which inflicted such heavy losses on Allied bombers. Used together, Carpet and chaff forced the Germans to divert much of their scientific talent from working on microwave systems to defeating the RRL's creations [56].

In the beginning, Terman and the RRL experienced difficulties in working on certain kinds of problems. This dilemma was first articulated by Ralph Bown in September, 1942, in a report prepared for the chairman of the Microwave Section. The RRL's "60 odd scientific workers" were engaged, Bown wrote, "primarily [in] an apparatus program aimed at producing designs of equipment to meet needs which have been specified to it by the Army, the Navy and the British RCM group." At the time, the laboratory tested long-wave and microwave radar systems to discover their susceptibility to jamming; the laboratory also developed long-range navigational aids. Though finding the investigations of electronic jamming techniques "well conceived" and the work on "apparatus development" progressing, Bown had some caveats. First, he thought the electronics engineers exhibited a "characteristic . . . tendency toward elaboration and perfectionizing" in their designs, instead of leaving final design characteristics to the manufacturers who would have to make changes anyway as they gained experience in production. Secondly and more seriously, he found amid the well-done "apparatus projects" that the engineers were ignoring the need for both simplicity and ruggedness in the field.

Bown's criticisms found fertile soil. The engineers of the Radio Research Laboratory learned to work with the manufacturers, forming liaisons and holding regular meetings. Staying close to the radar laboratory at MIT helped maintain the exchange of scientific and technical information and advice. The physicists and engineers of the RRL, like the scientists at other laboratories, learned also to design their apparatus with combat conditions in mind. As the laboratory's official administrative history explained the matter in 1946: The RRL's "responsibility for a development did not cease after that development left the Laboratory, but extended all the way through to the analysis and evaluation of its operation in the field." Bown's immediate concern had been for the use of a systems approach in the laboratory's developmental work, though, of course, such an encompassing approach had always been a part of effective engineering, whether in the design of an arc-lighting system or of microwave radar apparatus [57].

Terman best described the nature of the RRL's work in 1943, when he protected it from being diluted by Guy Suits, his division chief. Suits had asked Terman to produce four sets of a type of countermeasures apparatus that had been developed by the Naval Research Laboratory. Terman rejected the request. Even had his shop facilities been adequate, Terman explained to his Harvard superior, "it was not our job to be a manufacturer of other peoples' products." To accept such assignments would displace "just that much development and research work." After suggesting several alternatives, Terman reported that the RRL's project committee unanimously backed him, and its members were "somewhat excited about the idea of someone attempting to dump additional work" onto the laboratory. Had Suits been successful, "RRL could very quickly cease entirely to be a development organization, and become another model shop."

In his response to Suits, Terman had drawn lines around the type of work he thought important and thus appropriate to the laboratory. The RRL would engage in "development and research," he had written. His reversal of the phrase, moreover, was purposeful. Developmental work lay at the core of engineering work, Terman knew, entailing in the main design and testing. Engineering research at the RRL sought applicable outcomes, such as an invention or innovation, in anticipation of "the developmental phase of technological change." Terman did not deny other points on the engineering spectrum. Indeed, the laboratory was continously involved with production—though of products of its own design—and, as Bown had urged, its research associates had learned to consider the eventual use of the apparatus designed by the staff. Terman's point was that while the engineer's concerns might extend to actual production, the core of engineering work lay elsewhere [58].

Indeed, the Radio Research Laboratory spent the rest of the war producing supplies of chaff and adequate numbers of the various radar units designed by the laboratory. The staff had early and continuously worked on the problems of microwave radar as well. Given that Germany's radar technology remained in the long-wave category through most of the war, the countermeasures laboratory's long-range objectives in ultrahigh-frequency research amounted to basic engineering research. Still, the bulk of the laboratory's developmental activities fell in the high-frequency range. As late as 1946, a laboratory report referred to a "trend" toward "microwave equipment," and urged continued research in the postwar era, since radar in the ultrahigh-frequency range would be more difficult to jam. As Terman admitted, this circumstance had made the RRL's work much easier. However, in 1944, the RRL's "developmental program" actively sought to extend the "new microwave receivers . . . to still shorter wavelengths" and to design receiving systems "for setting jammers on frequency, with particular reference to the microwave range" [59].

Such a broad range of work required an equally broad range of talent, and this Terman had finally achieved. Though the OSRD instituted a program to convert some biologists and chemists into radar technicians, Terman had found both his physicists and his engineers. Of the 192 "research associates" in July of 1945, at the end of its "stable period," fifty-two were physicists and 116 were engineers. Industry and educational institutions had supplied over half (seventy and fifty respectively); yet, because of the problems of finding experienced professional workers, more than a third were hired directly out of graduate school or upon completion of undergraduate study. Because of the "difficulties of hiring service personnel," an RRL report explained in 1945, a policy had been adopted "of hiring . . . female applicants" or men "disqualified from military service by age or disability." In any case, the RRL had become a laboratory of both physicists and electrical engineers [60].

As the war wound down, Terman's staff began to write official histories of the Radio Research Laboratory. Drawing from evaluative and narrative reports prepared throughout the war, the laboratory entered upon what Terman described as "a formidable report writing program." Though Oswald Garrison Villard, Jr., did much of the actual writing, Terman was finally responsible. And indeed, the final histories reflected Terman's judgments. The RRL had served well in the general task of acting as "a clearing house for information of all kinds," both technical and tactical. Also, the laboratory had helped to force the Germans to concentrate on conventional radar rather than bending "their efforts in other, more fruitful directions, such as the development of microwave radar." Working mostly in the high-frequency range, the engineers had "obsoleted a whole class of radars" and had done so, the report explained, by being able "to make engineering decisions" on what ordinarily "would be considered insufficient information" [61].

Terman judged the laboratory's work positively, resting his judgments on the quality of the apparatus designed and produced, as well as on the number produced. Intelligence reports revealed that the laboratory's countermeasures devices had saved hundreds of planes and crews in the European theatre "that would otherwise have been shot down by German radar-controlled flak." And in the Pacific, countermeasures devices used in B-29's had "contributed to the strikingly low losses suffered by these planes during the final months of the war." With such a record, Terman declared, "the members of this laboratory have the satisfaction of knowing that they have contributed in an important way to the successful outcome of the war."

The engineers of the RRL stood at the center of the electrical engineer's varied experience of the war. Other engineers like William Wickenden, the president of the Case School of Applied Science and a prominent member of the AIEE, served in government agencies concerned with matters of industrial production. Wickenden had even turned down a request to head an emergency technical training program for the Federal Office of Education, insisting that, as Case president, he was contributing already in overseeing both the education of 1800 students and work on numerous defense contracts. Still, the research efforts in nuclear fission and electronics constituted the great legacies from what Wickenden called a "technological war" [62]. Their consequences went beyond the individual to the social. Nuclear energy introduced a major new fuel source, stirring great interest and activity within the electric power industry. Electronics rose to dominate the national industrial scene. But the full impact of nuclear energy and electronics would not come until the third quarter of the century. In the immediate postwar years, rather, the event that was reshaping the context of electrical engineering was the fashioning of a new American political economy in which the search for national security joined the more familiar quest for business prosperity.

The legacy of the war: A new world

It was "good to be tied to the tail of his kite," Frederick Terman later remarked of his former professor and wartime employer, Vannevar Bush. The image serves not only for viewing Terman's career but also for understanding the life of the profession during the postwar years. For a large portion of electrical engineers were similarly tied to a kite. Theirs was a much larger one, sent aloft by the powerful winds generated by the war. It rose on more than the ambitions of a single individual, but rather soared before the flight of what, fifteen years later, President Dwight David Eisenhower characterized in his Farewell Address as a "military-industrial complex."

But although that complex presented a new world to electrical engineers, much remained that was familiar. A dynamic industrial order still occupied the terrain, built upon a century of engineering achievements in electric power, manufacturing, communications, and electronics. And although small companies and independent laboratories remained a part of this private order, large national firms continued to dominate. There were other familiar elements. Large university programs at schools like MIT, the University of Illinois, and Stanford presided over a national educational establishment that was populated by scores of smaller college and university departments of electrical engineering. Governmental agencies like the Federal Communications Commission and the Bureau of Standards or federal laboratories like the Naval Research Laboratory and NACA's Langley Laboratory had become, by 1940, known elements in the engineering context. Though less prominent in the life of most engineers, industry associations like the Western Electronics Manufacturer's Association, the Association of Edison Illuminating Companies, and the Edison Electric Institute — organized in the 1930's, shortly after the demise of National Electric Light Association — were also well known [63].

Yet for all the familiar scenery, the organizational and technical innovations of the war were remaking the electrical engineering context. Besides the new technologies, there was the institutional novelty of scientific and technical advisers operating high in military and governmental circles. Unlike the demobilization which followed World War I, scientists and engineers like Vannevar Bush remained a leading voice in the realm of policy-making. In addition, wartime experts like Terman and Edward Bowles continued to advise the government and the military in specialized areas.

As early as the summer of 1944, Bush began to give attention to postwar research policy when he outlined a plan to transfer the work of the Office of Scientific Research and Development to the military at the end of the war. Though Bush and others later drew on the OSRD as a model for a civilian-run defense policy agency for science and technology, in 1944, Bush advised the military to administer its own R&D programs. Other than some "fundamental research work," which he did not believe "would . . . bear fruit during this war," Bush advocated that the "Services" take over all OSRD programs. This would include the most promising basic work, along with "weapons development" projects and "exploratory programs of research and development" [64].

Bush's advice to the military did not come from outside. Besides his OSRD directorship, in 1942, he had taken on the chairmanship of the Joint Committee on New Weapons and Equipment (JNW), established by the Joint Chiefs of Staff. Bush wore another hat, moreover: that of chairman of the Military Policy Committee, which oversaw the Manhattan Project. Although the agencies that were set up by the 1946 Atomic Energy Act took over the nuclear fission research, Bush's OSRD and JNW roles were combined after the war into the chairmanship of the Joint Research and Development Board (JRDB), also organized under the Joint Chiefs of Staff. The remainder of this five-member board represented the Army and Navy. Bush's holding of multiple positions in the war was not uncommon. Terman did also; but as might be expected, his extra-OSRD service focused on technical matters, such as the work of the Vacuum Tube Development Committee he chaired for the Army and Navy's Joint Communications Board. When the Committee was reorganized after the war, Terman remained to help provide a "medium through which the Services will cooperate with one another and industry on vacuum tube problems" [65].

Bowles' involvement similarly expanded beyond his role as technical advisor to the Secretary of War. In April, 1946, he took the initiative of drafting a memorandum on "Scientific and Technological Resources as Military Assets" and sent it to General Dwight D. Eisenhower. Eisenhower thought it "splendid" and, with some changes made by his own staff, issued it under his signature on April 30, 1946. It served as the General's position paper on postwar research and development policy. In it, he recognized that "scientists and business men contributed techniques and weapons which enabled us to outwit and overwhelm the enemy." Eisenhower wanted the peacetime military to continue using the nation's scientific and technical resources in the spirit of the OSRD, where "scientists and industrialists [were] given the greatest possible freedom to carry out their research." Vannevar Bush liked the document so much that he circulated a thousand copies to the top scientists and engineers who had been associated with the OSRD, along with the expressed hope that "the cordial co-operation of the past six years [will] continue in the peace" [66].

When the National Security Act of 1947 reorganized the military and separated the Air Force from the Army, combining the services under a Department of Defense, engineers continued to operate at the center of the national arena in which R&D policy came together. Bowles assumed chairmanship of the powerful Air Force Science Advisory Board, which was charged with keeping the Air Force current with scientific and technical advances. The National Security Act also replaced the earlier Joint Board with the Research and Development Board (RDB), of which Bush was named head.

Like the Joint Board, Bush wanted the RDB to adapt the work of the wartime agencies "to peacetime purposes" and perpetuate them "as part of the national security program." The work seemed critical to Bush for a number of reasons. Chiefly, he hoped that it would help maintain "the technological preeminence" of the country by having the "military look to the inventors and the technicians for new methods in warfare." In this spirit, the RDB represented a "master plan of R&D for military purposes" [67].

The encompassing purpose of national security had been invoked previously in matters greatly affecting the profession. In combating the moves for public control of radio both during World War I and, again, when the Federal Communications Commission was established in 1933, David Sarnoff insisted that an effective national communications system would be achieved only under private management. The greater efficiency of private enterprise, he had argued, made it best able to protect the nation. But never did the term carry so much weight as it did after World War II, when the notion of how to secure the nation was thought to require more than a prosperous private economy. No less than a partnership between the military, industry, and university was called for.

This was the objective of Edward Bowles when, in 1946, in the *Proceedings of the IRE,* he described a "mechanism" on which to build "a program of national security." Rather than becoming distracted by the work of "disposing of the vanquished," the nation needed to consolidate "our gains" at home. This was made possible by the wartime integration of "our three basic resources—the professional military, industrial and educational assets." The "barriers" that had previously stood "between civilian and military" were now removed, easing the way to establish a "peacetime counterpart pattern." With the impetus of the war, the nation's "educational and industrial areas" could be brought "as close to the military in peacetime" as they had been during the war [68].

That same impetus impelled Bush forward through the 1940's, during the war and in the strained peace of the postwar standoff between the United States and Russia. He aimed to revise the old order, bringing the government into the world of engineering. Bush's opportunity to influence the postwar settlement came when President Roosevelt invited him to recommend how this might be done. In a letter to Bush in November 1944, Roosevelt regretted that the OSRD's work had to be conducted in "the utmost secrecy." He promised that "some day the full story . . . can be told." For now, the President saw "no reason" why "the information, the techniques, and the research experience developed" by scientists and engineers in the OSRD, industry, and the universities could not be "used in the days of peace ahead." From the wartime work could come "the improvement of the national health, the creation of new enterprises bringing new jobs, and the betterment of the national standard of living" [69].

Bush responded with alacrity, submitting within seven months a 184-page report that dealt with the several issues raised by Roosevelt: the diffusion of scientific knowledge, medical research and disease, government aid to public and private research organizations, and developing the talent of the nation's youth. But throughout was Bush's insistent message about the importance of basic research. It was implicit in the title, "Science — the Endless Frontier," but Bush made it explicit: "Progress in the war against disease" and "defense against aggression" required a continued "flow of new scientific knowledge." And that could be obtained "only through basic scientific research" [70].

Efforts by Bush and others during the next several years sought to fulfill these objectives by, in essence, continuing the work of the OSRD. Besides the research and development projects taken over by the Atomic Energy Commission and the agencies of the newly organized Department of Defense, scientists and engineers actively sought to establish a nonmilitary agency which would sponsor basic research. This came at mid-century with a body much like that Bush had called for in his report to the President. It was named the National Science Foundation.

Bush also wanted an advisory group within the government which, again like the OSRD, would be independent of the military. While chairman of the RDB in 1948, he had established a committee to explore the need for "mobilization of the civilian scientific effort in the event of an emergency." The resulting report called for a body similar to the OSRD. The recommendation went to the White House, where, prompted by U.S. entry into the war in Korea, in 1951, President Harry S. Truman established the Science Advisory Committee (SAC) to provide "independent advice on scientific matters." Because its purpose was to aid defense planning, Truman organized the Committee under the White House's Office of Defense Mobilization.

SAC members came from the leading bodies of scientists and engineers, including the chairman of the RDB, the president of the National Academy of Sciences, and the director of the year-old National Science Foundation. Individual members were J. Robert Oppenheimer and university presidents Lee DuBridge of the California Institute of Technology, James B. Conant of Harvard, and James R. Killian of MIT. Though all were influential individuals, as part-time civilians without authority over research funds, several years would pass before the Scientific Advisory Committee gained influence in the making of R&D policy [71].

The Committee's work during the first year entailed assessing the nation's resources in the area of scientific and engineering manpower and utilization. As its first task, the Committee examined the pattern of military research and development expenditures since 1940. This, the members reasoned, would indicate the demands being made on the scientific and technical communities to meet both military and industrial manpower needs. Using data drawn from the Research and Development Board and the Bureau of the Budget, the committee reported not only an increase in Department of Defense expenditures committed to its own research installations, but also a tremendous growth since 1940 of defense funds going to industrial companies and universities. After spending had leveled off in 1945, the committee found a "precipitous rise" in obligations to federal laboratories incurred after 1948. Overall, however, the nearly $450 million assigned in 1948 to industry, federal R&D laboratories, universities, and nonprofit institutions, rose by little more than $50 million in 1950. With the Korean war, this sum jumped to $1.3 billion two years later. But the war represented no anomaly, serving rather to presage the large increases of the 1950's, when federal spending on R&D grew by over 16 percent a year [72].

Although allowance had to be made for the curtailment of ordinary industrial activity in the face of military demand, the impact of these rapidly growing expenditures on the engineering community was immense. To clarify this point for the Science Advisory Committee, the Committee's executive secretary, recent IRE president and Bell engineer Frederick B. Llewellyn, translated the "dollars into . . . the number of engineers required at industrial, university, and federal laboratories" [73].

The numbers alone indicated much about the shape of the engineering profession — and the place of the electrical engineer in it. In 1954, for example, the National Science Foundation — in part "a central clearinghouse for information about scientific and technical personnel" — estimated that the United States contained 200,000 scientists and 650,000 engineers. Nearly 200,000 of the engineers worked in research and development, about which the NSF report pointed out two distinctive characteristics. First, the heaviest concentrations in manufacturing were R&D workers for the aviation companies (21 percent) and the producers of electrical machinery (18 percent). Both were key areas of employment for the electronics engineer. Second, the report called attention to the shift in research personnel between 1941 and 1953 from government and industry to the university.

To be sure, the shift in funding signaled no revolutionary upheaval. By 1953, industry's numbers had dropped from 71 to 68 percent, leaving it the chief employer of engineering researchers. The government's loss was more significant, dropping from 20 to 17 percent as engineers employed by educational and nonprofit institutions rose from 9 to 15 percent [74]. The picture painted by the numbers, in brief, was that in a country of 150 million inhabitants, more than one in 320 were engineers. And of those, nearly one in four worked in research and development.

The changes represented in the shift to university-based research indicated not only an increase in basic research but also a significant addition to the realm of the research laboratory. During the 1920's and 1930's, the state of research in the country remained much as it had been earlier in the century, during the first era of industrial research. The founding of Bell Laboratories in 1925 as a separate entity in American Telephone & Telegraph substantially advanced this tradition. However, as historian Leonard Reich has shown, the competitive struggle for patents often led companies to pursue "research for monopoly control of markets" [75]. Equally subject to such policy, for example, were engineering scientist Irving Langmuir at General Electric and H. D. Arnold of AT&T as they worked to develop high-vacuum triode tubes in the 1910's.

Between the wars, when it was RCA that the telephone company competed with in attempting to establish national, popular broadcasting systems, Bell Laboratories spent a substantial portion of its resources on developmental research to help the company in its policy of "covering the field" through the accumulation of patents. As Reich sums up the situation for the researcher: though some large laboratories gave "relatively free reign to a few of their best scientists, . . . the majority of men and resources [pursued] those types of patents useful for market control."

Basic research did exist in the United States before 1940, in independent, nonprofit research organizations, in some of the industrial laboratories, and in a few places in the government. But in the universities, where research generally meant fundamental work, research continued to follow the traditional style of the solitary investigator [76]. Thus, the shift to university research after 1940 suggested that the newer tradition of laboratory research had found an additional home. Though their research budgets scarcely equaled the R&D funds spent in industry or in the 700 federal laboratories which existed by the early 1980's, many universities found themselves with more money for research programs than ever before [77].

Some of the new research organizations came directly out of the laboratories established by the OSRD during the war. An example was the establishment of the MIT Research Laboratory for Electronics in 1946 within the facilities of the Radiation Laboratory. New laboratories sprang up in many schools. Some, like the Hudson Laboratories at Columbia University, were relatively small. It had an annual budget from the Department of Defense (the Navy) of $4.7 million and a staff of 381 before it was merged with the Naval Research Laboratory. At the other end in size was the DOD-funded Stanford Research Institute. When sold to its own board in January 1970 for $25 million, SRI had an annual budget of $30.6 million and 2200 employees [78].

MIT's participation in the new research opportunities, while extreme, was not unusual. The Institute received more funding from governmental sources than any other university. Of $98 million, nearly $68 million came from the DOD and the National Aeronautics and Space Administration. The total represented 41 percent of the school's entire budget. Two of the laboratories at MIT were havens for electronics engineers. The Institute established the Lincoln Laboratory in 1954 at the request of the Air Force. Through the 1960's, both Lincoln, with its budget of $40 million and the Instrumentation

Laboratory, which received its annual $20 million from the DOD and NASA, developed advanced military and space hardware systems. Although basic research went on at a good many of the government-funded university laboratories, the Instrumentation Laboratory engaged mainly, as one writer has explained, in "advanced engineering development of hardware systems," often in close cooperation "with government personnel and industrial contractors." Nearly 38 percent of its research staff were electrical engineers (the next largest being mechanical engineers, making up 15 percent).

As the new pattern of spending expanded research and development activity and the demand for engineers, it contributed to the tremendous growth of the engineering profession. This could not have surprised the leaders of the burgeoning field of electronics. Like many others, in 1945, the California electronics engineer, Haraden Pratt, saw the "coming industrial revolution" in "the wondrous achievements made in weapons, machines, and basic science applications." Frederick Terman predicted that, as a direct result of "wartime electronic research," the radio industry would undergo an unprecedented expansion through intensively exploiting higher frequencies and developing new communication systems. "The return to society in the next two decades," Terman wrote in 1947, would be many times the cost of the "wartime electronic research program" [79].

The demands being made by the drive for national security only underscored the military presence in the new engineering world. But the changes brought by the war to the electrical engineering profession went beyond an active, peacetime military, as had been immediately apparent to IRE members like Terman and Pratt. Industrial growth and technical innovation, expanded educational programs and swollen enrollments, all were changing the profession.

In the middle of this new world were the engineering societies, struggling to come to terms with the postwar order. The challenge was greatest for the Institute of Radio Engineers, which, after 1940, emerged the clear winner in the competition for the electronics engineer. While growth in numbers came to the American Institute of Electrical Engineers, the IRE's swelling roster far outpaced the older society. But more significantly, the waxing field of electronics produced a host of new specialities. To meet this challenge, the Institute transformed its organizational structure, simultaneously sealing the fate of the AIEE and solving the professional society's age-old problem of splinter groups.

7/THE GROWTH OF ELECTRONICS AND THE PATH TO MERGER

It is interesting to observe how [the AIEE] is one of industry, rather than electrical engineering in its more scientific, inventive or academic aspect. Radio is a newer field and the IRE is as yet closer to the physics of things. But, we are fast approaching the age of great operational and manufacturing responsibility which is bound to put a similar predominant industrial tinge on IRE. As this condition arises, it will be time to reorganize the field of the electrical societies, and this time is rapidly approaching.

Lloyd Espenschied, 1946 [1]

The electrical societies and the "great growth industry"

Although, on the surface, it may have been accurate to describe the postwar AIEE as commercial and the IRE as technical, the two societies actually had much in common. They had often met on the shared ground of the traditional communications field and of the newer field of electronics. Just as the AIEE strove to attract engineers from the nonpower fields, the widening application of electronic controls in the power industry linked the IRE, in part, to power interests. In addition, the radio engineering society had shared basic concerns with industry, whether an RCA or an entrepreneurial firm.

Industry's use of electronic components, for example, made its needs central to the concerns of electronics engineering educators like Frederick Terman and John D. Ryder. While acting department head at Iowa State after the war, Ryder deftly tied curriculum to industrial need. Following the principle that "industry finds the school that will supply the kind of men they want," he installed a pulse techniques course to supply "pulse people" for a Minneapolis electronics firm. Active in promoting an indigenous electronics industry in his native state, Terman even described his disciplinary field as a "great growth industry" [2].

An "industrial tinge" had, of course, always colored the engineering mainstream in modern America. In his study of American engineering from the late nineteenth century to 1940, historian Edwin T. Layton, Jr., concluded that the engineer was "the hybrid offspring of a union between science and business" [3]. This same notion informed Lloyd Espenchied's vision that the industrial interests of both the AIEE and the IRE would lead to a future realignment of the electrical societies. To say that both the IRE and the AIEE displayed an industrial tinge, therefore, is only to say that each was an American engineering society.

But something did distinguish the two societies. Leading radio engineers understood their engineering calling in a manner essentially different from that of the AIEE's predominantly executive leadership. The conflicting orientations appeared concisely in the contrasts between the societies' presidents. It was not just that the presidents of the electronics engineering society were ten years younger on the average. The differences had to do rather with the substance of their engineering interests and careers. What was true of the two presidents who served on the eve of World War II stood for most of the presidents in office during the following quarter-century. On the one hand, Raymond A. Heising, who presided over the IRE in 1939, spent his career in research and development. During the 1930's, he headed Bell Laboratories' research on ultrahigh frequencies. On the other hand, AIEE president John C. Parker was a vice president for Consolidated Edison of New York. Though a former professor, his presidential biography noted that he held "many executive and advisory positions outside" Consolidated Edison.

The closest to an engineer-manager who held the IRE presidency after 1940 was W. R. G. Baker, the GE vice president. Although Baker had been active in standards work for a number of years, he attained the presidency in 1947, only after Terman and Haraden Pratt received assurance that he would make no radical changes in the Institute of Radio Engineers.

Differences between presidents went beyond individual preferences. Presidents came from the industry that each society represented. And unlike the steady state of traditional electrical engineering in the power industry, electronics engineering was undergoing rapid and fundamental change in both content and context. New specialties were emerging amid a private economic scene in which entrepreneurial firms shared the territory with the larger communications companies. Military interests dominated but shared a market composed also of industrial and consumer sectors. Moreover, innovations continued to appear in such areas as electron devices, and large new specialized groups developed around subfields like military electronics, the electronic computer, and information theory.

Given these conditions, the electronics engineers' choice of the IRE over the AIEE for their professional society made all the difference for the futures of the two societies. At the end of World War II, the AIEE stood supreme among the engineering societies. With a membership of 21,400, it had 1500 more members than the American Society of Civil Engineers and was larger by a fourth than the American Society of Mechanical Engineers. In contrast, the Institute of Radio Engineers ranked fifth. Only half the size of the AIEE, the IRE had 150 members more than the Society of Automotive Engineers and was slightly smaller than the Institute of Mining and Metallurgical Engineers. But already underway was a pace of growth that would leave the IRE, on the eve of its merger with the AIEE in 1963, with 27,000 more members than the older society and a surplus in its treasury of $3 million.

The rapidity of that growth was made even more apparent in 1954 when Haraden Pratt, then secretary of the IRE, charted the society's growth over forty years. From its founding in 1912 to the mid-twenties, the IRE had slowly grown to include 2500 members. And though a rapid increase carried the society to 7000 members by 1931, that figure dropped by several thousand in the next few years, settling at around 6000 in the late 1930's. Then, in 1940, the explosion occurred, showing up in Pratt's chart when the horizontal line representing the plateau of the late 1930's turned abruptly up. From 6000 in 1940, the IRE had climbed to over 35,000 a dozen years later and the line of growth still pointed upward [6].

These changes in numbers and technical emphases affected especially the internal organization of the IRE and nudged modifications from the AIEE as well. But while the AIEE struggled simply to recognize the field of electronics, the IRE worked to absorb the fast-appearing specialties issuing from the wartime R&D programs. Their internal responses to disciplinary expansion were chiefly organizational, yet behind them was a technological revolution within a revolution: the rise of microelectronics within the broader field of electronics.

Out of these momentous technological events, moreover, there emerged a broad movement for educational reform. The reforms sought in engineering education after World War II affected the entire engineering profession. Electronics engineers sought to change electrical engineering education both from within and in relation to the full constellation of engineering fields. In doing so, they worked for reforms that would cut as deeply as those MIT engineering professor Dugald C. Jackson had sought earlier in the century. This time, instead of curriculums geared to industrial practice, the reformers sought an education for engineers rooted in fundamental science, theory, and mathematics. Like the ideal engineering course envisioned at the turn of the century by GE engineering scientist Charles Proteus Steinmetz, the movement led by professors like Terman and Ryder did not ignore the needs of industry. It merely responded to those needs at what its adherents deemed a more basic level. Finally, the rapid growth during this era would alter the relations of the electrical engineering societies, just as Espenschied foresaw. This would lead, in the 1950's, to the AIEE's collapse before the same force that thrust the IRE into a position of great strength: the continuing transformation of electronics.

The organizational response

The responses of IRE and AIEE leaders to the rapid expansion of the profession roughly approximated the differing rates of growth of the two societies. The radio engineers, whose faster increase was apparent to the AIEE leadership, radically restructured their organization just two years after the war. But the AIEE did not act to alter the structure of its organization until 1950.

At mid-century, the AIEE's Board of Directors responded to the doubling of the number of technical committees since the war by creating five technical divisions with a Technical Advisory Committee to oversee their activities. Prior to 1950, the AIEE had sought to meet the demands of "rapid developments in technical activities" through holding special technical conferences and adding more general and district meetings. The Technical Advisory Committee seemed "an excellent foundation for still further decentralization." Later moves could give technical divisions and committees "the greatest possible opportunities to advance technical developments in their fields and to supply the corresponding information to members" [7].

That AIEE officers perceived the changes of 1950 as a decentralizing step—though they did not alter the society's hierarchical committee structure—perhaps reflected a suppressed desire to emulate the more substantial moves toward decentralization made by the IRE some two years earlier. Having nearly tripled in size during the war, IRE membership climbed from 17,000 in 1945 to 30,000 by mid-century. Responding to this increase, in 1947, several members urged the need for a new organizational form to handle the growth. The idea originated with ex-IRE president, Raymond Heising, who was joined by William Everitt, IRE president just two years before. Their efforts led to the appointment of an ad hoc Committee on Professional Groups, of which Everitt was named chairman [8].

Although it was true that the IRE responded to the society's rapid growth, actually, it reacted to the large blocs of members clustering around newly emerging specialized fields in electronics. Thus, the initial step came in the face of a movement to organize an audio engineering society outside the IRE. "We were afraid," Everitt later remembered, "that this might do the same thing to the IRE that we had done to the AIEE in its day" [9].

Everitt's committee reported to the Board in March 1948 with a plan that sought mainly to integrate the various technical interests within semi-autonomous national groups. In approving a Professional Group System, as *Proceedings'* editor Alfred Goldsmith described it, the Board had taken "another forward step, yea verily, a leap." Each Group was intended to insure adequate coverage of its special field in the Institute's publications, to hold meetings and special conferences, and to organize technical sessions at national and regional conventions. The right to publish papers given at their meetings along with other papers submitted independently, with a subsidy from Institute headquarters, followed soon after. This scheme of governance made the point of the Professional Groups even clearer, granting to each a large measure of autonomy with IRE headquarters providing policy, coordination, and general services. Group members elected their own officers, with chairmen serving on the Institute's standing Committee on Professional Groups. In turn, the chairman of this committee sat on the IRE Executive Board [10].

W. R. G. Baker, who held the presidency when Heising proposed the Group System, thought it especially useful to the IRE since electronics provided such an "ideal . . . climate to develop splinters." (But in spite of the formation of the Audio Group, a separate Audio Engineering Society was organized outside the IRE.) The tendency to fragmentation, Baker explained, began when "radio turned into electronics" as "the mutuality of interest based on one particular application was replaced by thousands of products and many, many different types of services." The IRE, as a "large parent body" with a "great diversity of interests," thus required an organizational form which would prevent the natural growth of "splinter organizations" [11].

Though recognizing that the American Society of Mechanical Engineers and the American Institute of Physics had served as models for the innovation, the GE vice president compared the Professional Group System with "decentralization as applied by industry." Just as a company might create divisions around engineering, manufacturing, and distribution, the Institute had created its semiautonomous groups to serve specific technical areas. To Baker, the strength of the system lay in allowing "the parent organization to assign the authority and fix the responsibility, and permits the men concerned to accept the accountability."

After the chairmanship of the Committee on Professional Groups passed from Everitt to Baker in 1950, Baker so actively promoted the system that many members tended to credit him with initiating the Professional Groups. Yet he insisted that "I had nothing to do with originating the idea or, in fact, with the initial planning." Baker credited Heising with the idea and Everitt with assisting him; in 1956, in fact, the Institute gave Heising the Founder's Award for having founded the system. Baker contributed, rather, by actively promoting the system. Besides writing a pair of articles for the *Proceedings* on the Professional Group System, Baker consistently encouraged the formation of new Groups [12].

Baker was in a strong position to promote the new system. Returning to the Board of Directors in 1946—he left the Board in 1940, when he lost to Terman in his bid for the presidency—Baker remained on the Board beyond mid-century, assuming, in 1953, the position of treasurer as well. When John Ryder became active in the national IRE in the early 1950's, he perceived Baker as a "king-maker." He remembered that "Baker's policy for the Groups was freedom." In 1957, when Ryder suggested that a Professional Group on Education be formed, Baker answered immediately: "O.K., if you will be the Chairman." Thus promoting the system by putting the weight of his prestige behind it, Baker could report in 1951 that eight new Groups had been added to the ten that existed a year earlier.

The subject fields of these new Groups suggested, not just an eighty-percent numerical increase but the maturing of a number of technical areas that would dominate the profession a decade later. The ten that existed in 1950 represented traditional areas of radio engineering, such as audio, broadcast transmission and receiving systems, and instrumentation, and the older theoretical areas like antennas and propagation and circuit theory as well. A Professional Group on Nuclear Science represented the only new area formed between 1948 and 1950. But as early as 1948, when the Audio and Broadcast Groups formed, members had begun to plan Groups in areas like telemetering and electronic computers. Thus, the new Groups of 1951 documented an exploding field of electronics specialties, including radio telemetry and remote control, airborne electronics, information theory, electron devices, and electronic computers.

These new subfields would not only dominate the future of the profession. They had begun already to capture the high ground in postwar electrical engineering. This was made manifest in a chart Baker prepared to demonstrate the expansion of Professional Groups through 1951. Traditional interests commanded the largest numbers, with over 1000 members gathering around audio, broadcasting, and antennas and propagation. In a similar category were the two largest Groups: Circuit Theory and Instrumentation. Yet a telling statistic was buried in Baker's table of membership totals, number of meetings held, and attendance at conferences. The Professional Group on Electronic Computers, with only 201 members in its first year, attracted the largest number of engineers to its conference in Philadelphia. The 832 who gathered there were greater by 300 than those attending a conference sponsored by the 1700-member Instrumentation Group. Within two years, the Computer Group was third in size and, by the end of the decade, the largest.

With additional Groups like Microwave Electronics and Medical Electronics added in 1952, Military Electronics four years later, and, soon after, Space Electronics, the Group System simply depicted the contours of the maturing field. This was illustrated by the six largest Groups in 1961, as shown by Ernst Weber, an ex-IRE president and director of a microwave research center who was then chairman of the Professional Groups Committee. These Groups came directly out of the earlier configuration of the electronics field, four of them having begun in 1951 and 1952. Weber listed, in order of membership: Electronic Computers, Circuit Theory, Electron Devices, Microwave Theory and Techniques, Antennas and Propagation, and Information Theory [13].

Yet beyond the configuration of the IRE's disciplinary interests, Weber saw something else in his list. The half-dozen largest Professional Groups, Weber realized, constituted the "theoretical groups" among the twenty-two existing Groups. As such they confirmed to Weber what many understood to be the special quality of the IRE: that it had "always been" the society's character "to integrate science and engineering." To the achievement of that persistent mission, the Professional Group System had contributed a way of doing so within an expanding technical environment.

Nuclear fission, electronics, and the power industry

The success of the Professional Group System rested, as much as on anything, on the dynamic nature of electronics itself. The system had been fashioned to handle potential splinter groups whose character was not yet known, yet which, as W. R. G. Baker recognized, would come rapidly out of the field of electronics. But electronics had another, related characteristic: its tendency to broad applicability. The strength of that tendency not only marked the IRE but came bluntly home to the AIEE when, after mid-century, it joined with industrial associations to effect the transfer of nuclear energy technology from public to private hands. Once developmental work began to make nuclear power generation a practical reality, it became clear that most of the innovations contributed by electrical engineering came out of electronics. The ubiquitous character of electronics would prove decisive in sealing the fate of the older society. Ironically, it came as the sharp split between power and electronics engineering dissolved before the continuing revolution in the electronics field.

Though the first commercial nuclear power plant did not begin operating until the late 1950's, the possibility of adding nuclear fission to the electric power industry's array of energy sources had been recognized even before the war. The editors of *Electronics* had discussed the "potentialities [for] a true atomic fuel" in 1940. That same year, *Electrical Engineering* published an article by nuclear physicist Enrico Fermi on the subject. And in the middle of a string of predictions made in a speech at the opening of RCA's new laboratory in 1941, David Sarnoff had pronounced it "conceivable that even before present-day youngsters become old, we may learn a good deal more along practical lines about the release of atomic energy" [14].

Those practical lines, which began to converge during the war, came fully together within a decade. The convergence was signalled in 1953, when the Engineers Joint Council (EJC), a successor to the post-World War I American Engineering Council, appointed a committee to prepare a position paper on nuclear policy to present to the Joint Congressional Committee on Atomic Energy, then considering changes to the Atomic Energy Act of 1946. The AIEE actively joined the EJC in this attempt to hasten the transfer of nuclear technology to the private power industry. Support came readily from a society in which power interests continued to dominate. Among the twenty-five members of the AIEE Board of Directors in 1953, in fact, nine engineers represented power companies and three more members came from the related areas of cable manufacture and coal mining. Others were employed by manufacturing companies or were professors, and one member came from a communications company.

Most telling, the AIEE's president for 1952–1953, Donald A. Quarles, was the chief officer of a nuclear research and development organization. Quarles had worked for the Bell companies since 1919, most recently as vice president of the Laboratories and of Western Electric. He described the New Mexico-based Sandia Corporation he headed as a "Bell system subsidiary working for the Atomic Energy Commission in atomic weapons ordnance." On the strength of this experience with military R&D, Quarles would serve in the presidential administration of Dwight D. Eisenhower throughout the fifties, moving from Assistant Secretary of Defense to Secretary of the Air Force and finally to Deputy Secretary of Defense [15].

At the Engineers Joint Council's request, Quarles appointed a member of GE's Nucleonics Department to represent the AIEE on the panel. Then in April, 1953, Detroit Edison president Walker L. Cisler informed Quarles of the founding of the Atomic Industrial Forum. The idea for the specialized power-industry association had come from another electrical engineer, T. Keith Glennan, who had succeeded William Wickenden as president of Case School of Applied Science upon Wickenden's retirement in 1947. Glennan, who had been a member of the Atomic Energy Commission (AEC), wanted the Forum to present "an informed voice to be heard at government levels as new atomic energy policy is hammered out" [16].

Impressed by this "constructive move," Quarles announced it to the AIEE directors and membership at a meeting that month in Louisville, Kentucky. He told them that, although "a lot of new technology [needed] to be mastered," which might take anywhere from two to thirty years, it was time that the "commercial application of atomic energy be pushed." The time had come, in part, because of the national government's increasing tendency to favor such technological transfers, he explained. It involved the AIEE, moreover, because "private enterprise and electrical engineering [were] two of the main ingredients" needed to commercialize nuclear energy. Quarles moved decisively to gather support for this initiative by electric power interests. When he learned in a news story, for example, that North American Aviation had developed a design for a small, civilian nuclear power plant, he informed the company's president of the EJC's joint committee that had been organized to promote the "commercial applications of atomic power."

The AIEE's representative on the EJC committee circulated lengthy comments on the existing legislation to Institute leaders, drawing from Quarles a detailed response. With similar communications from the other member societies, in July, the Engineers Joint Council testified before the Joint Congressional Committee. After claiming to represent a large majority of the nation's half-million engineers, the council laid out the key points of the national nuclear-energy policy desired by the power industry: Fuel should be made available to "all licensed organizations," free exchange of information was necessary to engineering progress, and the Atomic Energy Commission should allow the industry to establish its own safety codes and standards. The Engineers Joint Council's nuclear policy statement had virtually the unanimous support of the AIEE Board. Yet from within the AIEE committee that Quarles had named came a strong cautionary statement from Philip Sporn, a power engineer who possessed a reputation for technical acumen. Sporn told Quarles that power engineers were "barely scratching the surface of the technology and economics of atomic energy." An engineer with the American Gas and Electric Service Corporation of New York, Sporn thought that until "industry. . . . legislators and . . . the public" better understood these issues, the proposed "broad scale revisions of the Act would be very bad" [17].

The promise of more economical energy production—especially in the area of transporting fuel which, in the case of fossil fuels, can account for one-half of the cost—proved hard to resist. Sporn's was a lone voice in the counsels of the Institute. More representative was Titus Leclair, an engineer with Commonwealth Edison of Chicago and a past president of the AIEE. In June, LeClair called for a "speed-up" in the search for "more economical designs." He judged the matter to be the same as continuing to use black and white television while engineers developed color television: "Progress is usually made by doing rather than theorizing" [18].

Quarles agreed and, at the fall meeting of the AIEE in 1954, summed up the recent achievements in the new field of "what electrical engineers have came to call 'nucleonics'." The "atomic power field" had been active that year. "A new Atomic Energy Act" had opened the technology of nuclear fission to "development by private industry in our free enterprise tradition." Earlier that year, the Navy's atomic-powered *U.S.S. Nautilus* submarine had been

The opening of the Shippingport Atomic Power Station in 1958 began a new era, at once expectant and troubled, for the privately owned electric power company.

launched. And on Labor Day, President Eisenhower had "waved a radioactive wand" near Pittsburgh to begin construction of the country's first large-scale nuclear generating plant. To Quarles, the new field was "now passing from the invention and exploratory phase to the development and consolidation phase of a great new electrical-generation industry." A commercial nuclear reactor was achieved, much closer to the two years Quarles had estimated than to the thirty-year mark when, in 1958, Eisenhower launched the first commercial plant—the Duquesne Light-AEC Shippingport Atomic Power Station.

For all the attention the AIEE gave to nuclear energy during these years, the role of the electrical engineer was only part of a complex developmental task undertaken by chemical, mechanical, and civil engineers. (Recognizing this, in the early 1950's, some engineering schools established separate programs to train engineers in the distinctive problems of nuclear engineering.) Physicists were largely out of the picture after the war, when the engineering research and development phase began. Electrical engineers aided chemical and mechanical engineers by devising electronic instruments for control, measurement, and detection. Automatic controls to shut down reactors in case of component failure, for example, were among the early problems assigned to the electrical engineer. Electronics engineers also developed pile simulators to "duplicate electronically the dynamic behavior of the reactor" by means of "servo mechanism, electronic amplifiers and amplidyne type motors." The developmental stage, in short, required that electrical engineers design instruments for detection and measuring, drawing on "experience and techniques used in the radio and radar industries" [19].

Except for these contributions from electronics to a multifaceted technology like nuclear power, major innovations in the power industry tended to be in mechanical engineering. In a study of technical innovation in electric power generation from 1950 to 1970, only one category among nine lay squarely in the field of electrical engineering. Thus, amid advances in boiler design, boiler feed pumps, pollution abatement equipment, and fuel handling, the contributions of the electrical engineer to conventional power technology were the same as in nuclear engineering: that of automatic control equipment. This included, for example, Louisiana Power and Light's pioneer use of transistorized computers in 1959 and, in 1961, the same company's step of fully automating its Little Gypsy Station generating plant [20].

Traditional power engineering concerns were still pertinent, especially in dealing with problems of electrical generation and distribution in such traditional areas as meters, instruments, and relays. Still, as made clear by the electrical engineer's contributions to both nuclear and conventional electric power generation, electronics pervaded even the power engineering field in the postwar era.

The situation demanded that the AIEE give close attention to the extension of electronics into its end of the profession. It had little choice since, as a member of the headquarters staff reported in 1952, special technical conferences on computers and electronic components drew the largest numbers. The strongest conference programs, he reported, came out of "the Science & Electronics Division where there are a lot of practical applications and lively new subjects to cover." However, the "Power Division is generally rather complacent." In 1956, AIEE president Morris D. Hooven made the point explicit in assessing the members' expressed desire for more articles on electronics in the Institute's publications. Though recognizing that the largest group retained a primary interest in meters, relays, and instruments, a recent questionnaire sent out by headquarters indicated "that the electrical engineer who was formerly interested mostly in wires, cables, and switchgear is now showing an even greater interest in electronic tubes, semiconductors, and electronic components" [21].

The promise of future developments in power engineering similarly involved the use of electronics. Writing in 1961, Lloyd Berkner, then president both of the IRE and of the Graduate Research Center of the Southwest in Dallas, Texas, described the place of electronics in a changing power field. He foresaw in "today's magnetohydrodynamics based on relativity and wave mechanics" the direct conversion of heat to electricity without the aid of mechanical parts and at greatly increased efficiencies. Though transistors had begun to replace vacuum tubes in numerous areas, they told the same story, leading Berkner to conclude that "because of their economy and extreme flexibility, electronic methods are destined to become the nerve system of industry, entering into every aspect of the industrial process" [22].

The technological landscape of the IRE

By 1960, electronics, indeed, was becoming the nervous system not only of industry but also of the nation's defense system and its efforts in space. Yet awareness of the direction of electronics had come to the fore during the Second World War. Institute president William Everitt reported to the membership in 1945 that engineering applications once restricted to the radio field were rapidly expanding "both in size and breadth." Therefore, he announced, the Institute of Radio Engineers, which had "grown with the electronics art," intended "to continue to grow to serve the entire family of engineers who are working in this field" [23].

For the IRE, that intention meant, besides responding organizationally to the new disciplinary conditions, advancing the profession's knowledge-base through the bread-and-butter activities of any engineering society: conventions and publications. Wartime restrictions had diminished conference attendance and barred the publication of scientific and technical articles based on wartime research. However, as normal professional society activities got fully underway again, a plethora of papers and information was released. Additionally, Institute leaders canvassed the engineers returning to universities and sought out industrial contributions to society programs.

Swollen IRE conventions presented the most dramatic picture of growth as each major meeting set new attendance records. At the first postwar Winter Technical Meeting, held in New York during January 1946, over 7000 engineers attended. Not only was it the largest meeting in the society's history, but the convention's attendance far exceeded the 3000 who attended the year before. When, in January 1949, over 7000 showed up at the IRE West Coast Convention, its organizers accurately claimed the meeting as "one of the largest electronic events ever held." However, two months later, the standard was raised in a spectacular fashion when over 16,000 "engineers, physicists, and technicians" gathered in New York for the National Meeting. As the *Proceedings* editor wrote in his report on the convention, it was "the largest in the thirty-seven years of the Institute's existence" [24].

But the most substantial event lay behind the numbers and was to be found in the technical papers and reports. These traced the development of the field and, also, the pressing technical concerns of the era. Serious attention went early to the new developments in nuclear physics. At the 1946 meeting, Major General Leslie R. Groves, who had headed the Manhattan District, spoke to a joint meeting of the AIEE and the IRE on "Some Electrical, Engineering, and General Aspects of the Atomic-Bomb Project."

In a similar vein, attendees to the 1947 West Coast meeting visited the new 184-inch cyclotron at the University of California at Berkeley. At the same meeting two years later, "atomic bomb control was the principal topic of the convention." Physicist Robert A. Millikan read a succinct paper on "The Release and Utilization of Atomic Energy," which argued, in part, that the bomb's "greatest service" was to have clarified the "necessity for finding a substitute for war in international relations." Before the end of the meeting, an ex-IRE president received wide press comment when he asserted that the

Russian scientist's "fear . . . [of making] mistakes would bring victory to the United States in the event of war."

Such peripheral events alone indicated much about the larger technical scene which surrounded the conventions. Taken with the content of the formal programs during the five years after the war, convention activities revealed a field taking shape. At the 1947 meeting in San Francisco, besides the visit to the cyclotron at Berkeley, groups of engineers inspected the wind tunnel constructed at Moffet Field during the war by the National Advisory Committee on Aeronautics, the Electronics Laboratory at Stanford, and electronic manufacturing plants in the Bay Area. Some of the conference goers accepted the invitation of the West Coast Electronic Manufacturers Association to attend its annual show being held simultaneously in San Francisco.

Ample signs of the past and future of the profession existed. The chair of the IRE's convention committee for the 1947 meeting was Stanford professor Karl Spangenberg, who had headed the ultrahigh-frequency research at the Radio Research Laboratory during the war. RRL director, Frederick Terman, was himself the guest speaker at the convention banquet. A year earlier, 170 exhibits by 130 companies had included displays on peacetime uses for radar, vacuum tubes, magnetic recording, remote-control devices, and electronic navigation and direction-finding instruments. Control devices were prominently displayed, including systems for ground-tuning radio equipment in aircraft, pulse timers for automatic control equipment, and other items spinning out of wartime research.

At the West Coast meeting in 1949, engineers learned of Army experiments going on in the Los Angeles area on "guided missiles," and, that same year, at the record-shattering National Meeting in New York, symposia were held on network theory and electronic computers. The nucleonics session, it was reported in the *Proceedings,* "received more attention from radio men this year than ever before." From the radio laboratories, the report explained, were "coming newer and better instruments indispensable to the production, control, and utilization of fissionable materials." Among the most revealing elements at the 1949 National Meeting was a display by the Army's Signal Corps. As an example of its "'miniaturization' program," the Corps displayed a radio receiver and transmitter "so small that it fits into a king-size cigarette package."

Fitting electronics components into compact spaces was already a prime objective of industry and the national government during these years. But the Army Corps' example was not an accurate representation. Its pocket-sized radio contained miniature vacuum tubes and miniaturization would be fully realized only with the transistor. The critical breakthrough in semiconductors, moreover, would occur in an industrial setting, with no direct assistance from the military.

Working at the Bell Laboratories in New Jersey between 1945 and 1947, John Bardeen and Walter Brattain experimented with chips of germanium in their quest to make a semiconductor work as an amplifier. Though only indirectly involved in this project, William Shockley led the semiconductor team set up by Bell in 1945. His theoretical contributions made him, in effect, a partner in Bardeen and Brittain's work. All three were physicist-engineers. Shockley, a Californian, received his Ph.D. from MIT in 1936 before going to work for Bell Laboratories. Bardeen and Brattain came out of the Midwest. After getting a bachelor's and master's degree in electrical engineering at the University of Wisconsin, Bardeen earned a Ph.D. in physics at Princeton. Brattain's doctorate in physics came from the University of Minnesota in 1929, the year he joined Bell Laboratories [25].

From this trio came, first, Bardeen and Brattain's point contact transistor in December 1947 and, in 1950, Shockley's junction transistor. The junction transistor took the field, as engineers assumed the developmental work of solving the basic problems in design and manufacture. This meant, in the early years, mostly scientists and engineers at Bell, since the company initially maintained secrecy around the project.

Bell's dominance ended in 1951, when the military inaugurated a substantial R&D program on the transistor. Anticipating the impact of transistors on the youthful field of military electronics, the Department of Defense's Research and Development Board established an Ad Hoc Group on Transistors under its Committee on Electronics. Yet the research extended to the solid-state field generally, seeking knowledge of circuit applications as well as

Bell's trio of researchers — (from left) John Bardeen, William Shockley, and Walter Brattain — transformed the capabilities of electronics when they invented the transistor.

improved transistors. Efforts were made also to increase available facilities for the manufacture of transistors. As an Army observer reported, "the military services are presently supporting substantial contractual programs." They were specifically aimed at "transistors with higher power ratings, higher frequency response, lower noise, and the ability to operate satisfactorily over wide ranges of temperatures" [26].

The work already underway by both Bell and the military received a large boost in April 1952 when Bell ended the company's policy of secrecy by holding a symposium on the transistor for representatives of thirty-five electronics manufacturers. Charging each firm $25,000, Bell promised to deduct this later from licensing fees since its aim was to get more companies involved in working on the problems of the transistor.

Though a fundamental scientific and technical achievement, in a sense, the transistor represented just the newest piece of hardware in the story of postwar electronics. Miniaturization, which, as a concrete and widely applicable technical concept, underlay the explosion of electronics after the war, began earlier. Before the advent of the transistor, as demonstrated by the Army Corps' pocket-sized radio, the search for smaller and smaller components had concentrated on the vacuum tube. Especially at the demand of the military, whose funding carried a significant portion of the research and development costs, miniature and then subminiature tubes were developed.

However, the conception and development of the junction transistor around 1950 promised a level of miniaturization and dependability undreamed of with the tube. Even before Jack Kilby's invention of the integrated circuit at Texas Instruments in 1959, one million transistors were being concentrated within a cubic foot. With its low energy requirements, absence of heat-producing elements like the filament, and potential for permanence, the transistor's small size substantially defined its significance and that of other semiconductor devices. This cluster of factors, in short, came together in the transistor during the 1950's and 1960's to make possible developments in weapons systems and space technology that otherwise would have been impossible.

Above all, the transistor enabled the electronic computer to become, like the transistor itself, a versatile and widely applicable machine. The first electronic computers were scarcely older than Bell's transistor program. Constructed with thousands of vacuum tubes, they issued from wartime R&D projects at Harvard, Bell Laboratories, and the University of Pennsylvania, whose ENIAC (Electronic Numeric Integrator and Computer) has been credited with being the first large-scale, programmable electronic computer.

Physicist John Mauchley and engineer J. Presper Eckert built the ENIAC for the Army at the Moore School of Electrical Engineering, designing it to compute firing tables. However, when completed in 1945, the ENIAC's first assignment came from the Manhattan Project, with a request to determine the feasibility of a hydrogen bomb. The uses of the computer had expanded only slightly by the early 1950's, when the Department of Commerce estimated that only 100 large (that is, digital) computers would be needed to handle the nation's needs. But by 1963, it had become the largest single user

The Electronic Numeric Integrator and Computer (ENIAC) was a member of the small, select first generation of electronic computers built during the mid-1940's.

of transistors. Still an industrial item, the computer used more transistors than in all consumer products, including the radio [27].

Like the transistor, the computer came of age during the 1950's. The major contours of their stories were concisely drawn in a half-dozen of "Special Issues" published in the *Proceedings of the IRE* between 1952 and 1961 [28]. Two each appeared on the transistor and computer with single issues on solid-state electronics in 1955 and on space electronics in 1960. Together, they drew a picture that amply illustrated Terman's characterization of electronics as a "great growth industry." For between the first issues on the transistor and the computer in 1952 and 1953 and the second issues in 1958 and 1961, the technology and industry of electronics grew from infancy to maturity.

When the first transistor issue appeared in late 1952, just six months after Bell had offered the new device to the nation's electronics firms, the leading question was "how to transform the transistor from a laboratory oddity to a practical device which could be manufactured uniformly and in quantity." Already, the junction transistor was receiving the most attention but there was interest also in different semiconductor materials, in photocells, and in the problem of noise that plagued early transistors.

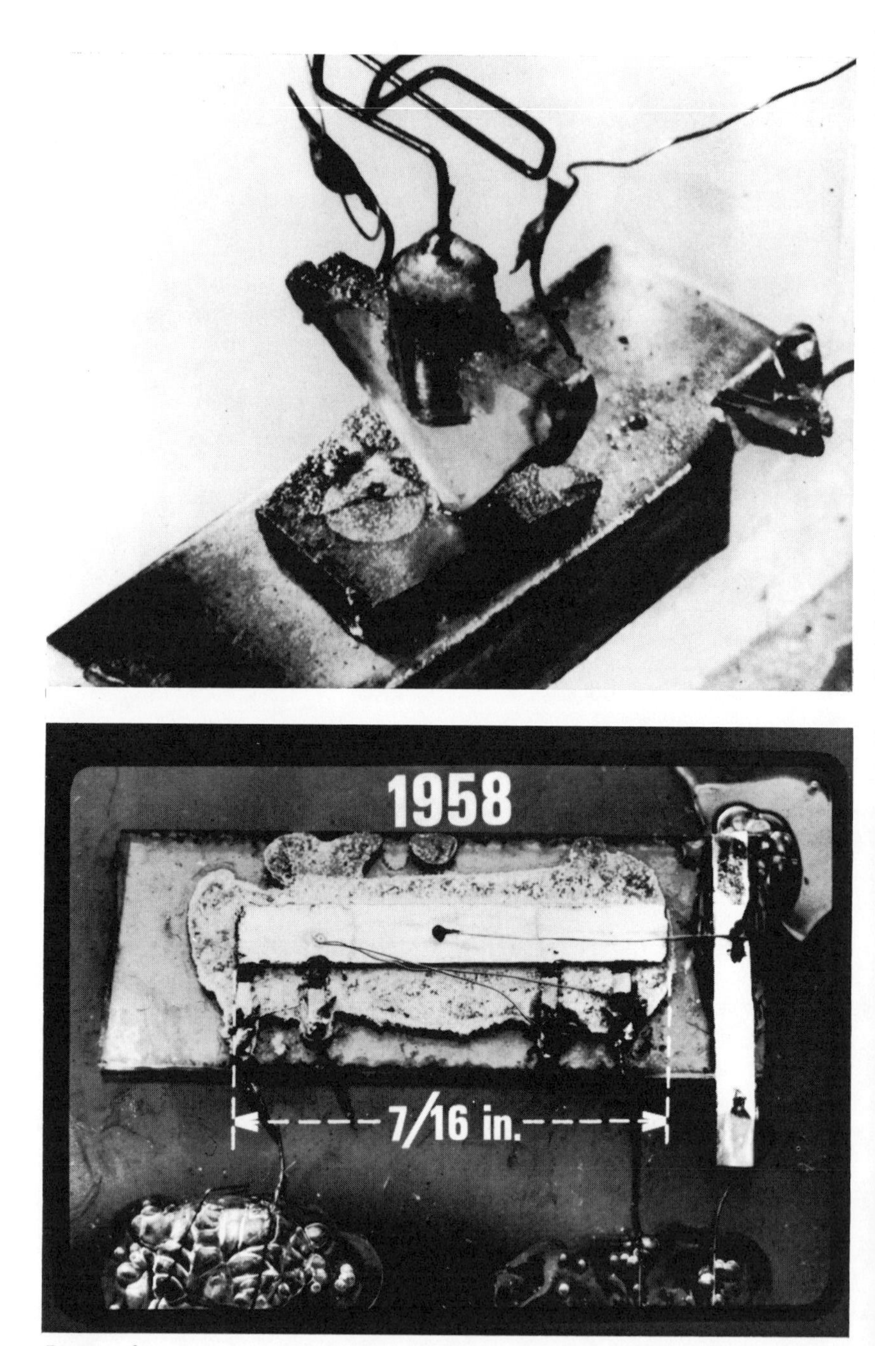

Between the announcement of the point contact transistor by Bell in 1948 (above) and the development of the integrated circuit by Texas Instruments a decade later (below), the foundation of the microelectronics revolution was put in place.

Bell wanted to replace the vacuum tube in computers and servomechanisms with transistors and to apply transistors in the Bell system "where vacuum tubes are not yet widely used." For the military, the transistor offered the potential advantage of both compactness and dependability. Specific applications desired by the military branches at this time ranged from wrist-watch radios for the Army, light electronic equipment for Air Force bombers, and Navy torpedoes with "'transistorized' electronic brains" [29].

At the outset, engineers hoped that the transistor would "close somewhat the great gap that now exists between our best electronics 'brains' and the human brain." It might also hasten the "industrial use of complex automata." In 1951, digital computers were giving "stiff competition" to the pioneer analog computers. With this development, the military already had begun to contemplate the use of "small compact 'transistorized' computers" in applications where the size and power needs of vacuum tubes had made their use impractical.

When the first computer issue appeared in 1953—edited by IBM engineer Werner Buchholz for the Professional Group on Electron Computers—several specific computers received most of the attention. Besides IBM's Type 701 Computer, papers were included on the SWAC, SEAC, the Navy's NAREC, and the University of Illinois' ILLIAC. Buchholz explained, the computer had become indispensable in two areas: that of "engineering calculations in defense industries, particularly in the aircraft industry" and "accounting applications to cope with the problems of an economy of ever-growing complexity." Buchholz predicted that the electronic computer might "soon affect the average person as much, if not as obviously, as radio and television."

By the time a second transistor issue appeared in 1958, Bell transistors, as the editor pointed out, were "orbiting around the earth in the Explorer and Vanguard satellites." Additionally, a great number of semiconductor devices besides the transistor and diode had been developed. Diffusion techniques in silicon and germanium were greatly advancing the fabrication of semiconductor devices. Though recognizing the importance of the computer as a source for the use of transistors, it was still seen as something "which will become more important as time progresses."

The editor of the second computer issue three years later had far more to report. Since the first issue of 1953, the computer field had been expanding at "an astonishing rate," making the computer "a factor in the lives of most electronics engineers." Computers had advanced on the fronts which earlier had been seen as of primary significance, leading to greater internal operating speeds and to new devices for internal memory or storage (especially thin-film memory and disks). While their use in control systems had predictably increased, computers were used for the first time as tools in circuit design and in simulation roles as an aid in engineering research.

These accomplishments of the formative years of microelectronics dramatically came together in the space program during the late fifties. The editor of the special issue on space electronics in 1960 predicted that "the new space environment will cause design and operating changes in practically

The Jupiter-C rocket used in this launch at Cape Canaveral, Florida, in 1958 rested upon the military research and development programs of the 1950's. But its payload, Explorer I—America's first satellite—looked to the 1960's and the growth of a massive aerospace industry.

every branch of electronics." He recognized, nonetheless, that the changes would rest on work already done in navigation and guidance, communication, telemetry, and instruments and measurements.

Indeed, the electronic equipment and techniques used in the space research and satellite programs had been adapted from that used in "earlier rocket experiments," conducted first by the Germans during World War II, then, afterwards (and often with the same scientists) "through the military efforts of the United States and Russia." Earlier in the decade, Simon Ramo, director of operations for Hughes Aircraft Company, had clearly seen the shape and thrust of the work on guided missiles when he claimed it as "a new field" for electronics engineers. Drawing together system considerations with specific electronics areas like communications, servomechanisms, device development, and computers, the "guided missile field," Ramo wrote, "is second to none . . . in its probable ultimate effect on the broad development of electronics."

Like the space program itself, the space electronics issue demonstrated the technical continuity of the era. But it also showed that the framework of expectations constructed by the postwar engineering community continued to bolster engineering ambition. When the editor asserted that the achievements of the late fifties rested on the "certain" belief that "human ingenuity will triumph . . . in the colonization of other bodies in space," he simply attached new goals to the technical achievements of the field that Ramo had announced in 1952. The profession, in large part, was already a part of Edward Bowles' "peacetime counterpart pattern" to the World War II R&D order. Electrical engineers were serving, as Donald Quarles told the AIEE in 1952, as the "shock troops of the Cold War." In 1960, electronics engineering joined a quest to build space vehicles as well.

In these achievements in military and space technology, the engineers were building a technology of radically new dimensions, not only in technical terms but in its social and economic aspects. Giving rise to one of the nation's largest industries, microelectronics made possible systems both unprecedented and increasingly pervasive. The IRE served this revolution essentially by transmitting technical knowledge through meetings and publications. In the papers before its meetings and in the articles in its journals (including the *Transactions* of the Professional Groups as well as the *Proceedings*), the IRE thus provided a technical record of the major achievements of the field and the scope of its uses as well. Among, the thousands of pages, moreover, there appeared at times coherent parts that described the whole. These were the Special Issues.

But besides the numerous Special Issues, there occurred a far smaller publishing event that also captured the whole. It was a conceptual event, having to do with defining the field of electronics, taking place in 1952 when William Everitt wrote a brief article for the *Proceedings* entitled "Let Us Re-Define Electronics" [30]. Everitt judged the older definition to be "too narrow" in defining electronics as "the science and technology of systems using devices in which electrons flow in a gas, no matter how dense or tenuous."

Not only did this limit the field to a device — to the electron tube, in fact — but the definition failed to recognize that "control" was only a part of electronics. Actually, the direction of electronics was far broader, Everitt insisted, aiming to extend "man's senses in space, as by the radio . . . ; in acuity, as by the electron microscope; in visual or audible range . . . ; and in speed, as by computers." Finally, Everitt explained, the field sought to supplement "man's brain," even to "making comparisons and judgments," as in servomechanisms and by solving mathematical problems.

Thus, the definition — at least for now, Everitt cautioned, since it would change again with the field — should be:

> Electronics is the science and technology which deals primarily with the supplanting of man's senses and his brain power by devices which collect and process information, transmit it to the point needed, and there either control machines or present the processed information to human beings for their direct use.

Some engineers protested, rejecting the broad, social cast of Everitt's definition. In a letter to the *Proceedings* editor, one engineer insisted that the new definition, like the old, should be restricted to devices but extended to include semiconductors. He rejected the defining of electronics "broadly by the ultimate service which it renders to mankind." Another writer disagreed with Everitt's introduction of the concept of supplementing human senses, confining the definition rather to "the control of electron movement." But Everitt insisted that engineers "must consider in their design the nature of the receptors which make use of the signals," whether human or machine; "the control of electron movement is only part of a complex picture."

Educating the engineer in the postwar world

It was the sheer complexity of the new electronics that raised the major professional issue of the postwar era: that of the education of the engineer. Though more pressing than the search for a definition of electronics, it was not alien to the spirit of that search. For the question of educating the engineer involved, at its core, defining the nature of the engineer and the content of engineering knowledge.

Though raised anew after the war, the issue of what constituted a proper engineering education had a long history within the profession. Following William Wickenden's epochal investigation of the 1920's, the Society for the Promotion of Engineering Education (SPEE) published a report on "Scope and Aims" in 1940 and followed it in 1944 with a study looking to the postwar situation. In the early 1950's, the growing concern over the state of engineering education climaxed. Again, as in the 1920's, the effort made by what was now called the American Society for Engineering Education (ASEE) would be a major one and, as never before, would move the engineering profession to confront fully the issues bared by the Wickenden report. Indeed, the 1950's eruption hearkened back to the turn of the century, when Charles Steinmetz similarly challenged the profession. For again, the place of engineering science in the curricula of electrical engineering education became a burning issue for the profession.

The immediate source of the sense of crisis came from the war. The differences Frederick Terman observed in the work of physicists and engineers in the Radiation Laboratory and the Radio Research Laboratory had raised basic questions about the quality of electrical engineering education to many engineers. But besides bringing understanding, the war spurred action. The idea that, as John Ryder put it, the "more abstract forms of engineering education" were necessary to the electronics field had emerged at numerous meetings even before the war [31]. At a 1939 West Coast AIEE conference on teaching communications and electronics, for example, a lengthy discussion revealed that no one disagreed on the need to study fundamental principles. A professor from the California Institute of Technology reduced that general notion to a specific prescription. Because "the commercial development of the ultrahigh frequency field" made the need for advanced study greater, the education of the engineer required more rigor. "Our circuit equations are only approximations," he explained, making it increasingly "necessary to resort to fundamental equations of Maxwell to obtain a rigorous and correct understanding of what is happening at these very high frequencies" [32].

How far the prewar generation was from acting on the need for knowledge of fundamentals became clear through other events in 1939. Most conspicuous was the AIEE's award of the Edison Medal to Dugald C. Jackson for a half century of combining "engineering education with engineering practice." Yet again that year, Wickenden had joined with the Society for the Promotion of Engineering Education to issue another general report. Even allowing that no outside funds were used and that no surveys were conducted, the 1940 report on the "Aims and Scope of Engineering Curricula" nonetheless avoided the high conceptual ground of the earlier study. It began boldly by charging that the broad aims of conventional programs, whose subject areas ranged from "science" to "commerce and finance," diluted basic engineering education and, along with it, student ambitions. But the report failed to discuss the place of science and mathematics in the curriculum, instead recommending "scientific-technological and . . . humanistic-social sequences of engineering education." It was these investigators, moreover, who acquiesced in the American student's avoidance of advanced study, asking only for sound training in the essentials of good practice [33].

But the war changed all this. In 1945, the future necessity of learning Maxwell's equations had become a present need. Terman made this a constant theme after he returned from the war to serve as dean of the School of Engineering at Stanford. He wanted also to train students in additional specialties so they could "exercise leadership in business activities." However, he established advanced studies to help students imbibe "technical knowledge . . . and thereby aim toward leadership in scientific and technological activities." Terman's aims were not narrowly academic. He wanted to aid Californians who had "long dreamed of an indigenous industry" of sufficient scope "to balance [the state's] agricultural resources." The war had advanced those hopes; yet as he wrote in his 1947 report to Stanford's president, the time had come to realize a "great new era of industrialization." Because advanced graduate training would be needed, he concluded with a warning: "Industrial activity that depends on imported brains and second-hand ideas cannot hope to be more than a parasite that pays tribute to its hosts, and is permanently condemned to an inferior competitive position" [34].

Besides the influx of German physicists to America after the war, Terman had in mind the historic reliance of his field on the physical sciences. His ideas here, too, rested on experience more than speculation. "It was quite clear," he later recalled, "that the war showed that the training of engineers was inadequate, that they didn't measure up to the needs of the war. Most of the major advances in electronics were made by physicists and people of that type of training rather than the engineers." Engineers typically terminated their training with a bachelor's or master's degree, Terman explained, following it with "practical experience" instead of pursuing an "understanding of the . . . fundamentals." The physicists, however, "were converted into extremely good engineers very quickly." The reason: "they were able to understand the complex technology that was involved with their superior training, whereas many engineers or most of them didn't quite understand these new phenomena" [35].

The 1944 report on engineering education showed that others shared Terman's judgment. That the profession perceived the need to move beyond the concentration on practice appeared clearly in the sharp break that the Society for the Promotion of Engineering Education made from its 1940 report. Though some committee members served again — Wickenden and the chairman, H. P. Hammond, for example — a sense of urgency suggested that the committee did not simply ride the rising tide of graduate education in engineering that followed the war. Its stress on training in science to produce critical engineers in the Steinmetz tradition had taken up the challenges appearing in the conclusions of the Wickenden report. Wickenden's survey had led, in 1932, to the founding of the Engineers Council for Professional Development (ECPD) as an accrediting agency, but that step had failed to reach the depths explored in the Wickenden analysis. Further, there was something in the 1944 committee member's experience of the war — they believed the war's "policy of expediency" had degraded the educational process — that pushed the new SPEE committee to deeper levels [36].

So while accepting that many students wanted a four-year college degree, the members also devised a body of educational principles intended to strengthen "engineering education at its base" and to develop creative engineers. To meet the requirements of "Engineering Education After the War," the SPEE committee urged two basic changes in engineering departments: increased training in applying the engineering method to the "management and operations of industry" and a quick response to the "greatly increased need for engineers prepared to practice at high scientific and creative levels." Three tracks included teaching science fundamentals to "regular students," introducing problems of production and operation to an "industrial group," and erecting programs to prepare students for "highly scientific and creative engineering work." This latter program, moreover, aimed at "broader and more fundamental preparation in scientific principles and method" for a longer period of study.

Rather than receding with the passing years, the strong urge of 1944 to "keep pace with the rapid developments in science and technology" intensified. The new crisis was forcefully raised in 1951 in a committee report of the ECPD. It affirmed both the continuing pertinence of the Wickenden analysis and the profession's disregard of its recommendations. The ASEE's new investigation flowed directly from the ECPD document and, like it, the society's "Evaluation of Engineering Education," published in its final form in 1955, broke fresh ground. The "Evaluation" also brought Frederick Terman once again onto the national engineering scene to promote engineering science.

Though the phrase itself was absent, a desire to establish *engineering science* in engineering education was apparent in the ECPD's incisive report on the "adequacy and standards of engineering education." Devoting half of its report to an historical review, the committee arrived at rather dismal assessments of the situation at mid-century. The curriculum of 1890 had so closely expressed the interests of self-educated "practicing engineers," by 1910, the committee could find "no elevating influences" in the traditional fields. The only exceptions resided in the curricula of the newer fields of chemical and metallurgical engineering and, after World War I, of electronics.

The ECPD's report soundly condemned the inroads made by commercial engineering during the period before the First World War. It had led colleges of engineering to develop various courses in "fringe areas," such as accounting and business. "There can be no question that these courses of reduced technical content were an aid to certain persons," the committee concluded. However, "the serious doubt was whether they should have been considered engineering courses." This dilution of engineering curricula, the report argued, underlay the constant threat to the modern engineer of "obsolescence." In the committee's opinion, engineers protect themselves best when they undergo rigorous instruction in the basic sciences, especially in mathematics. These were the "sustaining parts of the curriculum," enabling engineers to continue to learn as their "interest and need may expand." In short, the committee wanted the "science" of engineering taught in the schools, leaving "much of the art" to be "acquired in the field." Most remarkable was its conclusion with regard to the thirty-year-old Wickenden report: "Your committee is struck with the potency of the statement to today's situation, and the extent of the failure of the schools to implement much of the recommendations" [37].

With explicit reference to the ECPD report, the ASEE appointed a committee in 1952 to prepare the "Evaluation of Engineering Education." Nearly 50 percent longer than the 1944 report, the final document issued from elaborate surveys and from a series of conferences supported by the National Science Foundation. The resulting "Evaluation" was a watershed for a half-century of attempts to bring scientific rigor to the study and practice of engineering. Though much of it concentrated on the need for the educational establishment to put its own house in order, the report made an unprecedented effort to clarify the proper subject matter of engineering. It broke new ground in the area of curricula by spelling out the content of a more scientific core [38].

Given this central thrust, it was not surprising that the "Evaluation" emerged as a major statement on the "engineering sciences." It began with general definitions: By training in engineering science, the committee meant the study of "basic scientific principles as related to . . . engineering problems and situations." There followed a list of ten means to implement a "scientifically oriented engineering curricula." In mathematics, faculty competence needed to be developed to teach additional mathematics for students interested in research, development, and "the higher phases of analysis and design." Chemistry would aid engineers in their studies of such areas as properties of materials, corrosion, and industrial chemical processes. Chemistry, moreover, should be carefully coordinated with the teaching of "modern physics."

In the critical area of physics, the report elaborated, calling for a reorientation of the content in engineering courses in physics. Undergraduate engineering curricula should include modern physics, meaning "nuclear or solid-state physics." More fundamentally, in urging greater attention to engineering physics, the committee wanted the introductory physics course to be "redirected to place much greater emphasis upon sub-microscopic phenomena and the conservation principles, with virtual elimination of semi-engineering examples." In addition, it would strip basic physics courses of "the engineering sciences of mechanics, thermodynamics, and electricity," thus recapturing areas in classical physics that duplicated the engineering sciences. Following its definition of both basic and engineering science, the committee specified the six engineering science fields. All stemmed from the main branches of mechanical and electrical phenomena, leading to the mechanics of solids, fluid mechanics, thermodynamics, transfer and rate mechanisms (heat, mass, and momentum transfer), electrical theory (including fields, circuits, and electronics), and the nature and properties of materials.

It was this act of specification that raised the ire of the electronics engineers. Terman could have found nothing intrinsically abhorrent in the basic ideas and proposals of the report. For too long, engineering science had stood on the fringe of most electrical engineering programs, a position Terman had, for many years, wanted to change. During the thirties and forties, he had won a footing for advanced training at Stanford and helped found an electronics industry in the area south of San Francisco, what came to be called Silicon Valley. Returning to Stanford after the war, Terman helped establish the Stanford Electronics Laboratories and assumed the deanship of the School of Engineering. But engineering science, to Terman, was largely electronics. Yet in its discussion of engineering science and especially in its list of the engineering sciences, the committee had omitted "electronics." Terman was incensed. Parenthetical inclusion under electrical theory missed the mark.

To the IRE Board as well, that omission represented a colossal failure to comprehend the engineering scene at mid-century. Terman complained that the report came from a "46 man committee" that held "the compromised viewpoints of a progressive but middle-aged to senior statesman group of engineering educators, whose center of gravity is in traditional engineering."

Though the IRE Board contained engineers who might have protested on their own, Terman acted as a catalyst. The final ASEE report came in September 1955, just three months before Terman published his own views on the fragile state of electrical engineering education in the *IRE Student Quarterly*. In an article entitled, "Electrical Engineers are Going Back to Science!" Terman passionately summoned electrical and electronics engineers to seek a renewed understanding of the scientific character of engineering. Within the universities, he wanted a central curricular position for engineering science. But Terman went further. In a single, italicized paragraph standing in the middle of the brief essay, he urged engineering educators to accept areas like electronics "that lie between pure science and traditional engineering as being engineering." If they did not, "Colleges of Applied Science will develop on the campus and insulate engineering from pure science while taking over the interesting and creative areas," leaving "engineering to concentrate primarily on dull trade school subjects. I cannot see engineers allowing this to happen — too many engineers are ambitious for their profession" [39].

Terman's article quickly became a manifesto for electronics as the dominant disciplinary component of electrical engineering. *Proceedings* editor Don Fink reprinted the article so that, he told Terman, it could reach 50,000 readers instead of 7000. In the meantime, the IRE Board had begun to develop a statement of educational policy. University of Pennsylvania Professor John G. Brainerd, a member of what John Ryder called the Board's "educational fraternity," first raised the issue. He pointed out that Terman's article was "substantially at variance" with the ASEE report, which made it necessary for the IRE Board to discuss the matter. Brainerd worried that the report's use as a guideline for accrediting undergraduate programs might have an "adverse effect on electronics education." The problem was made more acute with the IRE unrepresented on the ECPD, which had adopted the "Evaluation" as a guide for its accrediting work [40].

Brainerd's comments initiated a year-long response to the slight. That spring, reprints of Terman's paper went out to the deans of all schools possessing IRE student branches. By early 1957, John Ryder had drafted a formal response to the ASEE report. Ryder's role in shaping the IRE response was strengthened by his position during the early 1950's as chairman of the education committees of both the AIEE and the IRE. Twice in the past five years, moreover, Ryder had urged the need to define "what kind of engineers" the profession needed. In 1949, Ryder urged that the electronics engineer be prepared to undertake "pure research" and, three years later, that the electrical engineer's education be "more fundamental in nature" [41].

Like Terman, he firmly supported the centrality of engineering science. In the draft resolution, he called for more emphasis on science and mathematics and less "educational time spent on mere skills and practical technologies." Ryder wanted the "Evaluation" to "be considered only as a photograph of the present, rather than as a blueprint for the future." The six engineering science fields listed in the report would fit modern students to "the needs of the past," he thought, rather than to the needs "of the future." In defining the "modern engineer," Ryder transcended the conventional boundaries of the field. He included as part of engineering work "the development of basic ideas and the discovery of new knowledge of nature, as well as the design and the building of 'things' based on those discoveries." Though Terman did not sit on the IRE Board, Ryder sent him a copy of the draft resolution. Terman agreed with its views but thought it too timid.

> In particular, I feel that the IRE in a polite and dignified way should raise a little bit of hell over the fact that electronics, the great growth industry of engineering, was given very little consideration, and apparently was only imperfectly understood in the ASEE study on engineering education [42].

In the end, the IRE Board took Ryder's conceptions and Terman's suggestion. Thus, the final statement adopted in April, 1957, while complimenting the ASEE report for raising the level of engineering education, sought to suppress its use as a guide for the Engineers Council for Professional Development in its accrediting work.

The ASEE report and the spirited response of the IRE leadership led to results beyond what could have been achieved by the "Evaluation" alone. For one, it drew the IRE into the ECPD. Following a rare meeting in Houston in 1957, at which the IRE Board met as a committee of the whole to discuss "the educational trends and problems of our profession," the directors decided to apply for membership in the ECPD. Four years later (following procedural delays), the IRE accepted an invitation from the ECPD. Beyond this, the conflict engendered by the ASEE "Evaluation" enlivened discussion and debate over electrical engineering education within the IRE. Though punctuated by defenses of "manual work" by engineers, the reformers' attacks on "technician training" and calls for study of "the fundamental sciences of nature" and for emphasis on "mathematics in a computer age" dominated the debate [43].

Thus did the issues reverberate through the electrical engineering community, leading to conferences such as the one held in Worcester, Massachusetts, in 1959. There, a speaker defined the work of "research scientists and engineers" as dealing with "the frontiers of science and technology." Each helped to "accelerate significant discovery," he said, both by advancing "new scientific knowledge" and by translating "science into new technologies." This was one discussion among many that helped to unfold the meaning of engineering science to the profession. During the late 1950's and early 1960's, to cite one of the more ambitious undertakings, with the support of the National Science Foundation, the IRE and the AIEE joined the ASEE and the IRE's Professional Group of Education to sponsor a series of International Conferences on Electrical Engineering Education at Syracuse University and Sagamore, New York. They focused on such topics as "Undergraduate Physics and Mathematics in Electrical Engineering" and "Electrical Engineering Education." The conferences often served the dual purpose of both exploring the new disciplinary content and celebrating the advances being won. Though the discussions touched on several types of engineering work, ranging from the routine to the creative, engineering science was a major topic at the meetings [44].

The movement to establish "engineering science" as a major emphasis in engineering education was yet another legacy from the war. The Sagamore Conferences demonstrated the widespread concern for an education that would train engineers capable of doing creative work comparable with that done by the physicists during the war. Certainly, after the mid-fifties, the number of graduate degrees awarded in electrical engineering shot dramatically upward. Two decades later, even after noting a decline in doctoral degrees among engineers, Terman proclaimed the revolution in engineering education a resounding qualitative success. "Never again," he wrote during the nation's bicentennial year, "will electrical engineering have to turn to men trained in other scientific and technical disciplines when there is important work to be done in electrical engineering" [45].

In judging the ultimate outcome of the 1950's revolution in engineering science, Terman once again recalled his war experience. Yet his comment looked more to the thirty years following the war. Not only had educational innovations amounting to a revolution been won, but also, as Terman believed would happen, the electronics field had come to loom over the engineering and industrial terrain.

But the results of that growth were mixed. Out of it would come conflict for the electrical engineering societies, and eventual merger.

The path to merger: The founding of the Institute of Electrical and Electronics Engineers

Numerous incidents gave rise to tension in the relations between the AIEE and the IRE during the twenty years before the two societies merged in 1963. The most persistent source of friction came out of the shrinking and swelling of membership numbers. Soon after 1940, following the great takeoff of the IRE membership, an AIEE member proposed the return of the radio engineers. Hearing of the proposed "return," RCA engineer Arthur Van Dyck wrote to Frederick Terman: "Fred . . . expect to have some fun." Similarly aware of the drift of students after the war to the electronics field, Terman saw in "the appeal of radio to students . . . a fight that would last for years." While admitting that the AIEE had done an "excellent job of building good will" among engineering faculty, nonetheless, Terman found "the power people . . . over conservative" and blind to "the hand writing on the wall" [46].

Into the 1950's, some AIEE leaders clung to the belief that, somehow, the radio engineers would return to the parent body. In 1953, for example, an AIEE board member again advocated that all "electrical professional people" unite under the older society. Though he admitted that the new organization might have a "different name than AIEE," the IRE would, in any case, come in as a "division" [47].

By 1948, the decline in student membership had brought the AIEE's situation painfully home to its leaders. But although faced with a disturbance so fundamentally threatening, instead of complaints, collective leadership responded with formal, organizational moves. For it was not that, after fifty years, Charles F. Scott's idea that student branches would increase membership no longer held. The cause of the disturbing trend was simpler: Most students were entering electronics and joining the IRE's branches. The AIEE's efforts at capturing electronics having failed, the two societies sought a peaceful solution and formed, in 1948, an AIEE-IRE Joint Student Branches Committee to deal with conflicts between the two societies.

Precisely at this point, the balance in the student branches shifted dramatically in favor of the IRE: Whereas, in 1948, the AIEE had 127 student branches and the IRE 37, in 1951, the AIEE had added just five more and the

IRE's branches had risen to 108. Though explained, in part, by the AIEE having saturated the ECPD-accredited schools by 1948, the IRE's faster rate of growth continued to diminish the AIEE's status in the schools. Between 1951 and 1953, in fact, the number of annual student applications for AIEE membership declined from over 14,000 to around 6000 [48].

The total AIEE student membership still remained higher than the IRE's. Yet within a few years, even that situation was reversed, helped by such events as the appearance of the *IRE Student Quarterly* in 1954. But the magazine served, rather than produced, the expanding body of electronics engineering students. The IRE's takeoff had begun fifteen years before and the consequences for the older AIEE were becoming ever clearer. Finally, in 1956, IRE student membership passed the AIEE's and the total membership of the IRE moved ahead the next year.

Amid statistics and tensions, then, the two electrical engineering societies moved toward merger. The first direct step toward organizational merger was the Joint AIEE-IRE Coordination Committee in 1952. The logic of this broad-ranging joint committee did not escape the AIEE Board. When appointing the two Board members and its secretary to the committee, as agreed, the directors declared that "Should this lead to a situation in which the two organizations can be merged, such merger would be favored by the A.I.E.E." [49].

Competitive feelings continued to shape the actions of individual engineers. Early in the decade, one of the AIEE appointees to the Joint Committee charged that the IRE's new Professional Group on Communication was a "further invasion of A.I.E.E. territory, just as the establishment of I.R.E. many years ago was an initial invasion." But the official reaction to the new Professional Group was more characteristic. Donald Quarles advised that "no direct action" could be taken. He promised instead to appeal to the "cooperative spirit" that existed between the two organizations and call the matter to the attention of the IRE president, Bell Laboratories' vice president, James W. McRae. With the controversy heating up, Morris Hooven, who had a reputation in the AIEE as a "gentleman philosopher," advised Quarles to "avoid undue competition" by meeting with McRae for a "soul-to-soul talk on cooperation" [50].

The talk took place in December 1953 between the new AIEE president, Elgin B. Robertson (a consulting engineer based in Dallas, Texas), and representatives of the IRE. Though Quarles could not make it, McRae and IRE Board member, William Hewlett, who were both traveling east for an IRE executive committee meeting, met with Robertson at Love Field, the Dallas airport. The three agreed, Robertson reported, that it was "perfectly silly that the AIEE and IRE cannot live in the closest harmony" [51].

Throughout the fifties, by means of both ad hoc meetings like the Love Field discussion and formal arrangements like the Joint AIEE-IRE Coordination Committee, the two societies grew closer. John Ryder used the opportunity of his IRE presidency in 1955 to promote a joint membership policy: "if a man was a member of one Institute at a certain membership level, he was to be immediately acceptable at a comparable level in the other Institute."

Ryder quickly won over AIEE president Hooven, but the AIEE Board declined to go along [52].

Then, in 1958, after joint groups had for some years coordinated activities in such areas as student branches and standards, IRE president Donald Fink moved the discussion to a new level. He proposed to the IRE Board that it seek "closer cooperation" with the AIEE on four fronts: appointment of joint technical committees in common areas would dissolve the long-time, vexing problem of duplication; agreement on a joint student branch policy might heal that wound to the AIEE ego; encouragement of joint section meetings would move toward the harmony of the Love Field meeting; and allowing automatic entrance to equivalent grades would fulfill Ryder's suggestion by moving toward merger in a sensitive area. By the end of the decade, each society had appointed task groups to act as a liaison committee. At an early meeting, the liaison committee devised a joint membership policy on the basis of both institutes being "professional societies with professional stature" and, thus, except for the Fellow rank, admission to a membership grade in one equalled the same grade in the other [53].

At this point, with the sentiment for merger apparent, it remained only to work out the organizational changes that would lead to the Institute of Electrical and Electronics Engineers (IEEE). In October 1961, the Boards of Directors of the AIEE and the IRE resolved to move toward amalgamation. A committee of eight drawn from both Institutes assumed the task of determining the feasibility of consolidation, then planning it. From this Joint Committee on Consolidation came an additional nine committees to develop guidelines in areas ranging from finances to standardization to student branches [54]. From the work of these committees came the "Principles of Consolidation," which were adopted by the two boards in March 1962 and by the membership within a few months.

Numerous decisions remained to be made once consolidation was approved. For this work, a fourteen-member merger committee began to plan for merger by January 1, 1963. In the case of the AIEE's membership magazine, *Electrical Engineering,* which had an outside editor at the time of the merger, the transition committee reestablished technical control. Ryder was given the editorship of *Electrical Engineering* in 1963 and, additionally, the headship of an editorial policy committee. The committee decided, in the main, to establish a new membership magazine and to retain the *Proceedings.* The *IEEE Spectrum,* as the IEEE's membership magazine was called, began publication in 1964 and quickly became a major magazine in electronics and in the diverse technical areas that gathered under the new engineering society [55].

But the essential choice was between the IRE's decentralized Professional Group System and the hierarchical AIEE Technical Committee Structure. Both had been designed to handle what an AIEE member had earlier described as the "problems of bigness and expansion of activity, coupled with a continually increasing technological development." He had obtained the idea from William Hewlett's 1953 article in the *Proceedings of the IRE.* Hewlett, soon to assume the presidency of the IRE, discussed the problems faced even under the Professional Group System. Fundamentally, he prescribed the need for new generations of engineers to periodically bring fresh energy to the Professional Group System and, thus, to the challenges of specialization in the expanding electronics field [56].

As it turned out, the IRE system provided the flexibility and energy which would be necessary if the new society were to grow. IRE leaders had long before come to see the Professional Group System as the ideal form for engineering societies. This was in evidence in 1959 when the IRE president asked Frederick Terman his opinion about the merger. Terman professed indifference. Within twenty years, he thought, "the AIEE will probably become still more a specialized society representing a particular segment of the industry . . . like that of the present Society of Illumination Engineers." By then, the older society would represent "little more than a Professional Group in the broad field of electronics" [57].

The merger came far sooner, of course, and a large part of the AIEE later became the Power Society — that is, a Professional Group within the Institute of Electrical and Electronics Engineers. Ironically, Terman was invited to be the speaker at the first annual banquet of the IEEE in 1963. In it, he stated the advantages of the "unlikely marriage":

> Through the technical committee approach, the corporate body of the IEEE possesses the means of developing a technical conscience. At the same time, the professional group system provides an almost unlimited outlet for individual participation and initiative. The professional group system also provides a flexibility, and an opportunity for experimentation with a minimum of regimentation from above, that will serve effectively as a cement for holding the IEEE together as a coherent organization [58].

It was true, of course, that the new engineering society owed much to the AIEE. As one of the founder societies, it had pioneered in the area of organized professional engineering. Still, the AIEE's four-tiered system in which committees formed divisions whose chairs became the Technical Operations Committee reporting to the Board of Directors gave way before the momentum of the largely self-governing Professional Group System of the IRE.

In short, Terman had spoken politely. The Technical Committees pioneered in 1903 by AIEE president Charles Scott, along with the full AIEE structure, had long been part of the inheritance of the IRE, whose founders, in 1912, frankly borrowed organizational forms from the older society. In 1961, there was little left to borrow. What prevailed instead was the system that had proved so responsive to the IRE's commitment to an engineering professionalism based on technical knowledge and guided by technical concerns. This commitment had shaped the Professional Group System, which became, with only subtle changes, the organizational foundation stone of the IEEE.

The word "professional" joined the AIEE's use of "technical" to form the title, Professional Technical Group. Some of the AIEE's Technical Committees were immediately absorbed into Groups; others retained their committee status, to be later merged into the Group system. The author of an editorial in the magazine of the Boston section of the IRE aptly concluded: "What has the IRE to lose [from the merger]? The AIEE has agreed to an organization patterned closely after that which has worked so well for IRE." Some dissent did exist. One member accused the national leadership of pursuing a "missile-age timetable." Also, the Society of Broadcast Engineers was founded in protest. However, by the summer of 1962, the AIEE and IRE memberships had voted by impressive margins to approve the merger [59]. The next year, the Institute of Electrical and Electronics Engineers began its official existence.

The makers of the IEEE had drawn copiously on its tangible past and, so, in a real sense, this new national engineering society was formed, not founded. Its technical fields, publications, and convention habits were only the most obvious components of a rich and detailed inheritance. These were standard programs to be found in any engineering society.

Yet neither so standard nor so tangible was the baggage of engineering values that the profession carried into the new Institute. That inheritance, as much as the organizational forms, would define the new group.

But another critical factor remained: the changing social and political economic context through which the IEEE moved. Before the end of the sixties, events would occur within that context that would modify but not divert the pure course of the IEEE.

8/THE IEEE AND THE NEW PROFESSIONALISM

We engineers are not, most of us, at ease in the presence of the conflicting forces of politics and the maneuvers of power, forces by which the outlook for electronics — as for every type of work — is ultimately shaped. Nevertheless, we must face them.

Donald G. Fink, 1972 [1]

Looking ahead

Looking into the future is an old habit of engineers. After World War II, the tendency became especially strong in an electronics community driven by rapid economic growth and an expansive military-space establishment. Technical advances came so thickly during the postwar years that predicting future trends was a precarious undertaking. How precarious was demonstrated by a "guess at the future in electronics" made in a 1952 article by General Electric Research Laboratory engineer and IRE Fellow, William C. White.

White recognized the breadth of the field, observing that the country had entered a technological "age of communication and control." He thought that this "modern industrial revolution" would "devalue the human brain" just as the earlier transformation had diminished the place of the skilled worker. But there were obstacles to this new revolution, especially in the difficulties faced by builders of systems in the area of "computers and military applications." It would prove difficult, White predicted, to assemble a "large number of simultaneously used tubes and circuits" because of the "most discussed problem" of the day: "reliability" [2].

On these matters, White's vision was clear. But the closer he got to the actual world of technical innovations, his vision blurred. In his prescription for overcoming the unreliability of the tube, White missed what, literally, lay under his nose. Standing at the point at which the infant technology of transistors was about to overwhelm the mature technology of electron tubes, White placed the transistor last in a list of five "significant" trends, after ceramics. He admitted that the recent achievements in the semiconductor field (the crystal diode and the transistor) would "undoubtedly replace tubes for many applications," but, as he saw it, they represented "no new trend."

The transistor, of course, was a "new trend." But only as the discipline absorbed the advances in semiconductor technology and the systems engineering concepts which were devised during the 1940's and 1950's did the central place of the transistor and the integrated circuit become apparent.

Indeed, the basic devices of the microelectronics revolution developed so rapidly that, by 1962, when a group of distinguished electronics engineers were looking ahead on the occasion of the IRE's 50th Anniversary, they ignored the subject. Instead, they focused on the computer and exulted in the promise of advanced automata.

Among those who looked closely at automata were leaders in the field, including one of the engineers who had helped invent the ENIAC at the University of Pennsylvania during World War II, the research director of the Navy's major laboratory, and the chief advisor on electronics to the Department of Defense. They were struck by the potential for what computer inventor J. Presper Eckert called "a self-reproducing automata which can improve itself!" By the early 1960's, "memory, eyes, ears, hands, and logic" had become "about as good" in automata as in human beings. Still, he admitted, "recognition ability, certain types of information retrieval, and the ability to taste and smell" were areas where humans still excelled. But even here, Eckert promised, the millions spent annually by industry in these areas would bring success in fifty years [3].

To R. M. Page of the Naval Research Laboratory, the main advance of the age had been the automation of "human operations," but "mechanization" was invading "the functions of the human brain." Similarly, J. M. Bridges, chief of electronics for the Office of the Director of Defense Research and Engineering (the successor to the postwar Research and Development Board), saw "adaptive computers" in the future. These "design machines" would perform the tasks of systems analysis, layout, and equipment design. Further, they would configure the system and then feed the information to an automatic assembly machine. These feedback arrangements would discover errors so that "the design will be optimized" [4].

As in these visions of the technical future, the educators invited in 1962 to look into the next fifty years concentrated on the most exciting happening in their midst: the rise of engineering science. To Frederick Terman, a vice president and provost at Stanford, the "electronic scientist" would replace the electrical engineer. The curriculum would emphasize mathematics, classical physics, and chemistry, with much of the current work in microwaves being standardized as "handbook stuff." Advances in controls and computers were included in Terman's vision. For alongside the rise of the electronics scientist, who designed the technology, would have grown up a new field in which the "logic" concepts of "thinking machines" would have developed into a new discipline. Though without the specificity of Terman's view, William L. Everitt assumed the continuing prominence of engineering science. An engineering dean at the University of Illinois, he predicted that universities would concentrate on both "training"—the teaching of practice—and "education"—"a guided enlargement of creative ability and understanding." As in the twentieth century, however, "education" would receive the greatest attention, Everitt predicted [5].

In one critical respect, these engineers were accurate. Each assumed continued advances in electronics and, except for Everitt, specifically mentioned the most conspicuous and promising technical achievement of their time: the computer. The integrated circuit, though only three years old, had already won acceptance and large-scale integration (LSI) lay close ahead. But, just as certainly, something blurred their vision. It could be found in what they did not discuss. The trajectory followed by the advancing computer and the increasing numbers of electronic scientists, like the space vehicles their work made possible, moved through a frictionless atmosphere. In the engineers' predictions of the future of electronics technology, they took social, political, and economic issues as constants in their predictive formulas — rather than as the dynamic factors they would prove to be.

Attempts were made to bring social questions into the equation, especially by Franz Tank, a retired professor from the Swiss Federal Institute of Technology in Zurich who had been IRE vice president in 1955. The social issues appeared in the form of disquieting notions about the limits to innovation in the electronics field. Tank thought that the "technics" based on the work in solid-state physics would have become "classic," with amplifiers, transmitters, servomechanisms, and computers attaining "a certain settlement," a "final form." However, by then, electronics would face problems that went beyond technology. "New social problems" would emerge amid a world of scarce raw materials. Further advances would require huge efforts and, in any case, would bring neither "more power nor more happiness" [6].

Though such dour thoughts did not appear elsewhere in their future visions, the engineers drew negative conclusions about the impact of electronics developments. Eckert predicted that unemployment would follow the spread of automata, though it would affect "skilled workers" only. However, Bridges, the Defense Department's director of electronics, predicted that the future role of "adaptive computers" in devising complex electronic systems would replace "the thousands of electronics engineers" then required.

But all these anniversary seers missed the crucial, professional event which was to take place during the IEEE's first twenty years. True, Bridges came close in foreseeing unemployment among electronics engineers, but in his vision, its source was technical, not political and economic, as would be the actual case. Even Tank's "social problems" had gone unillustrated as he shifted to talk of resource depletion. Indeed, these engineers appeared myopic as they looked ahead, peering through a tunnel constructed of technical components only.

They failed to foresee that, in the late 1960's, political and economic storms would blow a host of socio-technical issues into this problems-oriented profession. Within a few years, the storms would lead to a redefinition of professionalism within the IEEE and to major organizational changes to accommodate it. These two elements — professionalism and organization — were mirror images of each other. So long as the profession rode the crest of the postwar economic growth, the structure and purposes of the IEEE would continue to reflect pure technical interests. But as the stormy half-decade between 1969 and 1973 would demonstrate, if one changed the other would surely follow.

The IEEE and the "engineering mission" of the 1960's

If the IEEE's first half-decade indicated nothing else, it proved the merger to have been more a procedural act than a formative event. To follow the Institute of Radio Engineers during the two decades before 1963, then to observe the IEEE from 1963 to 1969, reveals continuity in both organizational activity and technical purposes. The merger, in short, formally consolidated what had been accomplished by the juggernaut of electronics.

This appeared in the persistent emphasis on technical communication as the society's prime mission and in the steady reliance on the Professional Group System and publications. Even so, the tradition of organizational innovation continued to transform what, under the IEEE, came to be called Professional Technical Groups. Within a year of the merger, the Institute's leadership moved to restructure the Group System, leading them to create boards in an on-going attempt to meet technical changes with organizational innovation.

A pattern of moving from the technical to the organizational, the one naturally following from the other, had emerged. The natural order of this pattern could be seen in the IEEE Constitution adopted in 1963. In Article I, on purposes, advancement of the "electrical, electronics, and related fields" in their "scientific and educational" aspects constituted the Institute's commitment. The remainder of the Constitution described an organizational structure and method of governance to support it.

This constitutional order accurately forecast IEEE activities during the 1960's. Basking in the warm rays of an expanding technical field and the strength of the Group System, the society's leadership worked, as it had since the end of World War II, to meet the challenge of rapid growth. When the 92,000 members of the IRE in 1961 became, with the merger and continued growth, 150,000 in 1963 (140,000 in the U.S.), it was clear that more would be required than simply adding groups.

Handling this massive growth required managerial skills, something the IEEE's leadership was especially rich in during the decade. By the early 1970's, complaints that the IEEE was "manager-dominated" led one president to explain that, since the Board of Directors' function was "to manage the affairs of the Institute, it is quite natural that it should include a majority of those skilled and trained in the art of management" [7].

Natural or not, the charge especially fit the IEEE's presidents during these years. After the inaugural presidency of professor and research scientist Ernst Weber in 1963, between 1964 and 1969 a string of corporate vice presidents held the office. From the industrial world, there was Clarence H. Linder of General Electric, Bernard M. Oliver of Hewlett-Packard, Walter K. MacAdam of American Telephone and Telegraph, and Seymour W. Herwald of Westinghouse. That string was broken only by William G. Shepherd, in 1966, a vice president at the University of Minnesota. At the end of the period, in 1969, came F. Karl Willenbrock, provost at the State University of New York at Buffalo [8].

Their experience in managing large, multifaceted institutions was reflected in the major organizational innovations undertaken during these years. The first step came in September 1964 when the Executive Committee combined the AIEE's Technical Operations Committee (TOC) with the IRE's Groups Organization to form the Technical Activities Board (TAB). Seeking to consolidate the technical activities—what the Constitution called the "scientific" purpose—TAB assumed TOC's six general committees. This included Standards, for example, which itself was a major agency with 145 committees at several levels. In addition, TOC had six technical divisions with 69 technical committees and their 320 subcommittees. Joining the hundreds of AIEE committees with the IRE's Professional Group System formed a massive agency. But in giving the IEEE's central technical directorate the stature of a "board," a new organizational form had been created which had the authority requisite to its tasks. The import of this act became clear in 1967, when an Educational Activities Board was established, creating for the first time a major interest area nominally equal to the technical work of the society. The next year, a Regional Activities Board was formed to give geographical representation [9]. What had actually been created was a way of giving stature to the fundamental commitments of the Institute.

That the new boards' status was nominal could be seen in the official statements of IEEE leaders. Their assertions formed a statement of directions and purpose to guide the IEEE leadership. When, in 1965, President Bernard Oliver discussed the immediate challenges to the Institute, he pointed to the essential fact that "new technologies appear, grow, flourish, and then either reach a semistatic maturity or gradually wane." To remain organizationally current with the changing technical scene, Oliver called for attention to the principle membership services of publications and meetings and to the need for the Institute to develop "modern methods of abstracting and information retrieval" [10].

During his presidency in 1966, William Shepherd described the IEEE's central purpose as an "engineering mission." It entailed giving support to the technical triumverate of engineering technology, engineering practice, and engineering science. The engineering society must serve the full range of engineering, recognizing the validity of both "abstract contributions" and "useful devices." Seldom was the technical character of the IEEE put so succinctly as when President Seymour W. Herwald explained the Institute's mission in 1968: "Our underlying purpose in getting together is to enhance communication about our field. Our goal is to advance the technology that we all, in one way or another, depend on" [11].

This concentration on the technical led, in 1969, to the clustering of TAB's thirty-one specialized groups into six divisions. The objectives were developed by Institute President James B. Mulligan, Jr. as a means of decentralizing "management responsibility" and building "good management–member communications links." Mulligan's idea was to "decrease the communications span within the IEEE" and increase "information flow." Before the divisions were formed, a vice president for technical activities had full responsibility for the thirty-one groups. Under the new structure, the directors of the six divisions plus a seventh "interdivisional" director would share the labor. All would serve on the IEEE Board of Directors. The divisional structure was intended to give the group system the "flexibility" needed to keep pace in a time of "rapid technological expansion" [12].

Clustering the technical groups would also help members find their technical niche within the complex of committees, groups, meetings, and publications. Not only would groups with similar interests be pulled closer, but declining groups could be phased out more easily, shifting specialized interests and members into contiguous groups.

While the divisions were being formed, the Institute's Technical Planning Committee, chaired by RCA engineer Edward W. Herold, reviewed the Institute's technical activity to identify important areas not being covered. The committee then established committees to cover the areas — including, for example, cable television, plasmas and magnetohydrodynamics, cryogenics, social systems, and the history of electrical engineering. The committees helped consolidate membership interest in their specialties. Should the response suggest more permanent arrangements, then "long-term solutions" would be sought in the Group System [13].

Similar innovations continued to alter the IEEE organizational structure as society members worked to keep organizational vitality equal to the rate of technical change. In early 1970, the Board of Directors approved the formation of Societies — beginning with the Power Engineering, Computer, and Electronic Controls Societies. The new agencies, the Board explained in its policy statement, would provide the means for merging closely related or declining groups and to bring non-IEEE societies into the IEEE.

J. V. N. Granger, who served as president in 1970, when Societies were created, thought the innovation was the "most significant . . . since the "Professional Group concept" entered the IRE twenty years before. Still, he attributed it to the same forces that had led to the Group System. Both Society and Group, Granger explained, derived from the great expansion of electronics engineering in World War II which had "continued, virtually unabated, ever since" [14].

War, jobs, and the "road to professionalism"

As the technical exuberance of electronics continued unabated, the smooth organizational evolution of professional electrical engineering was being disturbed. Indeed, Granger's remarks on the technical mission of the society came in January 1971, a year in which a constitutional amendment would go before IEEE members in an attempt to make nontechnical purposes primary in the organization. Already, in 1969 a major step had been taken to open up the *IEEE Spectrum* to social and economic material. By 1970, the Board had appointed IEEE general manager Donald G. Fink to begin a study of the financial and legal implications of absorbing within the organization the social and economic concerns that had entered the IEEE's flagstaff publication.

Among the sophisticated weapons systems in place during the 1960's that relied heavily on the work of electronics engineers was the SAGE (Semi-Automatic Ground Environment) system (operators shown above). Nike Hercules missiles, controlled by SAGE and able to carry conventional or nuclear warheads, were designed to destroy entire fleets of manned enemy aircraft.

A desire to expand the purposes of the Institute had never been entirely absent. In a 1957 article on the future of the Institute of Radio Engineers, Fink had advised that the society would have to find "some effective and acceptable means of increasing by concerted action the economic and social rewards of a career in electronic science" [15]. But only rarely did such sentiments prescribe new, nontechnical functions. In 1912, Dugald Jackson had used his presidential address to remind engineers of their obligation to support the political goals of the privately owned electric utilities. Twenty years later, Lee deForest had urged IRE engineers to prevent the large companies from controlling radio technology at the same time that Admiral Stanford Hooper was asking IRE members to appoint an engineering spokesman to defend the social and economic interests of the profession before governmental bodies.

That such issues were again entering the society appeared in 1968 in a membership attitude survey sponsored by the Institute. According to the survey, taken by an opinion research organization in Princeton, New Jersey, three fourths of the membership found the association "fairly valuable" or "very valuable." When asked what they thought the objectives of the Institute should be, only one, the dissemination of information, was mentioned by a majority (53 percent). But the next largest response asked the IEEE to work for the "upgrading of the profession." More specifically, slightly over half of the members favored "placing more emphasis on economic and career aspects" in *Spectrum* [16].

Increasingly, electrical engineers were redefining professionalism to include economic concerns. It is doubtful, however, that any of them would have expected that the way would be prepared by the intrusion of moral and political issues into the technical exclusivity of the IEEE.

Indeed, the first signs of the invasion escaped even the most perceptive observers. In an article in *Spectrum* in 1973, Leo Young designated the early 1970's as the time when the society had started down the "road to professionalism." Board member Young had recently left a position as a research engineer at the Stanford Research Institute to join the Naval Research Laboratory. He explained that the Institute had extended its traditional concern with "engineering" to include an interest in the "engineer." Young traced the source of the new interest to the "crisis in electrical engineering" brought on in the late sixties by reduced expenditures for military and space research and development programs. From this economic blow, he wrote, the issue of "whether IEEE should engage in professional activities" had "caught fire in 1971 and 1972" [17].

Apparently, Young remembered just the second conflagration that engulfed the profession. At that time, the loss of jobs that resulted from the near collapse of the aerospace industry and the sharp decline in military R&D expenditures had jolted a profession which had known only growth for three decades. However, the initial force that drove the Institute down the road to professionalism had actually come in 1969, borne by the moral and political reaction to the Vietnam War and to the reliance of electronics engineering on the design and development of military weaponry.

These issues first arose in the spring of 1969 over a piece of engineering hardware: the antiballistic missile (ABM) system. It was called Safeguard by President Richard M. Nixon's administration, which had requested $800 million from Congress to begin deploying it. The development of the system had dominated military electronics during the sixties, much as the development of the intercontinental ballistics missile (ICBM) had dominated the fifties. Both systems were armed with nuclear warheads. The ABM was first proposed a few years before by President Lyndon B. Johnson as Sentinel. Like Johnson's, Nixon's Safeguard system would defend against attacks on the nation's ICBM sites located in the West [18].

As with the antiwar protests, which had spread across the nation by the late 1960's, the protest to this program of the Nixon administration entered the Institute by way of a university, ironically, one that was preeminent within the military-university complex: the Massachusetts Institute of Technology. Its medium was a letter to the editor of *Spectrum* from the recently organized Union of Concerned Scientists at MIT. The letter called for a day of speeches and protest at MIT on March 4, 1969. It also strongly condemned the actions of the United States government in both developing and using sophisticated weapons.

The letter's first sentence set the tone, asserting that the "misuse of scientific and technical knowledge presents a major threat to the existence of mankind." Not only had its misuse in Vietnam shaken confidence in the American government, but there was "disquieting evidence of an intention to enlarge further our immense destructive capability." The letter named one specific technical artifact: the ABM. Arguing that the scientific community was "hopelessly fragmented," the letter called "on scientists and engineers at MIT, and throughout the country, to unite for concerted action and leadership." The group wanted, among other matters, research to be turned away from "the present emphasis on military technology" with more attention given to "pressing environmental and social problems" [19].

The letters from IEEE members that flowed into *Spectrum's* offices made it immediately clear that something extraordinary had happened within the society as well as within the pages of the IEEE's magazine. "This is to protest in the strongest possible terms" the inclusion of the letter from the MIT group, wrote a Canadian engineer. He hoped "such offensive articles" would not appear again. A California engineer thought the "unsigned article . . . an improper intrusion of politics into the pages of a professional journal." "Technologists are free to decline work," he pointed out, and, besides, opposing numbers of engineers were equally concerned that the "survival of the United States is at stake." Another engineer questioned the omission of the signers of the letter and posed a series of questions which collectively asked "whether the IEEE is still a scientific and professional society" [20].

A conflict clearly had begun. But it was not to be between the Union of Concerned Scientists and the protesting members. It was between those who resented this editorial intrusion into the pages of a technical magazine and the volunteer editor of *Spectrum*, Dr. J. J. G. McCue, a research engineer at the

Lincoln Laboratory of MIT. But more than McCue stood behind the *Spectrum* article. As E. W. Herold, a member of the magazine's Editorial Board, explained in 1970, the Board had backed McCue throughout [21].

But McCue's choices were also personal. He acted out of the world in which he was involved, a wide-ranging university-community containing leading social scientists and scholars in the humanities as well the scientists and engineers who naturally dominated the institution. Lincoln Laboratory had been founded by MIT in 1951 at the request of the Air Force. Its work, as McCue described it, had been in "electronics for defense and space." But in 1969, he explained, it and similar university-based military R&D laboratories, like the Instrumentation Laboratory at MIT and the Stanford Research Institute, had come "under fire" [22].

If he had not taken up a rifle, McCue was, nonetheless, determined to involve his fellow electrical engineers as part of his responsibility as editor of *Spectrum.* He was reporting, as he later explained, a movement working for a "change in the sponsorship for research in electronics" [23]. Other, similar editorial decisions followed [24]. In June, McCue published a speech made at the March 4th meeting at MIT by Harvard Professor George Wald, a Nobel Prize winner in physiology. McCue described it as an "address on the question of what is worrying the young." At the end of the summer, McCue published a lengthy report on the ABM by a *Spectrum* staff writer plus pro and con statements on the ABM. Their authors were two established arms advisers to the government and military; the papers had been read recently before a meeting of the American Institute of Physics.

It was appropriate that the final "controversial" item of the three published by McCue in the spring and summer of 1969 was the report on the ABM with its pair of opposing viewpoints. The character of this group of articles, entitled, "The Antiballistic Missile," captured best the form in which *Spectrum* would continue an open policy of publishing sociotechnical issues that related to electrical engineering. That stance was expressed in the editor's introduction when he wrote: "Not all of these facts are technical ones, but technical men—and others—must assess all of them in any case . . ." [25].

McCue had extended the notion of what information was necessary to "technical men," and, in doing so, had initiated not only the debate over military R&D but also one over editorial policy. Theoretically, *Spectrum* had been founded to serve all the engineering interests of the membership. This had been proposed by the Editorial Policy Committee established by the Merger Committee in October 1962. Headed by John Ryder, the Policy Committee had the task of formulating a policy for the entire range of publications. When Ryder explained this policy in 1964, moreover, he described a distinctive role for *Spectrum.* Whereas the *Proceedings* would remain a "broad research and development-oriented journal," *Spectrum* would move into new areas [26].

Innovations appeared at *Spectrum* on several levels. Among the technical magazines of professional engineering societies, *Spectrum* pioneered with a commercial magazine format, utilizing color and bold illustrations. Though an outside consultant suggested these innovations to the committee, the engineers on the committee determined the editorial policies. Besides Ryder, these included men like Joseph M. Pettit, who had been with Terman at the Radio Research Laboratory and was dean of engineering at Stanford; Bernard Oliver, Hewlett-Packard's vice president for research and development; Thomas F. Jones, Jr., president of the University of South Carolina; Seymour W. Herwald, a Westinghouse group vice president; and Donald Sinclair, General Radio Company president. Besides Oliver and Herwald, who served as IEEE presidents during the decade, Sinclair had been IRE president in 1952.

In the technical area, the committee prescribed that *Spectrum* should cover "all frequencies of the electromagnetic spectrum." Beyond that, the magazine would "include news of the IEEE and of IEEE people, news of the profession and of education, and letters to the Editor on topics of general interest." The official statement of editorial policy was even more explicit: The *IEEE Spectrum* was "the core publication" and would include "news of political and social interest to the profession."

But, as McCue explained in an October 1969 editorial, that formal declaration of policy had, "until the past few months, . . . been pretty much of a dead letter." He insisted that if the policy was to be "taken seriously in these times of questioned values, it must result in the treatment of controversial themes." It was this conviction that had led McCue, during the months between the letter of "Concerned Scientists" in April and the October editorial, to fully air the "questioned values" and "controversial themes" of the day [27].

Of the trio of volunteers who edited *Spectrum* between the mid-sixties and 1972, McCue was not the only one to incorporate what would be called sociotechnological issues. After assuming the editorship in the early fall of 1970, McCue's successor, IBM engineer David DeWitt on several occasions defended and advanced McCue's position. In his inaugural editoral, DeWitt announced his intention of continuing a questioning posture. Besides serving *Spectrum's* "prime purpose" of displaying the "wide range of electrical engineering" to give "access to all of the tools of our profession," DeWitt wrote, there existed a "second purpose of *Spectrum*": that of including "studies and proposals where the tools and insights of our profession are applied to social and international problems and the fine arts" [28].

His personal beliefs regarding the controversy over the uses of electronics technology were also close to McCue's. The "major nations" of the world, he wrote, were engaged in "defensive and mutually reactive games that waste vast resources and use as their ultimate means the infliction of suffering, death, and destruction." Its relevance to the engineers: "Those nations necessarily employ many of us in their defense." DeWitt even went so far as to characterize as a form of "Hitlerism" the tendency of the same nations to claim as their "basic purpose the achievement of a world of peace."

For the IEEE, the personal testaments and policy discussions of 1969 and 1970 translated into organizational change over the next three years. Al-

The Apollo 11 flight of 1969 which put astronaut Edwin E. "Buzz" Aldrin, Jr., on the moon represented the high point of the aerospace program of the 1960's. For electonics engineers, however, it marked the end of three decades of steady growth.

though the issues that induced them provided the background for the IEEE's transformation, the organizational modifications marked the actual trail of change.

In September, Institute President Karl Willenbrock asked for responses from the membership on a proposed policy statement to govern "the presentation of controversial material." Four months later, after two-thirds of nearly 300 respondents opted for the new policy and the Board had engaged in "extended discussion," a policy statement was approved [29].

In the formal statement, the wording was refined, becoming a statement of policy for "the Presentation of Socio-Technical Material." McCue had previously stated the basic points, but the Board's unanimous action made them official: the new subject matter, whether social, technical, or economic, should be "relevant to the field of electrical and electronics engineering and to its relationship to the needs of society." Moreover, "reasonable efforts" would be made to air different viewpoints. In any case, the "opinions expressed" would be the author's, not those of "the Institute, its officers, or its members."

But as significant as was this act of policymaking — specifically, its defense of *Spectrum's* coverage of the protests to the war and to the military's heavy involvement in electronics — another matter arose that cut a much wider swath across the member-ranks of the IEEE. It had to do with engineering jobs and certain problems related to the shifting interests of the national governmental-military R&D complex.

In the first place, the budget of the National Air and Space Administration dropped considerably following the contraction of the aerospace industry around the time of the landing of Apollo 11 on the moon in 1969. This was accompanied by a drop in military expenditures on R&D. In September 1970, for example, a Defense Department official announced that 367,000 jobs in the defense industry "already had been lost because of military spending cutbacks." And he expected "600,000 additional layoffs" within the next ten months. How serious this was, McCue made clear when he explained that the military "had funded a large fraction of the employment in electronics in the U.S. for nearly a generation, and has sponsored a large fraction of the research" [30].

Given the great numbers of electronics engineers employed in the aerospace industry and military R&D, the size of the economic reductions created disturbances beyond those who lost their jobs. The context in which this was happening and its meaning to the Institute became clearer in the report of an IEEE ad hoc committee appointed "to assess U.S. economic conditions in the electrical, electronics, and related industries." Seeking to produce a timely report, the ad hoc committee mined its data from studies already done by agencies like the Department of Labor, the Engineering Manpower Commission, the Stanford Research Institute, and the National Science Foundation [31].

The committee quickly found that the gross data it drew on failed to "pick up 'minor' perturbations." Yet these could "cause significant dislocations for a small percentage of the population." As a result, the committee admitted, "we did not see a dip in employment of engineers around 1970."

But the dip was there. Donald Fink, who had served the IEEE as general manager since the merger, used data from a 1971 National Science Foundation study to determine the level of unemployment among IEEE members. Though the figures were lower than had been expected, the data indicated that unemployment among electrical engineers had risen by ten-fold in recent years: from 0.4 percent during the "good years" of 1965 to 1968 to 4 percent in 1970 and 1971. In short, Fink found that "the employing institutions — the offices and laboratories of industry, government, and the universities — are suffering an upheaval of unemployment as well as a cessation of the productive work on which the future depends" [32].

The problems that confronted the electrical engineers were to be found precisely in the short-lived dips. Though slight by gross standards, these instances of economic dislocation were systemic. Engineers continuously moved around within the aerospace industry as contracts ended and workers were laid off. The shifting priorities of national policymakers were partly responsible. As described by a California engineer in 1970, aerospace engineers in the 1950's "went from aircraft design to missile design," then in the early 1960's "from missile into space vehicle design," and, "recently . . . back to airframes and V/STOL aircraft design" [33].

Among the responses to this first year of economic upheaval were calls for portable pensions and pamphlets on career development. Yet beneath the concrete responses to the social and economic protests of the early 1970's ran by an undercurrent of discussion about the meaning of professionalism. Whether, in the words of IEEE President J. V. N. Granger, it was a question of "the undesired social effects of technology" or, as McCue wrote, of the Institute's responsibility for the economic well-being of its members, the wrenching issues of the late 1960's led to the broader question of the IEEE's role [34]. In short, aside discussions of the meaning of engineering professionalism, there arose the necessity of a programmatic response from the IEEE leadership.

The programmatic response

From the first year of controversy, discussions of professionalism appeared regularly alongside the organizational initiatives. This could be seen in an editorial in the July, 1969 *Spectrum,* in which McCue examined a "spectrum of professions." But after locating the law, divinity, and medicine at one end, he found engineering, "alas, somewhere near the other end." Nonetheless, McCue believed that the fundamental professional criteria of "discipline, devotion to an ideal, and complete dedication of self" were there to be claimed by the engineer.

This was in spite of the serious contradiction that existed between the nature of engineering employment and that basic criterion of professionalism: independence. Engineers generally were "salaried," in service to "company management," McCue argued, and, thus, lacking the "freedom of action" characteristic of professionals at the high end of the spectrum. More than any other factor, McCue reasoned, this explained "why engineering is not universally regarded as a profession" [35].

Though he saw "no common ground between unionism and professionalism"—given that "individual merit," rather than "the welfare of the hive," marked the professional—the point could "not be dismissed summarily." Many engineers who talked of unions were in the "U.S. aerospace industry" and were thus frequently "hired by the hundred when a contract comes in, and laid off the same way when the contract terminates." So even if relatively few electrical engineers advocated unions, McCue advised, "many would like to see a strong drive—spearheaded, perhaps, by IEEE—toward organizing the profession" along new lines.

McCue's ideas struck a chord in the membership. The responses to his editorial rivaled in number and passion those won by the three anti-ABM documents published that year. The percentage of respondents favoring McCue's call for a new mission constituted the first of many decisive votes for an expanded professionalism. The first to appear was a brief letter from a California engineer telling McCue that, "after all these years, it is heartening to see that IEEE is addressing itself to professionalism." Many of his colleagues, he wrote, had shared with him a concern "with our status in life," which often led them to feel like "migrant workers" [36].

Throughout the fall, engineers wrote—and McCue published their remarks—to say that they, too, felt like "nomads following the contracts and trying to keep one step ahead of the layoffs." Because this Connecticut engineer did not believe engineers could be professionals in the same sense as doctors and lawyers, he thought that IEEE members should admit that "engineers are workers subject to the whims of management." He admitted that might lead to a national organization which would work for higher salary levels, fringe benefits, and pension plans "so that engineers can keep pace with the plumbers, electricians, and masons" [37].

Within a few months of the Institute's first encounters with these social and economic issues, a number of critical steps had been taken: an analysis and iteration of the issues had been made, a constituency had begun to form, and the IEEE had been suggested as the proper vehicle to achieve these new goals. Although the debate over professionalism would eventually be codified, during 1970 and 1971, talk predominated over organizational action. But in late 1971 the response moved significantly beyond discussion. New committees began to spring up throughout the society and membership ranks were invaded by surveys, referenda, and constitutional amendments.

Some of these activities had begun early, as had the informal poll on the new editorial policy for *Spectrum*. That step had been followed in 1970 by the Technical Activities Board's appointment of an ad hoc Committee on the Application of Electro-Technology to Social Problems. Two years later, a number of members joined together to establish a Professional Group on the "Social Implications of Technology."

But the economic issues easily dominated. As IEEE President Robert H. Tanner, an engineering administrator with Bell-Northern Research in Canada, told the membership that year: "Probably no single situation faced by members of IEEE since its formation has been as traumatic in nature, or caused as much soul-searching, as the present U.S. unemployment situation" [38].

Taking Institute initiatives as a gauge, the truth of Tanner's assertion was evident. Evidence enough existed during 1971, as the IEEE moved on a number of fronts to deal with the job crisis. They did this, as Tanner recognized, even though constitutional limitations prevented the Institute from acting outright. Most of the requests came from the "many [U.S.] members" who wanted the IEEE "to influence legislation." Skirting the constitution, therefore, in January, the Board arranged with the National Society of Professional Engineers (NSPE) to serve the profession's needs in legislative and economic matters.

With an office in Washington, the NSPE, which was founded during the depression of the 1930's, specifically lobbied for legislative relief from unemployment caused by "reductions in defense and aerospace production." Senator Edward Kennedy and Representative Robert Giaimo had already introduced one such bill in Congress. The Conversion Research and Education Act of 1970 aimed, as NSPE president and Mississippi State University engineering dean Harry C. Simrall explained, to reorient the work of engineers toward "serious social problems" in the areas of housing, transportation, pollution, health, and crime [39].

The IEEE continued this work during the first half of 1971. The Institute helped get a $750,000 contract from the Department of Labor to learn how to transfer the skills of "aerospace and defense engineers and scientists" into new areas. A joint project of a half-dozen engineering societies, the NSPE served as the contracting agency for the study. In addition, the IEEE joined with the American Institute of Aeronautics and Astronautics (AIAA) to develop a program in which volunteer engineers established committees to find and fill available engineering jobs with the unemployed. By 1972, President Tanner was able to report that, since 1970, when the IEEE and the AIAA conducted the first employment workshop in Baltimore, 170 workshops had been held in forty-three cities for 15,000 engineers. Additional initiatives were taken by the IEEE in the areas of career development (with a new staff member added to administer programs) and salary standards [40].

The job programs were capped with meetings attended by IEEE officers in 1971 and 1972, first in California with President Nixon and other officials on aerospace unemployment, and, later, with the heads of several engineering societies and the Secretary of Labor [41]. These largely ceremonial events emphasized the external character of a good portion of the Institute's early undertakings. More profound, however, were the changes — a committee here, a programmatic initiative there — underway inside the Institute. These were the seeds that would grow into an expanded mission within the field of electrical and electronics engineering — and, in the meanwhile, extend the IEEE into new areas.

The seeds germinated within the Institute's Boards. Their generic name suggested a sort of equality with the Board of Directors. They acted accordingly. Within the Technical Activities Board, for example, the Washington, D.C. section started an experimental program to examine the work of "technological forecasting and assessment." TAB hoped to "point the way toward methods" capable of obtaining "a clearer picture of the future implications of . . . current activities." When the attempt to establish a professional group on the social implications of technology failed the next year, TAB directors established an ad hoc committee to carry out that work. TAB also responded to environmental concerns with a Committee on Environmental Quality which, among other activities, cosponsored a successful conference on "technology and governance in achieving environmental quality" [42].

Similarly, in 1971, the Educational Activities Board (EAB) moved beyond on-going activities in the areas of accreditation, precollege guidance, and continuing education to establish a new committee on Career Development. It took over this activity from the IEEE Board of Directors. EAB leaders created programs to be implemented by a joint Regional Activities Board/EAB Institute Career Development project. At this time, also, committees were established within the EAB to look after the concerns of "women in engineering" and "minority groups" [43].

But the crowning innovation came in response to forces outside the leadership. The instigating act came when members, acting in opposition to the Board, attempted to overturn the primacy of the Institute's technical mission. It was unprecedented in the nearly ninety years of organized electrical engineering. The attempt came in the spring of 1971 in the form of a petition seeking to amend the Constitution. The amendment proposed that "the primary purpose of the IEEE is to promote and improve the economic well-being of the membership. . . ." Its "secondary purposes" would be "scientific, literary and educational" [44].

With 71 percent of the international membership voting no, the amendment was defeated. The move to relegate the society's technical mission to secondary status had failed to gain two-thirds of the votes cast. In the United States, however, the move to make the economic well-being of the engineer the prime purpose of the Institute garnered 52 percent of the vote.

The petition amendment did not fail in its essential purposes. At the November meeting of the Board of Directors—just three weeks after the results of the vote were revealed—the Board took several actions that responded to the concern for the economic well-being of members. Because the balloting on the amendment was considered subject to several interpretations, the Board decided that a fuller survey of membership opinion would be needed. But Board members did not dispute what the amendment vote had made dramatically clear: that when Institute activities "go beyond the dissemination of technical information, . . . members in different countries may have substantially different interests and desires for action" [45].

As a result of this last, persuasive fact, the Board took a step which necessarily modified the international status sought by the IEEE's makers. It asked the Directors of the six United States Regions to recommend "actions desirable to meet member needs in the area of professionalism.'" Given the name of United States Activities Committee, USAC was a committee of the Regional Activities Board. It possessed only a portion of the social and economic programs then appearing throughout the Institute. However, it had the authorization to establish an Institute office in Washington, D.C., so as to aid in the "exchange of information" with Congress and the executive branch. With these actions, a coherent response to the social and economic unheavals could be seen taking shape within the profession [46].

Innovation did not cease at the end of 1971; it had only begun. In March of the following year, Institute executive director, Donald Fink, submitted a report on "IEEE participation in socioeconomic programs." Reflecting a "wide variety" of views and suggestions, Fink's report led the president to call him the "chief distiller." At the same meeting, as Fink later wrote, "a major milestone" was reached in IEEE history when the Board voted to recommend changes in the Constitution that would permit IEEE entry into "non-technical professional activities." Further, the U.S. questionnaire initiated in November had been completed by 40 percent of U.S. members. By a margin of more than two to one, they opted for the Institute to become "more active in economic and political matters" [47].

From such information and with knowledge gained from two years of agitation, the Board devised a constitutional amendment to guide the Institute down the road to an expanded professionalism. It left intact the primary goal of promoting scientific and educational activities for the "advancement of the theory and practice of electrical engineering, electronics, radio and the allied branches of engineering. . . ." Instead, additional clauses were added that extended IEEE purposes to those that were "professional, directed toward the advancement of the standing of the members of the professions it serves."

Later, Institute leaders repeatedly described the vote on the amendments as "overwhelming" and, indeed, 86 percent of the members ratified the Board of Director's proposed constitutional amendment. The import of the amendments—which took effect in January 1973—was straightforward. Fink explained them in a "Blueprint for Change," a document prepared to reflect the views of the Board. Until the amendment, the Institute could go only so far down these new paths. Although many of the programs instituted in 1971 had dealt with the social implications of technology and the crisis over unemployment, the leadership was "limited by . . . the present IEEE Constitution to an insubstantial fraction of the Institute's efforts and resources." To take more "substantial steps," the Board had sought a constitutional amendment.

The new version of the "purposes clause," Article 1 of the Constitution, covered three realms of engineering activity: the technical, the economic, and the social [48]. Since it retained the primacy of the technical mission, the first purpose of the society remained the task of advancing the "scientific and educational" interests of the electrical field through meetings and publications.

The second purpose was new, seeking to advance the "standing of the members" by "professional" means. This section also discussed means, although, as in the technical part, the leadership was not limited to these. To advance the engineers' professional standing, the Institute could conduct surveys and prepare reports on "matters of professional concern." In addition, the society could collaborate with "public bodies and with other societies" for the benefit of the profession. Finally, "standards of qualification and ethical conduct" could be established. Only one negative note entered this section, as a statement of what was not allowed in the quest for professional status: "The IEEE shall not engage in collective bargaining on such matters as salaries, wages, benefits, and working conditions, customarily dealt with by labor unions."

The third section on social responsibility reflected the issues that arose in the first year of upheaval. It instructed the IEEE to "strive to enhance the quality of life for all people throughout the world through the constructive application of technology in its fields of competence." Further, "it shall endeavor to promote understanding of the influence of such technology on the public welfare."

The agency for seeking these goals had been created in 1971 when the United States Activities Committee was formed under RAB. That it was not then seen as the umbrella agency it would become, the Board demonstrated when, in 1972, it created a Professional Activities Committee. But that step, too, left most professional activities scattered throughout the Institute, since the Board assigned to the new committee only the two areas of manpower planning and government relations [49].

Early in 1973, however, these activities plus the other "professional programs" were consolidated under USAC. USAC became a nine-member committee with three Divisional Directors to complement the Directors from the six United States Regions. The committees established under USAC demonstrated what a regional director termed "a new focus for IEEE's professional goals." To achieve this focus, USAC formed subcommittees on government relations, pensions, employment practices, member employment, and manpower planning. It also took on a monitoring function for both TAB's committee on technology forecasting and assessment and the Educational Activities Board's committees on career development and continuing education, surveys, and public relations. The following year, USAC was elevated to board status, thus placing social issues at the highest level of authority within the Institute [50].

The new professionalism

A new definition of professionalism had congealed within the society. The amended Constitution and the United States Activities Board (USAB) reflected it and Leo Young affirmed it in his 1973 assessment of the steps taken toward professionalism. Writing as the amendments were being approved and professional activities were being consolidated under USAB, he discussed not only the crisis over professionalism but also the events that had precipitated and nursed the changes. Young concluded with a list of the "distinct attributes" of professionalism: possession of a systematic body of knowledge and useful skills, recognition of the engineer's authority by the profession's clients, sanction of engineering authority by the community, possession of a code by which to judge ethical performance and conduct, and achievement of a sense of group identity [51].

During these years, when concern over sophisticated weapons systems and lost engineering jobs jostled the traditional technical mission of the IEEE, the direction of change frequently received such analyses as Young's. The year before, an editorial appeared in *Spectrum* that both defined the movement and gave it a name. The editorialist was Donald Christiansen, a Cornell University-educated electrical engineer who had recently been editor-in-chief of *Electronics* magazine. In January, 1972, he became *Spectrum*'s first staff editor. As Christiansen would do frequently throughout the next decade, he summed up for his readers the nature of the issues that were before the Institute.

The basic components of professionalism had been familiar for some time, Christiansen wrote. A professional required "specialized skills and knowledge," and "buried somewhere" in the notion of professionalism "were ethics" or "honest practice." Other attributes were a job well done, a concern for public approbation, and the need for "group action to protect our 'rights'" [52].

But recently, engineers had been buffeted by unemployment and misemployment and had even heard their professionalism questioned. Nonetheless, Christiansen believed, something good had come from this criticism. Even as engineers addressed the economic aspects of employment, they expressed concern for more than "bread alone." Engineers also wanted "a friendly environment" for their work and a "say as to the uses to which their expertise is applied." In short, they were interested

> in what might be termed the "new professionalism," professionalism that is based not only on traditional high standards of technical achievement but that embraces concern for the impact of technological developments on society as well.

Christiansen properly singled out the interest in the impact of technology as the fresh ingredient in the new professionalism. The half-decade of social upheaval had been codified in the Constitution. And except for the new ethics code adopted in 1974, the changes that followed were organizationally contained in USAB. But these acts of 1974 had gathered more than the organizational initiatives of the previous five years. USAB and the ethics code rested on a hundred years of attention to external social and economic conditions.

The events of the half decade from 1969 to 1973 had by no means transformed the Institute or the profession. Christiansen recognized this when he placed the social question after the technical mission in the hierarchy of values represented by the new professionalism. The society's continued growth during the decade that followed was to be seen, as always, in the technical arena. The basic signs of organizational expansion continued to appear in the increase of the number of technical publications, groups, and societies. It was this persistent emphasis that pushed the IEEE toward a membership of 260,000 by 1984.

Nonetheless, the tree of electronics had been dramatically revealed to be a hybrid growth, containing both technical and social branches. This was confirmed on the occasion of the IEEE's tenth anniversary in 1973. When asked to comment on the future of the profession, members urged attention to such problems as engineering education, the increasing difficulty of technological choices, the engineer's responsibility to the public, and the need to ensure the safety of complex technical systems such as nuclear generation plants and automated transportation networks.

In taking on these hybrid issues, the society assumed no easy task, as the MIT engineer, Gordon S. Brown, explained. Defining "software" as "the laws, the policies, the doctrines whereby we govern ourselves," Brown stated the main challenge to professional engineers as they neared the end of their first century: "We find it relatively easy to produce hardware, but it is in devising our software . . . that we encounter our most baffling problems" [53].

So although the changes precipitated by the upheavals of the late sixties failed to transform the IEEE, they added to the array of responses available to

IEEE CODE OF ETHICS FOR ENGINEERS

PREAMBLE

Engineers affect the quality of life for all people in our complex technological society. In the pursuit of their profession, therefore, it is vital that engineers conduct their work in an ethical manner so that they merit the confidence of colleagues, employers, clients and the public. This IEEE Code of Ethics is a standard of professional conduct for engineers.

ARTICLE I

Engineers shall maintain high standards of diligence, creativity and productivity, and shall:
1. Accept responsibility for their actions;
2. Be honest and realistic in stating claims or estimates from available data;
3. Undertake engineering tasks and accept responsibility only if qualified by training or experience, or after full disclosure to their employers or clients of pertinent qualifications;
4. Maintain their professional skills at the level of the state of the art, and recognize the importance of current events in their work;
5. Advance the integrity and prestige of the engineering profession by practicing in a dignified manner and for adequate compensation.

ARTICLE II

Engineers shall, in their work:
1. Treat fairly all colleagues and co-workers, regardless of race, religion, sex, age or national origin;
2. Report, publish and disseminate freely information to others, subject to legal and proprietary restraints;
3. Encourage colleagues and co-workers to act in accord with this Code and support them when they do so;
4. Seek, accept and offer honest criticism of work, and properly credit the contributions of others;
5. Support and participate in the activities of their professional societies;
6. Assist colleagues and co-workers in their professional development.

ARTICLE III

Engineers shall, in their relations with employers and clients:
1. Act as faithful agents or trustees for their employers or clients in professional and business matters, provided such actions conform with other parts of this Code;
2. Keep information on the business affairs or technical processes of an employer or client in confidence while employed, and later, until such information is properly released, provided such actions conform with other parts of this Code;
3. Inform their employers, clients, professional societies or public agencies or private agencies of which they are members or to which they may make presentations, of any circumstance that could lead to a conflict of interest;
4. Neither give nor accept, directly or indirectly, any gift, payment or service of more than nominal value to or from those having business relationships with their employers or clients;
5. Assist and advise their employers or clients in anticipating the possible consequences, direct and indirect, immediate or remote, of the projects, work or plans of which they have knowledge.

ARTICLE IV

Engineers shall, in fulfilling their responsibilities to the community:
1. Protect the safety, health and welfare of the public and speak out against abuses in these areas affecting the public interest;
2. Contribute professional advice, as appropriate, to civic, charitable or other non-profit organizations;
3. Seek to extend public knowledge and appreciation of the engineering profession and its achievements.

A product of committees of the United States Activities Board and the IEEE Board of Directors, the first new ethics code since 1912 was approved in December 1974. IEEE President Arthur P. Stern, described it as "a significant step forward in the area of professionalism."

electrical and electronics engineers as they make their way in a changing world. These historic responses and the events that precipitated them constitute the essence of the involvement of the IEEE and its predecessor societies in the story of electrical engineering in America: as such, they reveal the making of a profession. In addition, they form the IEEE's inheritance: a legacy of engineering values for the second century.

REFERENCES

Chapter 1

[1] E. W. Houston, "Inaugural Address," *AIEE Trans.*, 11 (1894), p. 276.
[2] Franklin Institute, *Official Catalogue of the International Electrical Exhibition: Philadelphia* (Philadelphia 1884), p. 5; *Elec. World,* III (April 5, 1884), pp. 111–112; *Operator,* 15 (September 15, 1884), p. 175.
[3] *AIEE Trans.*, I (1885); see also Harris J. Ryan, "A Generation of the American Institute of Electrical Engineers, 1884–1924," ibid., XLIII (June, 1924), pp. 740–744.
[4] *Report of the Electrical Conference at Philadelphia, September, 1884* (Washington, D.C., Government Printing Office, 1886).
[5] "Scientific Theory and Practice," *The Operator,* 15 (October 1, 1884), p. 180.
[6] Thomas Parke Hughes, *Networks of Power: Electrification in Western Society, 1880–1930* (Baltimore, Md.: Johns Hopkins University Press, 1983).
[7] Daniel J. Kevles, *The Physicists: The History of a Scientific Community in Modern America* (New York: Alfred A. Knopf, 1978).
[8] W. James King, *The Development of Electrical Technology in the Nineteenth Century: 1. The Electrochemical Cell and the Electromagnet* (Contributions from the Museum of History and Technology, U.S. National Museum, Bulletin 228, Smithsonian Institution: Washington, D.C., 1962), pp. 281–284, 294–297.
[9] Matthew Josephson, *Edison: A Biography* (New York: McGraw-Hill Book Co., Inc., 1959), pp. 42ff., 61–63, 78, contains a detailed discussion of Edison's early inventive work in the telegraph industry; *Dictionary of American Biography (DAB),* Supplement 10.
[10] For Pope, see *DAB,* XV.
[11] See R. L. Thompson, *Wiring a Continent: The History of the Telegraph Industry in the United States, 1832–1866* (Princeton, N.J., 1947), pp. 245–246, for a discussion and illustration of the typical telegraph office; for Prescott, see *DAB,* VIII.
[12] Prescott, *History, Theory, and Practice of the Electric Telegraph* (Boston, 1875), pp. 1–2.
[13] King, 2. *The Telegraph and the Telephone,* pp. 289, 292–293, 294, 300–302.
[14] Ibid., p. 306.
[15] For Bell, see *DAB,* II; King, *Telegraph and Telephone,* pp. 312, 318, 319. For a fuller recounting of the encounter with Hemholtz's work, see David A. Hounshell, "Bell and Gray: Contrasts in Style, Politics, and Etiquette," *Proc. IEEE,* 64 (September, 1976), pp. 1306–1307.
[16] For Gray, see *DAB,* VII; see also David A. Hounshell, "Elisha Gray and the Telephone: On the Disadvantages of Being an Expert," *Technology and Culture,* V. 16 (April, 1975), 133–161.
[17] The following account draws on Rosario J. Tosiello's *The Birth and Early Years of the Bell Telephone System, 1876–1880* (New York: Arno Press, 1979), pp. 22–31. This is a facsimile publication of a dissertation written under Robert V. Bruce whose *Bell: Alexander Graham Bell and the Conquest of Solitude* (Boston, 1973) is the most recent biography of the inventor.

[18] In "Gray and the Telephone," fn. 67, p. 155, Hounshell suggests the opposite of Thomson when he makes the point that the committee viewed Bell's achievement as simply a "marvel." In his later article on "Bell and Gray," p. 1312, he implies a different attitude when he relates that Thomson immediately became "the herald of Bell and the telephone in Europe." The Thomson quotation is from Tosiello, p. 31.
[19] Alfred D. Chandler, Jr., *The Visible Hand: The Managerial Revolution in American Business* (Cambridge, Mass.: Harvard University Press, 1977), p. 197.
[20] Thompson, ibid.; Committee on Science and the Arts, Case No. 165, "Electro-Magnetic Telegraph," Report dated February 8, 1838, with correspondence, Franklin Institute Archives. The earlier report on "visible systems" was Science and Arts Case No. 144 ½, Report dated April 18, 1837.
[21] Ibid., Case 165.
[22] Henry O'Rielly to Amos Kendall, December 25, 1845, O'Rielly Manuscript Collection, New York Historical Society, quoted in Thompson, p. 49; Thompson, p. 31.
[23] Ibid., p. 24.
[24] Ibid., pp. vii–viii.
[25] Tosiello, p. 81; Josephson, p. 141; King, *Telegraph and the Telephone*, pp. 325–329; William H. Forbes to Bell, March 5, 1879, AT&T Archives quoted in Tosiello, p. 397; C. E. Hubbard to Gardner Hubbard, January 7, 1878, ibid., p. 176.
[26] For Green, see *DAB*, VII, and Lester G. Lindley, "Norvin Green and the Telegraph Consolidation Movement," *Filson Club Historical Quarterly*, 48 (July, 1974), pp. 253–264.
[27] *New York Times*, May 17, 18, 21, 1881; ibid., August 31 and September 1, 4, 1883; "The Government and the Telegraph," *North American Review* (November, 1883), passim.; *New York Times*, August 15, 1881.
[28] Tosiello, p. 176; for Vail, see *DAB*, XIX; ibid.; Chandler, *Invisible Hand*, p. 201.
[29] Ibid.
[30] Tosiello, 429–430; quoted in Leonard Reich, "Research, Patents, and the Struggle to Control Radio: A Study of Big Business and the Uses of Industrial Research," *Business History Review*, LI (Summer, 1977), p. 232.
[31] *Report of the 27th Exhibition*, p. 10.
[32] Ibid., pp. 17–29, 171.
[33] Lewis Mumford, *Technics and Civilization* (New York: Harcourt, Brace and World, 1934), p. 110; *Report of the 27th Exhibition*, pp. 5–6, 205, 227, 230.
[34] Ibid.
[35] King, *Development of Electrical Technology*, p. 254; *Report of the 27th Exhibition*, pp. 193–196.
[36] Ibid., pp. 248, 250, 254; Reese V. Jenkins, *Images and Enterprise: Technology and the American Photographic Industry, 1839–1925* (Baltimore, Md.: Johns Hopkins University Press, 1975), p. 22. On electrochemistry, see also David S. Landes, *The Unbound Prometheus: Technological Change and Industrial Development in Western Europe from 1750 to the Present* (London, Eng.: Cambridge University Press, 1972), p. 284.
[37] For Weston, see the *DAB*, Supplement 2; the quotation is from Passer, p. 32.
[38] For Thomson, see the *DAB*, Supplement 2.
[39] Houston's biography is also in the *DAB*, V. 9. Thomson's work and the early partnership with Houston are discussed in Passer, pp. 21–31; Elihu Thomson, "Electricity during the Nineteenth Century," *Smithsonian Institution, Annual*

Report, 1900, Part 1, (Washington, D.C., 1901), p. 339; for an account of the development of the Gramme dynamo, see King, "Early Arc Light and Generator," pp. 371–386; the Gramme machines at the Centennial are discussed in James E. Brittain, "B.A. Behrend and the Beginning of Electrical Engineering, 1870–1920," (Unpublished Dissertation: Case Western Reserve University, 1970), p. 20; Thomson, ibid., pp. 340–341.

[40] This and the following paragraphs on Brush rely on the *DAB,* Supplement 1, and Passer, pp. 14–21.

[41] "Report of the Committee on Dynamo-Electric Machines," *Journal of the Franklin Institute,* 3rd Series, CV (1878), pp. 289–302 and 361–378. Also see the account of Harold C. Passer, *The Electrical Manufacturers, 1875–1900* (Cambridge, Mass.: Harvard University Press, 1953), pp. 15–16; Passer incorrectly states that "no precedent" existed for the Institute tests, yet tests of European made dynamos had been conducted in England during the winter of 1876–1877, for which see King, Part 3, "The Early Arc Light and Generator," pp. 365–367.

[42] Thomas P. Hughes, "Edison's Method," *American Patent Law Association Bulletin,* (July-August, 1977), p. 436.

[43] Thomson to J. W. Hammond, July 25, 1927, quoted in Passer, p. 373, fn 29.

[44] Quoted in Hughes, "Edison's Method," p. 438. This is a concise statement of what Hughes describes as the "inventing, developing [engineering], and innovating process" and labels "entrepreneurship," p. 436; for an extensive treatment of the ways and purposes of the inventor-entrepreneur, see Hughes' *Elmer Sperry: Inventor and Engineer* (Baltimore Md.: Johns Hopkins Press, 1971). The following discussion of Edison's pre-1884 activities come from Passer, pp. 78 ff., and Matthew Josephson, *Edison: A Biography* (New York: McGraw-Hill Book Co., 1959).

[45] Ibid., p. 258; Passer, pp. 90, 19; Forrest McDonald, *Insull* (Chicago, Ill.: University of Chicago Press, 1962), pp. 26–27.

[46] Passer, p. 91; Josephson, pp. 262–263; quotation from ibid., p. 264.

[47] Josephson, pp. 248–249.

[48] Passer, pp. 15–16, 21–22.

[49] Josephson, pp. 127–130; Passer, pp. 27, 109–111.

[50] For Houston, see *DAB,* v. 9; Thomson to Houston, August 26, 1884, Elihu Thomson Papers, American Philosophical Society.

[51] *Report of the 27th Exhibition,* passim.

[52] *Elec. World,* III (April 5, 1884), pp. 111–112; ibid., (April 19, 1884), p. 130.

[53] *AIEE Trans.*, 1 (May-October, 1884). pp. 1, 3–4; refers to this and the following paragraphs.

Chapter 2

[1] Scott, "President's Address, June 29, 1903," *AIEE Trans.*, XXII (1903), pp. 3–15.

[2] *AIEE Trans.*, IV (1887), pp. 122–123.

[3] "Rules of the AIEE, 1884," *AIEE Trans.*, I, Supplement (1885), p. 3; Minutes, AIEE Council, February 5, 1889, Archives, Institute of Electrical and Electronics Engineers (IEEE); ibid., February 7, 1890; ibid., June 20, 1891.

[4] Ralph W. Pope, "Monthly Meetings of the Institute: Their Origins and Proposed Development," *AIEE Trans.*, X (1893), p. 533; *AIEE Trans.*, IV (1887), p. 123; Passer, *The Electrical Manufacturers,* p. 132.

[5] Pope, "Monthly Meetings," pp. 537 ff, 541; *AIEE Trans.*, XXI (1903), pp. 479–496.
[6] *AIEE Trans.*, I (1885), Appendix, pp. 1, 7; Minutes, October 21 and December 9, 1884, IEEE.
[7] Ibid., II (1886); Rules (December 8, 1885), printed in ibid., appendix.
[8] Minutes, May 18, 1887; April 21 and October 27, 1891; November 12 and December 3, 1889; February 7, 1888; December 3, 1889; and October 21, 1890, IEEE Archives.
[9] The Rules were printed as supplements in the *AIEE Trans.*, and in 1894, the *Handbook.* Copies of the Rules were also filed with the Minutes, IEEE.
[10] Alan Trachtenberg, *The Incorporation of America: Culture and Society in the Gilded Age* (New York: Hill & Wang, 1982), p. 3; Besides Chandler, *Visible Hand,* see Glenn Porter, *The Rise of Big Business, 1860–1910* (Arlington Heights, Ill.: AHM Publishing Corp., 1973); see also David F. Noble, *America by Design: Science, Technology, and the Rise of Corporate Capitalism* (New York: Alfred A. Knopf, 1977).
[11] Tosiello, "Birth . . . of Bell Company," p. 96; Agreement between Elihu Thomson, Edwin J. Houston, and the American Electric Company, July 9, 1880, Thomson Papers, American Philosophical Society; for Steinmetz, Kennelly, and others, see the List of Members and Associates in *AIEE Trans.*, XIV (1897).
[12] Report on the Electrical Conference at Philadelphia, p. 28; *Electrical World,* 13 (May 11, 1889), p. 268.
[13] *AIEE Trans.*, IX (1892), p. 490.
[14] *DAB,* XV; Passer, p. 133.
[15] Ibid, pp. 133, 137; for Stanley, see Hughes, *Networking Power,* pp. 98–104.
[16] Pope to John E. Hudson, June 21, 1895, American Telephone and Telegraph Archives (hereafter AT&T).
[17] Ibid.
[18] Hudson to Pope, July 25, 1895; Pope to Hudson, September 14, 1895; Hudson to Pope, September 18, 1895, AT&T; David O. Woodbury, *A Measure for Greatness: A Short Biography of Edward Weston* (New York, 1949), p. 141; Lockwood to C. P. Ware, January 24, 1907, AT&T.
[19] Woodbury, p. 141.
[20] For Thomson's speech, see *AIEE Trans.*, VI (1889), p. 483.
[21] Thomson to G. W. Stockley, May 31, 1882, Thomson Papers, American Philosophical Society.
[22] Ibid.
[23] Thomas Parke Hughes, *Elmer Sperry: Inventor and Entrepreneur* (Baltimore, Md.: Johns Hopkins Press, 1972), pp. 41–44, 51, 53, 68.
[24] James E. Brittain, "B. A. Behrend and the Beginnings of Electrical Engineering, 1870–1920," (Unpublished Dissertation: Case Western Reserve University, 1970), pp. 18–19, 52–53; Pupin, *From Immigrant to Inventor* (New York: Charles Scribners' Sons, 1925), pp. 279–280.
[25] For this and the following paragraph, see the Second Annual Circular (1882) and Catalogues of the Case School of Applied Science, Archives, Case Western Reserve University.
[26] See the Catalogues of the University of Texas and The Engineering Department: Civil, Sanitary, Electrical, Mining (University of Texas: Bulletin No. 40, 1904–1905), Archives, University of Texas, Eugene Barker History Center, Austin.

[27] *AIEE Trans.*, III (1887), p. 4; for Mailloux, see National Cyclopedia, XXVI; *Elec. World,* IV (September 20, 1884), p. 96; for Anthony, see *DAB,* Supplement 1.
[28] *AIEE Trans.*, VIII (1892), pp. 212–213, 224–225.
[29] Ibid.
[30] For Steinmetz, see *DAB,* XVII, and John H. Hammond, *Charles Proteus Steinmetz: A Biography* (Century Co., 1924).
[31] Brittain, pp. 78–80; Scott, "The Institute's First Half Century," *Electrical Engineering* (1934), p. 656; Philip L. Alger, *The Human Side of Engineering: Tales of General Electric Engineering over 80 Years (1972),* pp. 23–24.
[32] Brittain, pp. 7–8.
[33] Ibid., pp. 66 ff, 116; *AIEE Trans.*, XIII (1896), pp. 108 ff.
[34] Ibid., pp. 108–110; see Brittain, pp. 99, 134, for a comparison of Steinmetz's and Pupin's style of argument.
[35] Pupin, *From Immigrant to Inventor,* pp. 280–283.
[36] Benjamin Garver Lamme, *Electrical Engineer: An Autobiography* (New York: G. P. Putnam's Sons, 1926), p. 29.
[37] Hammond, Steinmetz, p. 219; Lamme, p. 87.
[38] Kendall Birr, *Pioneering in Industrial Research: The Story of the General Electric Research Laboratory* (Washington, D.C.: Public Affairs Press, 1957), pp. 2; for early industrial research at GE, see pp. 29 ff.; at Bell, see Leonard S. Reich, "Industrial Research and the Pursuit of Corporate Security; The Early Years of Bell Labs," *Business History Review,* LIV (Winter, 1980), pp. 504–529; see also Lillian Hoddeson, "The Emergence of Basic Research in the Bell Telephone System, 1875–1915," *Technology and Culture,* 22 (July, 1981), 512–544; at Westinghouse, see Lamme, *Autobiography,* pp. 28 ff.
[39] Federal Communications Commission, *Investigation of the Telephone Industry in the United States* (Washington, D.C., U.S. Government Printing Office) pp. 184, 186; Hubbard to Hudson, August 5, 1890(?), AT&T; George H. Clark, *The Life of John Stone Stone* (San Diego, Calif.: Frye & Smith Ltd., 1946), pp. 11–17.
[40] Clark, p. 23; Stone to Hayes, January 2, 1891; Hayes to Hudson, March 7, 1892, AT&T Archives.
[41] Clark, pp. 24–26.
[42] Hayes to Hudson, March 30, 1896; July 13, 1898, January 12, 1899, AT&T.
[43] Hoddeson, pp. 524–526; Hayes to Hudson, March 19, 1894; Davis to Hudson, January 30, 1895, and March 30, 1892, AT&T.
[44] Jacques to ? Madden, September 1, 1883; Hayes to Hudson, December 30, 1886, and January 2, 1890, AT&T.
[45] Ibid.
[46] Ibid.
[47] Hayes to Davis, January 29, 1895, and January 1, 1896, AT&T.
[48] T. B. Doolittle to Davis, January 22, 1897; Davis to Hudson, January 25, 1898; Davis to E. J. Hall, January 12, 1891; Davis to Hudson, March 18, 1899, AT&T.
[49] Charles F. Scott, "President's Address," *AIEE Trans.*, 22 (1903), pp. 6–7.
[50] *AIEE Trans.*, XVIII (1902); *AIEE Trans.*, XX (1907); see A. Michal McMahon, "Corporate Technology: The Social Origins of the American Institute of Electrical Engineers," *Proc. IEEE,* 64 (September, 1976), pp. 1383–1390 for comparisons of the AIEE leadership in 1884, 1900, and 1910.
[51] *AIEE Trans.*, XVIII (1902), p. 924.
[52] Minutes, Board of Directors, March 27, 1903, IEEE.

[53] Minutes, ibid., April 23, 1903, IEEE.
[54] Minutes, ibid., February 26, March 25, and April 22, 1904, IEEE; Rollo Appleyard, *The History of the Institution of Electrical Engineers (1871–1931)* (London: IEE, 1939), p. 167.
[55] Catalogue of the University of Texas, 1898–1899 and 1902–1903 (Austin, 1898 and 1902); Lamme, *Autobiography*, p. 91.

Chapter 3

[1] "Address of John J. Carty," AIEE Symposium on "The Status of the Engineer," *Proc. AIEE* (April, 1915), Pamphlet Reprint, AT&T.
[2] Frederick Leland Rhodes, *John J. Carty: An Appreciation* (New York, 1932), p. 69.
[3] Rushmore, "Relation of the Manufacturing Company to the Technical Graduate," *AIEE Trans.*, XXVII (1908), p. 1473; Ferguson, "Centralization of Power Supply," *AIEE Trans.*, XXVIII (1909), p. 355.
[4] Scott, "First Half-Century," *Elec. Eng.* (May, 1934), p. 659; "The Engineer in the Twentieth Century," *AIEE Trans.*, XX (1902), p. 305.
[5] Ibid.; Houston, "Review of the Progress of the AIEE," *AIEE Trans.*, XI (1894), pp. 274–275, 281, 285; Kennelly, "The Present Status of Electrical Engineering," *AIEE Trans.*, XV (1898), p. 279; Scott, "Proposed Developments of the Institute," *AIEE Trans.*, XX (1902), pp. 3–4.
[6] Ibid.; for Scott's technical contributions, see Thomas P. Hughes, "The Science-Technology Interaction: The Case of High-Voltage Power Transmission Systems," *Technology and Culture*, 17 (October, 1976), pp. 647 ff.
[7] For Martin, see *Elec. Eng.* (May, 1934), p. 789; Scott to Martin, September 15, 1902, Scott Papers.
[8] Martin, *Central Station Electric Light and Power, 1902* (Special Report, Bureau of the Census, Washington, D.C.: Government Printing Office, 1905), p. 3–4.
[9] "Proposed Developments," pp. 5–14.
[10] *AIEE Trans.*, IV (1888), p. 2; Houston, "The International Electrical Congress . . . of 1893," *AIEE Trans.*, X (1893), p. 472.
[11] Ibid., IX (1892), pp. 467, 477, 479, 492.
[12] Session on "Education of the Engineer," *AIEE Trans.*, XIX (1902) pp. 1145–1210.
[13] Hammond, *Steinmetz*, pp. 36 ff., 282 ff.
[14] *AIEE Trans.*, XXI (1903), p. 599.
[15] Ibid., IX (1892), p. 471; Owens, "A Course of Study in Electrical Engineering," *Proc. Society for the Promotion of Engineering Education (SPEE)*, V (1898), pp. 40–42.
[16] For Jackson, see the biographical sketch, *Elec. Eng.*, 53 (May 1934), p. 799; *AIEE Trans.*, XXI (1903), pp. 601–606.
[17] Ibid., pp. 587, 590, 592–593; Francis C. Caldwell, "A Comparative Study of the Electrical Engineering Courses Given at Different Institutes," *Proc. SPEE*, VII (1899), p. 128.
[18] Lamme, *Autobiography*, p. 153; Scott, "First Fifty Years," p. 657.
[19] Jackson to Henry S. Pritchett, August 14, 1906, Jackson Papers, MIT Archives. The GE-MIT cooperative course was frequently discussed in engineering periodicals during the twenty years following its first proposal in 1907. More recently, it has been discussed in Noble, *American by Design*, especially pp. 188 ff.

[20] For the Alexander paper and discussion, see *AIEE Trans.*, XXVII (1908), pp. 1459–1497.
[21] Brittain, "B. A. Behrend," pp. 206 ff.
[22] Wickenden to Maclaurin, April 6, 1918, Jackson Papers, MIT Archives.
[23] For a biography of Wickenden, see *Elec. Eng.* (March, 1945).
[24] Bush, *Pieces of the Action* (New York, 1970), pp. 252–255.
[25] Millikan to Henry S. Pritchett, Oct. 31, 1921, Millikan Papers, California Institute of Technology, quoted in Ronald Tobey, *The American Ideology of National Science, 1919–1930* (Pittsburgh, Pa., 1971), p. 218; Wickenden, "A Comparative Study of Engineering Education in the United States and in Europe" (Bulletin No. 16, "Investigation of Engineering Education," SPEE, June, 1929), Foreword, pp. 253–266.
[26] *Elec. Eng.*, 53 (May, 1934), pp. 681, 686; Kennelly, "The Work of the Institute in Standardization," ibid., p. 676.
[27] AIEE Council Minutes, June 3, 1884, and June 2, 1885, IEEE Archives.
[28] *AIEE Trans.*, II (1885), p. 19.
[29] Council Minutes, December 3, 1889; June 26 and September 22, 1891; April 9, 1892; November 15, 1893, IEEE Archives.
[30] Edwin Houston, "Historical Sketch of [the AIEE's] Organization and Work," *AIEE Trans.*, VIII (1891), pp. 601–608; "Memorial to the U.S. Congress," *AIEE Trans.*, XI (1894).
[31] "Annual Meeting, American Institute," *AIEE Trans.*, III (1886), pp. 13–14.
[32] Ibid.; Council Minutes, June 8, 1886, IEEE Archives; "Annual Meeting," p. 18.
[33] Ibid., p. 22. Council Minutes, November 16, 1886; March 8, 1887; December 3, 1889; February 7, March 18, and June 17, 1890; IEEE Archives; see the "Preliminary Report of the Standard Wiring Table Committee," *AIEE Trans.*, VII (1890), pp. 1–18; Final Report, pp. 407–411.
[34] "The Standardizing of Generators, Motors and Transformers (A Topical Discussion)," *AIEE Trans.*, XV (1898), p. 3.
[35] See the "Report of the Standardization Committee," *AIEE Trans.* (1899), pp. 255–268; "Standardization Rules of the AIEE," Appendix, *AIEE Trans.*, XXVI (1908), pp. 1826–1827.
[36] Scott to Steinmetz, September 23, 1902, Scott Papers; Scott, "President's Address," *AIEE Trans.*, XXI (1903), p. 9.
[37] Committee creation and name changes can be traced in the annual reports of the Board of Directors and, though the dates are not always reliable, in the Fiftieth Anniversary issue of *Elec. Eng.*, 53 (May, 1934), pp. 830–831.
[38] "Topical Discussion," pp. 4–5.
[39] *Elec. Eng.*, 53 (May, 1934), p. 796; "Topical Discussion," p. 13; Kennelly, "President's Address," p. 4.
[40] Insull, "Standardization, Cost System of Rates, and Public Control," *Central-Station Electric Service: Public Addresses (1897–1914)*, (Chicago, 1915), pp. 35–36, 39.
[41] Scott, "First Half Century," pp. 656–657; Scott to Steinmetz, September 23, 1902; Rice to Scott, October 29, 1902, Scott Papers.
[42] Scott to Mershon, October 1, 1902; Scott to Masson, November 15, 1902; Masson to Scott, November 26, 1902, Scott Papers.
[43] Scott to Masson, December 1, 1902; Masson to Scott, December 9, 1902, Scott Papers.
[44] Council Report, April 30, 1900, IEEE; Henry S. Pritchett, "The Story of the Establishment of the National Bureau of Standards," *Science*, XV (January-

June, 1902); Kevles, *The Physicists,* pp. 1–2, 9; Dupree, pp. 272–273, 275; John Perry, *The Story of Standards* (New York: Funk & Wagnalls Co., 1955), p. 130–131; *AIEE Trans.,* XXVI (June-December, 1907); Kevles, p. 67 fn; quotation from G. A. Weber, The Bureau of Standards (Institute for Government Research, Service Monographs No. 35, Baltimore, 1925); E. B. Rosa, "National Bureau of Standards and its Relations to Scientific and Technical Laboratories," *Science,* XXI (1905), p. 173, quoted in Dupree, p. 275.

[45] A. E. Kennelly, "The Work of the Institute in Standardization," *Elec. Eng.,* 53 (May, 1934), p. 679; Adams, "Industrial Standardization," AAPSS (1919), p. 298–299.

[46] Ibid.

[47] Adams, "Major Events," pp. 686–687 and "Industrial Standardization," p. 296; Le Maistre, "Standardization," *AIEE Trans.,* XXXV (1916), p. 490.

[48] "Discussion," ibid., pp. 497–499.

[49] Application to Signal Corps, Officer's Reserve Corps, April 26, 1917, AT&T; Frederick Leland Rhodes, *John J. Carty: An Appreciation* (New York, 1932), pp. 66–68.

[50] Carty to Jackson, November 28, 1910, Jackson Papers; "Address of J. J. Carty," op. cit.

[51] Steinmetz, "Industrial Efficiency and Political Waste," *Harper's Monthly,* CXXXIII (November, 1916), pp. 925–928; "The Individual and Corporate Development of Industry," *GE Review,* XVIII (August, 1915), pp. 813–816; and "Competition and Cooperation," *Collier's,* LVIII (September 23, 1916); *New Epoch,* "Introduction," n.p.; the last quoted phrase appears throughout the book.

[52] Carl E. Schorske, *German Social Democracy, 1905–1917* (New York, 1955), pp. 3, 17–18; Gary P. Steenson, "Not One Man! Not One Penny!" *German Social Democracy, 1863–1914* (Pittsburgh, Pa., 1981), p. 211; see also James Gilbert, *Designing the Industrial State* (Chicago, 1976), especially the chapter, "Charles Steinmetz and the Science of Industrial Organization," pp. 180–199. Gilbert is apparently unaware of the varieties of socialism and thus dismisses Steinmetz's socialism as casually as do his corporate biographers whom Gilbert criticizes. On Steinmetz's political activities during this period, see Hammond, *Steinmetz,* pp. 310 ff. and James Weinstein, *The Corporate Ideal in the Liberal State, 1900–1918* (Boston, 1968), p. 112.

[53] *New Epoch,* pp. 17–18.

[54] Ibid., "Introduction," pp. 42, 120–122, 127–128.

[55] Ibid., pp. 129–130, 153–155, 157.

[56] *New Epoch,* pp. 162–163. Steinmetz used the same passages on engineering standardization in the conclusion to "Industrial Efficiency and Political Waste," p. 928.

[57] *New Epoch,* pp. 80–81.

Chapter 4

[1] "Discussion," *AIEE Trans.,* 25 (1906), p. 266.

[2] For the proceedings of Insull's dinner, see "The De Ferranti Dinner," *Central-Station Electric Service,* pp. 215–233.

[3] AIEE Board Minutes, August 22 and October 13, 1911, IEEE.

[4] For Mitchell, see Sidney Alexander Mitchell, *S. Z. Mitchell and the Electrical Industry* (New York, 1960); Thomas P. Hughes, "The Electrification of Amer-

ica: The System Builders," *Technology and Culture,* 20 (January, 1979), pp. 153–159.

[5] For Wright's innovation, see Forrest McDonald, *Insull* (University of Chicago Press, 1962), pp. 67–69.

[6] James Weinstein, *The Corporate Ideal in Liberal America,* pp. 24–25, quotations from pp. 10 and 87; AIEE Board Minutes, December 16, 1909, and February 11, 1910, IEEE.

[7] For opposition to centralization, see the discussion following Insull, "Railroad Electrification," *AIEE Trans.* (1912).

[8] Kennelly, "Present Status of Electrical Engineering," pp. 273–275; Scott, "President's Address," p. 3.

[9] Scott, "Tendencies of Central Station Development," *AIEE Trans.*, XXI (1903), pp. 403–405.

[10] Insull, "Massing of Energy Production an Economic Necessity," p. 136, *Central-Station Electric Service;* McDonald, *Insull,* p. 58; Thomas P. Hughes, "Electrification of America," p. 141.

[11] McDonald, *Insull,* p. 70; Insull, "Stepping Stones of Central Station Development over Three Decades," *Central-Station Electric Service,* p. 354.

[12] For the number of stations, see Table 5, p. 10, in Martin, *Central Electric Light and Power Stations* (1902); for occupations, see Chap. IV, "Employees, Salaries, and Wages," pp. 56–59.

[13] "Discussion in Pittsburg," *AIEE Trans.*, XXII (1904), p. 763.

[14] McDonald, *Insull,* pp. 136–141; *Insull,* "The Production and Distribution of Energy," *Central-Station Electric Service,* p. 358.

[15] Martin, *Central Electric Light and Power Stations: 1912* (1915), p. 111, 112, 212 and *Central Station Light and Power Stations: 1907* (1910).

[16] For the role of the load dispatcher, see Hughes, "Electrification of America," pp. 147–148; Martin (1907), p. 92 and (1912), pp. 76–79.

[17] Insull, *Central-Station Electric Service,* pp. 151–152.

[18] Kennelly, "Work of the Institute," *Elec. Eng.*, 53 (May, 1934), p. 676; "Commercialized Engineering," *Eng. News,* LIII (January 5, 1905), pp. 20–21.

[19] Scott, "President's Address," *AIEE Trans.*, XXI (1903), p. 3; Tredgold, "Description of a Civil Engineer" (1828), *Minutes and Proceedings of the Institution of Civil Engineers,* XXVII (1867–1868), (London: The Institution, 1868), pp. 182–183; Stott, "The Evolution of Engineering, President's Address," *AIEE Trans.*, XXVII (1909).

[20] Dunn, "The Relation of Electrical Engineering to Other Professions: President's Address," *AIEE Trans.*, XXXI (1912), pp. 1027–1034.

[21] Day, "Engineering Education and Engineering Ethics," *Proc. AIEE,* XXV (1906), pp. 3–4; for Taylor and scientific management, see Daniel Nelson, *Managers and Workers: Origins of the New Factory System in the United States, 1880–1920* (Wisconsin: University of Wisconsin Press, 1975), p. 65; see also Noble, *America by Design,* pp. 82, 267.

[22] Ferguson, "Centralization of Power Supply," *AIEE Trans.*, XXVIII (1909), pp. 355–361.

[23] Jackson to Mailloux, April 13, 1908; Mailloux to Jackson, April 17, 1908; Steinmetz to Jackson, January, 1911; Carty to Jackson, November 28, 29, 1910, Jackson Papers.

[24] "The Status of the Engineer," A Symposium held in New York, February 17, 1915, *Proc. AIEE* (1916), pp. 635–674.

[25] Jackson, "Electrical Engineers and the Public: President's Address," *AIEE Trans.*, XXX (1912), pp. 1135–1142.

[26] Wheeler to Jackson, March 3 and April 8, 1910; Wheeler to Jackson, April 8, 1910; Steinmetz to Jackson, January 6, 1911, Jackson Papers; Committee on Code, Final Report, *Proc. AIEE* (1912), p. 149.

[27] "Report of Committee on Code of Principles of Professional Conduct," *Proc. AIEE* (April, 1912), p. 150; Schuyler Wheeler, "Engineering Honor," (President's Address), *AIEE Trans.*, XXV (1907), pp. 241–248.

[28] "Proposed Code of Ethics," AIEE Committee on Code of Ethics, submitted to the Board, May 17, 1907, *AIEE Trans.*, XXVI (1908), pp. 1421–1425; "Proposed Code of Ethics: Principles of Professional Conduct for the Guidance of the Electrical Engineer," Appendix, *AIEE Trans.*, XXVI (1908), pp. 1789–1793; Wheeler to Jackson, March 3, 1910; Jackson to Wheeler, October 24, 1907, Jackson Papers.

[29] *AIEE Trans.*, XXVI (1908), pp. 1426–1427.

[30] Ibid.

[31] "Code of Principles of Professional Conduct," *Proc. AIEE* (1912), pp. 447–449.

[32] AIEE Board Minutes, March 11, 1910; at the meeting on August 12, Stillwell reappointed the committee with Frank Sprague as chairman and himself and others as Members; Wheeler to Jackson, March 3, 1910; Wheeler to Jackson, March 3, 1910; Jackson to Wheeler, May 4, 1910; Wheeler to Jackson, May 24, 1910; Jackson to Wheeler, September 13, 1910, Jackson Papers.

[33] Jackson to Steinmetz, December 19, 1910; Steinmetz to Jackson, January 6, 1911 (misdated 1910), Jackson Papers; AIEE Board Minutes, June 27 and December 8, 1911.

[34] "Report of Committee on Code of Principles of Professional Conduct," *Proc. AIEE* (April, 1912), pp. 149–150; "Code of Principles of Professional Conduct," *Proc. AIEE* (1912), pp. 447–449; Newell, "Ethics of the Engineering Profession," *Annals of the American Academy of Political and Social Science,* 101 (March, 1922), pp. 76–85.

[35] Discussions of engineering codes of ethics generally rely on the final document and thus miss the complex of engineering concerns from which the few acceptable issues emerge: Besides Newell, ibid., see M. C. Lockwood, "A Hard Look at Engineering Ethics," *Civil Eng.*, 24 (May, 1954), p. 370; William G. Rothstein, "Engineers and the Functionalist Model of Professions," pp. 87–91, in Robert Perrucci and Joel E. Gerstl, eds., *The Engineers and the Social System,* New York: John Wiley & Sons, 1969; and Edwin T. Layton, Jr., *The Revolt of the Engineers, Social Responsibility and the American Engineering Profession* (Cleveland, Ohio: Case Western Reserve University Press, 1971), especially pp. 84–85.

[36] "Proposed Developments," p. 7; Insull, *Central-Station Electric Service,* "De Ferranti Dinner," p. 227–228. Layton, in *Revolt,* p. 84, criticizes the code as being "phrased as if the engineer were a consultant." Yet the code was written, not "as if" the engineer was a consultant, but specifically to control the conduct of the consulting engineer in the marketplace.

[37] John W. Lieb, Jr., "Presidential Address. The Organization and Administration of National Engineering Societies," *AIEE Trans.*, 24 (1905), pp. 282–296.

[38] Jackson to C. C. Chesney, May 20, 1910, Jackson Papers. Board Minutes, April 30, 1909; April 30, May 16, and December 8, 1911, IEEE Archives. Weaver to Jackson, August 20, 1910, Jackson Papers. *Proc. AIEE,* 31 (February, 1912), p. 41.

[39] "Constitutional Amendments," *Proc. AIEE,* XXXI (February, 1912), pp. 44–52, with introduction by Gano Dunn, pp. 41–44.

[40] Layton, *Revolt of the Engineers,* examines in detail the relation of the constitutional crisis to the "business counterrevolution," which occurred at this time in the AIEE, pp. 88–93. Yet, he errs in asserting, pp. 40 and 51, that the constitutional changes of 1912 granted "full membership" to executives. Under the 1912 amendments, however, full membership denoted the "Fellow" grade, a rank explicitly denied to executives. The AIEE had not constitutionally abandoned professionalism; instead, an historically contained event had taken place under a temporary proviso. Layton's fundamental argument is nonetheless correct: the AIEE had undergone "a gradual shift in orientation from professionalism toward business," p. 93; *Proc. AIEE,* 31 (February, 1912), p. 57.
[41] AIEE Board Minutes, August 8 and October 11, 1912, IEEE Archives.
[42] Ibid., November 8 and December 3, 13, 1912, IEEE Archives.
[43] Ibid., December 3 and January 10, 1913.
[44] Ibid., February 14, 1913.
[45] Thomas P. Hughes, "The Science-Technology Interaction: The Case of High-Voltage Power Transmission Systems," *Technology and Culture,* 17 (October, 1976), pp. 650 ff.; *Proc. AIEE,* 32 (March, 1913), p. 80; for the list, see ibid., (January, 1913), p. 8 ff.
[46] "Communication . . . from President Mershon," *Proc. AIEE,* 32 (April, 1913), pp. 130, 132–133.
[47] AIEE Board of Directors, Minutes, April 9, 1913; *Proc. AIEE,* 32 (September and October, 1913), pp. 342, 375; "Injunction Denied," ibid., 32 (July, 1913), p. 270.
[48] "Communication . . . from President Mershon," p. 137; "Communication, ... " p. 140.
[49] Ibid.
[50] "Electrical Engineering," *Proc. AIEE* (August, 1912), pp. 326–329.
[51] AIEE Board Minutes, March 12, May 18, November 12, 1909.
[52] Jackson to W. D. Weaver, Aug. 25, 1910, Jackson Papers.
[53] AIEE Board Minutes, November 8, 1912; May 20 and August 8, 1913; February 10 and April 14, 1911, IEEE Archives. Rushmore to Jackson, February 9, 1911, with abstract of Pupin's remarks. Rushmore to Board of Directors, March 6, 1911, Jackson Papers.
[54] Board Minutes, October 13, 1911; *Proc. AIEE* (February, 1912), p. 58.
[55] AIEE Board Minutes, August 8 and November 8, 1912, IEEE Archives.
[56] Mailloux, "The Evolution of the Institute and of its Members," *AIEE Trans.*, XXXIII (1914), pp. 819–838.
[57] Ivan S. Coggeshall, "IEEE's Endowment in Electronics from AIEE," *Elec. Eng.* (January, 1963).
[58] *Proc. AIEE* (November, 1909); Reginald A. Fessenden, "Wireless Telephony," *AIEE Trans.*, XXVII (1908), pp. 553–629; John V. L. Hogan, "What's Behind IRE?" *IRE Proc.* (April, 1951), pp. 340–341.
[59] See "The Genesis of the IRE," *Proc. IRE,* 40 (May, 1952), pp. 516–520; Hogan, "What's Behind IRE?" ibid., pp. 340–341.
[60] "Genesis," p. 517.
[61] Ibid., pp. 517–518.
[62] IRE Board Minutes, Feb. 10, 1915; C. T. Pannill to E. J. Nally, May 18 and 28; Nally to Pannill, May 22, 1914, IEEE Archives.
[63] IRE Board Minutes, January 13 and October 10, 1915, January 13, 1915, IEEE Archives.
[64] IRE Board Minutes, April 22, 1914, IEEE Archives.

Chapter 5

[1] Terman, President's Address, *Proc. IRE,* 29 (1941), pp. 406–408.

[2] *Electronics,* 4 (September, 1933); Hugh G. J. Aitken, *Syntony and Spark—the Origins of Radio* (New York: John Wiley & Sons, 1976), p. 284.

[3] Sarnoff to F. M. Sammis, January 7, 1914, Sarnoff Papers, RCA, Princeton.

[4] Quoted in Lessing, *Man of High Fidelity,* pp. 54–55.

[5] Lessing, pp. 54, 56, 60–62, 65.

[6] AIEE Board Minutes, March 27 and September 25, 1903, IEEE.

[7] Ibid., February 19, 1915, IEEE.

[8] Dupree, p. 303.

[9] Carty, "The Relation of Pure Science to Industrial Research," Number 14 in the *Reprint and Circular Series of the National Research Council* (Washington D. C.: n.d.), pp. 1–2.

[10] Scott, p. 38; Dupree, pp. 305, 308–310, 312.

[11] *New York Times,* January 19, 1936, and Godfrey, "Certain Considerations Concerned with the Council of National Defense," a manuscript copy of a speech, c. 1917, University Archives, Drexel Institute of Technology; Noble, 149–150.

[12] Dupree, p. 310, 314; Scott, p. 119; Pupin, *From Immigrant to Inventor,* pp. 367–368; on the Engineering Foundation, see Noble, 127.

[13] Scott, pp. 30, 116–117.

[14] *New York Times,* May 30, 1915; from Hughes, Sperry, p. 246.

[15] Daniels to Edison, July 7, 1915, in Lloyd N. Scott, *Naval Consulting Board of the United States* (Washington: GPO, 1920), Appendix, pp. 286–288.

[16] Scott, pp. 11–12; AIEE Board Minutes, August 10, 1915, IEEE.

[17] John C. Parker to AIEE Directors, Past Presidents, Chairmen of Sections, July 27, 1915, Jackson Papers; Jackson to John C. Parker, August 4, 1915, and Parker to Jackson, August 10, 1915, Jackson Papers.

[18] For a full list, see Scott, pp. 14–15.

[19] Hughes, Sperry, pp. 258–259.

[20] "The Submarine and Kindred Problems" (Bulletin No. 1, July 14, 1917), p. 252; "The Enemy Submarine" (Bulletin No. 2, May 1, 1918), p. 263, reprinted in Scott, NCB, pp. 252–274. "Problems of Aeroplane Improvement" was issued August 1, 1918, jointly with the War Committee of Technical Societies, in Scott, pp. 275–285; for an account of the work on the aerial torpeda, see Hughes, Sperry, pp. 258–274.

[21] Scott, p. 67.

[22] Scott, pp. 23, 27–30, 68–69, 235–238.

[23] Scott, pp. 70, 74.

[24] Scott, pp. 74–80; Kevles, *The Physicists,* pp. 122–123.

[25] Scott, pp. 80–81; Bush, *Pieces,* pp. 71–73.

[26] Scott, pp. 81–82.

[27] Bush, *Pieces,* p. 73; Scott, p. 83.

[28] Pupin, *From Immigrant to Inventor,* p. 376; Bush, *Pieces,* pp. 72–74; see the letters between Mershon and Bush and Lamme in the Records of the NCB, General Correspondance, B. G. Lamme, 1915–1921, NARS.

[29] Scott, pp. 222–223.

[30] IRE Board Minutes, August 17, 1915, Sarnoff Papers.

[31] De Forest, Father of Radio, 342, 344; Lessing, *Man of High Fidelity,* pp. 65 ff.

[32] Donald McNicol, President's Address, *Proc. IRE,* 15 (February, 1927), p. 73.

[33] Hugh G. J. Aitken, *Syntony and Spark — The Origins of Radio* (New York: John Wiley & Sons, 1976), p. 199, 274, 277, 279.
[34] Maclaurin, pp. 59–60.
[35] Maclaurin, 70–74 ff.; DeForest, *Father of Radio,* p. 215.
[36] G. Marconi to F. M. Sammis, January 22, 1914, Sarnoff Papers.
[37] Lessing, pp. 58–59, 61; Maclaurin, p. 90.
[38] Lessing, p. 58; John W. Griggs, President, "To the Stockholders of the Marconi Wireless Telegraph Company of America," October 22, 1919, Sarnoff Papers; see Sarnoff's memorandum on "Transoceanic Radio Communication" to E. J. Nally, February 16, 1920; ibid; and see Alfred Goldsmith, "World Communication," *J. AIEE,* 40 (1921), pp. 885–889, Maclaurin, p. 94–95.
[39] W. Rupert Maclaurin, *Invention and Innovation in the Radio Industry* (New York: Macmillan Company, 1949), pp. 33–34; Philip T. Rosen, *The Modern Stentors: Radio Broadcasters and the Federal Government* (Greenwood Press, 1980), p. 21.
[40] Quoted in Maclaurin, p. 99, 105; Rosen, pp. 22–23.
[41] "The Genesis of the IRE," *Proc. IRE,* 40 (May, 1952), p. 519 (That this training was of slight consequence is suggested in the 1920 memorandum to Nally on "Transoceanic Radio Communication," ibid. In establishing his credentials within the new organization in a detailed section of his "Record of Telegraph Experience," Sarnoff omitted mention of the Pratt Institute).
[42] Discussion following M. E. Parkman, "The Training of the Radio Operator," *Proc. IRE,* 3 (1915), pp. 338–342.
[43] *Proc. IRE,* 4 (December, 1916), p. 576; Minutes, IRE Board of Direction, November 9 and December 1, 1916, David Sarnoff Papers, David Sarnoff Library, Sarnoff Research Laboratory, RCA, Princeton. The following discussion of the IRE resolution comes from these minutes.
[44] David Sarnoff, ed., *Looking Ahead: The Papers of David Sarnoff* (New York: McGraw-Hill, 1968), pp. 11–13. See also the official biography by Eugene Lyons, *David Sarnoff, a Biography* (New York: Harper & Row, 1966).
[45] Noble, pp. 93–94; Rosen, p. 24; Gleason Archer, *History of Radio to 1926,* pp. 155, 161 ff.
[46] Rosen, p. 30; Noble, pp. 91, 93–94.
[47] RCA Annual Report, December 31, 1922, Sarnoff Papers; Gleason Archer, *History of Radio to 1926,* p. 17, 19; Maclaurin p. 148.
[48] IRE Board of Directors, Minutes, December 15, 1925, IEEE.
[49] IRE, Board Minutes, September 1, October 6, and December 6, 1926; ibid., July 27, 1927, and April 18, 1928, IEEE Archives.
[50] Goldsmith, "Cooperation Between the Institute of Radio Engineers and Manufacturers' Association," *Proc. IRE,* 16 (April, 1928), pp. 1065–1071.
[51] IRE Board Minutes, January 4 and June 6, 1928, IEEE; *Proc. IRE,* 17 (1929), p. 2108, and 18 (1930), p. 1777; see also in *Proc. IRE,* 17 (1929), the statement to the members on "Radio Standardization," pp. 2097–2100.
[52] "Radio Standardization," ibid.
[53] Beard, *The American Leviathan: The Republic in the Machine Age* (New York: The Macmillan Company, 1930), pp. 5–8, on radio; see also pp. 437–446.
[54] Beard, p. 6. Robert E. Cushman, *The Independent Regulatory Commissions* (New York: Oxford University Press, 1941), p. 297. Prepared for President Franklin D. Roosevelt's Committee on Administrative Management; Cushman dedicated his study to Charles Beard.
[55] Archer, *Radio to 1926,* p. 111.

[56] *Looking Ahead,* pp. 11–13; "General Memorandum: Radio Broadcasting Activities," April 5, 1923, Sarnoff Papers.

[57] "Progress Report Regarding Superpower Survey," Memorandum for the Press, Department of the Interior, Office of the Secretary, February 24, 1921, Scott Papers; for the evolution of the conservation committee, see *Proc. AIEE,* XXV (1906), p. 17 and AIEE Board Minutes, April 30, 1909, and August 22, 1911, IEEE.

[58] Cushman, *Independent Regulatory Commissions,* p. 277; "Progress Report," ibid.

[59] Murray, "Economical Power: The Strongest Agent for Maintaining Supremacy in World's Trade," *J. AIEE,* 39 (January, 1920), pp. 27–31; for a study of the idea as it developed in Pennsylvania, see Thomas P. Hughes, "Technology and Public Policy: The Failure of Giant Power," *Proc. IEEE,* 64 (September, 1976), pp. 1361–1371; AIEE Secretary to Committee on Super-Power members, April 3, 1920; Murray to Scott, February 28, 1921, Scott Papers.

[60] "Statement of Mr. David Sarnoff," Committee on Merchant Marine and Fisheries, House of Representatives, 69th Congress, 1st Session, Hearing on H. R. 5589, January 14, 1926, Sarnoff Papers; Sarnoff, "The Message of Radio," January 27, 1923, ibid.

[61] Sarnoff to H. P. Davis, March 19, 1924, with Guy E. Tripp, "A Plan of Super-Broadcasting to the Nation," Sarnoff Papers, RCA Princeton.

[62] Goldsmith, "Reduction of Interference in Broadcast Reception," *Proc. IRE,* 14 (October, 1926,) p. 575; Rosen, pp. 112–114.

[63] Arthur S. Link, *American Epoch* (3d ed., New York: Alfred A. Knopf, 1967), p. 265; H. P. Davis, "Early History of Broadcasting in the United States," *The Radio Industry* (Chicago, 1928), p. 220.

[64] For the relation of developments in advertising and economic growth, see A. Michal McMahon, "An American Courtship: Psychologists and Advertising Theory in the Progressive Era," *American Studies,* XIII (Fall, 1972), pp. 5–18, and Otis Pease, *The Responsibilities of American Advertising: Private Control and Public Influence, 1920–1940* (New Haven, 1958).

[65] Sarnoff, "General Memorandum: Radio Broadcasting Activities," April 5, 1923, Sarnoff Papers; Marriott, "United States Radio Broadcasting Development," *Proc. IRE,* 17 (1929), p. 1412; Raymond Williams, *Television: Technology and Cultural Form* (New York: Schocken Books, 1975), pp. 20–21.

[66] John D. Hicks, *Republican Ascendency, 1921–1933* (1960), pp. 26, 51.

[67] James E. Anderson, *The Emergence of the Regulatory State* (Washington, D.C., 1962), p. 145, quoted in Richard F. Fenno, Jr., *The President's Cabinet: An Analysis in the Period from Wilson to Eisenhower* (Cambridge, Mass.: Harvard University Press, 1959), p. 31.

[68] Rosen, p. 31.

[69] Gleason L. Archer, *Big Business and Radio* (New York, 1939), p. 33; Ray L. Wilbur and Arthur M. Hyde, eds., *The Hoover Policies* (New York: Charles Scribner's Sons, 1937), p. 208.

[70] Cushman, p. 299; Rosen, pp. 50, 52–53.

[71] Laurence F. Schmeckebier, *The Federal Radio Commission* (Washington: Brookings Institution, 1932), p. 6.; Cushman, p. 300.

[72] Rosen, pp. 68–69, 72–73; Archer, *Big Business and Radio,* pp. 230–231. For another view on Sarnoff and advertising, see pp. 31–32. However, this work and the companion volume, *History of Radio to 1926* (1938), were prepared with the close cooperation of Sarnoff and RCA.

[73] Quoted in Cushman, p. 299; Cushman, pp. 302–303. Quoted in Cushman, p. 313; Cushman, p. 320.

[74] Schmeckebier, "Federal Radio Commission," p. 66; *Proc. IRE,* 16 (1928), p. 237; Special Committee Report, Appended to IRE Board Minutes, May 2, 1928, November 7, 1928, *Proc. IRE,* 17 (1929), p. 411.
[75] Hoover Policies, p. 215.
[76] De Forest, *Father of Radio,* p. 374.
[77] Ibid., pp. 372–373; Inaugural Address (*Proc. AIEE*), 18 (July, 1930), p. 1123.
[78] *Electronics,* 7 (February, 1934), p. 55.
[79] "Retiring Address," *Proc. AIEE,* 19 (February, 1931), p. 169; *Proc. AIEE,* 18 (July, 1930), p. 1123; "Address of Welcome," *Proc. AIEE,* 18 (November, 1930), pp., 1791–1792.
[80] Ibid., p. 1792; *Proc. AIEE,* 18 (July, 1930), p. 1123.
[81] *Proc. AIEE,* 18 (November, 1930), p. 1792; *Proc. AIEE,* 18 (July, 1930), p. 1123; *Proc. AIEE,* 19 (February, 1931), pp. 170–171.
[82] *Proc. AIEE,* 19 (October, 1931), pp. 1894 and 20; (October, 1934), p. 1148. *Proc. AIEE,* 19 (October, 1931), pp. 1844–1846.
[83] Ibid., pp. 1847–1847.
[84] Ibid., pp. 1847–1848.
[85] *Proc. AIEE,* 18 (November, 1930), p. 1792; Rosen, p. 22.
[86] Rosen, pp. 172–175.
[87] "The relation of Engineering to the Radio Industrial Code," *Proc. AIEE,* 21 (September, 1933), p. 1233; for a biography of Hull, see *Proc. AIEE,* 17 (June, 1929), p. 912; William E. Leuchtenburg, *Franklin D. Roosevelt and the New Deal, 1932–1940* (New York: Harper and Row, 1963), pp. 64 ff.; Lessing, p. 178.
[88] *Electronics,* 3 (July, 1931), pp. 1, 4 and (May, 1932), p. 151. *Electronics,* 4 (May, 1932), p. 151.
[89] "The 'Industry Recovery Act'—and Radio," ibid., 6 (July, 1933), p. 179; "NRA Radio Industry Code," ibid. (August, 1933), pp. 208, 230.
[90] "This New Radio Prosperity," *Electronics,* 6 (November, 1933), p. 296. Wm. E. Leuchtenburg, *FDR and the New Deal,* p. 66 ff.
[91] Baker, "Work of Engineering Division of R.M.A.," *Electronics,* 8 (March, 1935), pp. 78.
[92] For a biography of Baker, see *Proc. IRE* (January, 1947); Lessing, pp. 162–163.
[93] Virgil M. Graham, "Work of Engineering Divison of R.M.A.," *Electronics,* 8 (March, 1935), p. 79; Donald G. Fink, "Perspectives on Television: The Role Played by the Two NTSC's in Preparing Television Service for the American Public," pp. 1323 ff.
[94] Terman Interview, Stanford University Archives.
[95] *Electronics,* 1 (May and June, 1930); Terman to Raymond F. Guy, April 8, 1946, Terman Papers.
[96] Pratt to Terman, August 2, 1946, Terman Papers.

Chapter 6

[1] "Radio Research in Wartime: Achievements of RCA Laboratories" (c. 1946), RCA Archives, New York.
[2] *Electronics,* 1 (April, 1930), p. 53.
[3] Ibid., 13 (April, 1940), p. 20; ibid., 6 (June, 1933), p. 148.
[4] *Proc. IRE,* 18 (June, 1930), p. 908; for the "tree," see *Electronics,* 1 (May, 1930) and 7 (May, 1934), p. 147.
[5] *Electronics,* 1 (April, 1930), pp. 8–9, 22.

[6] R. Hilsch and R. W. Pohl, "Zur Photochemie der Alkali—und Silberhalogenidkristalle," *Z. Phys.*, 64 (1930), cited in Ernest Braun and Stuart MacDonald, *Revolution in Miniature: The History and Impact of Semiconductor Electronics* (London: Cambridge University Press, 1978), p. 23.
[7] Ibid., 4 (October, 1932), p. 305.
[8] *Electronics,* 13 (April, 1940), pp. 21–24.
[9] Arthur L. Norberg, "The Origins of the Electronics Industry on the Pacific Coast," *Proc. IEEE,* 64 (September, 1976), p. 1316; Frederick E. Terman Interview, January 3, 1974, Stanford University Archives.
[10] Norberg, "Pacific Coast," pp. 1317–1319; Terman Interview, p. 54.
[11] Milton Moskowitz, Michael Katz, and Robert Levering, eds., *Everybody's Business: An Almanac* (New York: Harper & Row, 1980), p. 831.
[12] Norberg, "Pacific Coast," p. 1321; Terman Interview, pp. 56–57.
[13] Norberg, "Pacific Coast," p. 1321; Getting, "Radar," p. 61, in Carl F. J. Overhage, ed., *The Age of Electronics* (New York: McGraw-Hill, 1962).
[14] *Electronics,* 8 (September, 1935), p. 18–19.
[15] Terman Interview, pp. 10, 50.
[16] See the Insull file in Jackson Papers; Jackson to Joseph W. Stickney, May 12, 1922, Jackson Papers.
[17] "Electrical Engineering Department," October 13, (1918?), Jackson Papers; Terman Interview, Stanford Archives.
[18] Wildes, pp. 4–74; for the proposed radio laboratory, see Jackson Papers.
[19] Wildes, p. 4–60 ff.; Terman Interview, pp. 17–23; Terman, "Characteristics and Stability of Transmission Systems," (Unpublished Dissertation: MIT, 1924), pp. 1–2.
[20] Terman Interview, p. 17; Jane Morgan, *Electronics in the West: The First Fifty Years* (Palo Alto, Calif.: National Press Books, 1967), p. 89.
[21] Terman Interview, p. 11; *J. AIEE,* 40 (July, 1921), p. 587.
[22] Terman Interview, p. 11; SPEE, "Report," *J. Eng. Educ.* (March 30, 1940).
[23] Letter to H. P. Westmon, October 14, 1938, Terman Papers.
[24] E. J. O'Connell to Members of Communications Committee, AIEE, October 25, 1934, Terman Papers.
[25] *J. AIEE,* 40 (July, 1921), p. 587; AIEE, Committee on Communications, Minutes, March 23, October 25, November 2, 1934, Terman Papers.
[26] Minutes, Sub-Committee for Study of Electronics Papers and Electronics Committee Considerations, March 16, 1934, Terman Papers.
[27] *Trans. AIEE,* 29 (1910), pp. 1730 ff.; AIEE Board Minutes, August 8, 1912.
[28] Alger to W. S. Gorsuch, May 9, 1929; R. H. Park to Gorsuch, July 9, 1929; Jackson to Park, July 22, 1929, Jackson Papers; Parker to H. H. Henline, August, 15, 1938 and AIEE, Communications Committee Memorandum, September 17, 1937, Terman Papers.
[29] AIEE, "Communications Committee—Papers on Hand and Proposed," October 25, 1934, Terman Papers.
[30] AIEE, Communications Committee, Minutes, October 1, 1936; Terman to G. Ross Henninger, July 17, 1936, Terman Papers.
[31] G. Ross Henninger to Terman, October 8, 1934, and Terman to Henninger, November 6, 1934, Terman Papers.
[32] AIEE Board Minutes, January 21, 1935.
[33] Terman to E. J. O'Connell, September 13, 1937 and Terman to William L. Everitt, January 19, 1938, Terman Papers.
[34] Everitt to Terman, December 11, 1937, Terman Papers; AIEE, Commu-

nications Committee Minutes, October 4, 1935 and Draft of January 26 Minutes dated April 25, 1938, Terman Papers.

[35] Terman, "Contributions for Annual Report of Communication Committee," AIEE, n.d. and "Tentative Outline of and Status of Information for Report of Committees on Communication," April 12, 1940, Terman Papers.

[36] Terman to H. P. Westman, November 15, 1940; Terman to Arthur F. Van Dyck, July 27, 1942; Van Dyck to Terman, August 8, 1942, Ernst Weber to Terman, July 22, 1942; Terman to Weber, August 18, 1942; Terman Papers.

[37] Charles S. Rich, Secretary, Technical Programs Committee, to Members of Technical Committees, October 24, 1941, Terman Papers.

[38] IRE Proceedings Membership Survey, 1941; Van Dyke to Terman, August 8, 1942, Terman Papers.

[39] "Science Advisory Board," typescript of cover letter for Report of September 1, 1934, James R. Killian Correspondence, Papers of the Office of the President of MIT, 1930–1959, MIT Archives; Carroll W. Pursell, Jr., "The Anatomy of a Failure: The Science Advisory Board, 1933–1935," *Proc. Amer. Phil. Soc.*, 109 (December, 1965); Second Report of the Science Advisory Board, September 1, 1934 to August 31, 1935 (Washington, D.C., 1935), Killian Correspondence.

[40] Kevles, p. 297; Quoted in Pursell, "Science Agencies in World War II: The OSRD and Its Challengers," in *The Sciences in the American Context: New Perspectives,* edited by Nathan Reingold (Washington, D.C., 1979), p. 362.

[41] Dupree, p. 371; Quotation from Kevles, p. 300; see also Pursell, "Science Agencies in World War II," pp. 359–378.

[42] For challenges to the OSRD's role, see Pursell, "Science Agencies in World War II."

[43] James Phinney Baxter III, *Scientists Against Time* (Boston, 1968; originally 1948), pp. 423–424, 433.

[44] David Kite Allison, *New Eye for the Navy: The Origin of Radar at the Naval Research Laboratory* (Washington, D.C.: NRL, 1981), pp. 41–45.

[45] Baxter, p. 140; for a detailed account of radar research at the NRL between the wars, see Allison, *New Eye for the Navy.*

[46] Baxter, p. 140; Baxter, p. 141n.

[47] For this and the next paragraph, see Kevles, 309–310 and *Applied Physics: Electronics, Optics, Metallurgy,* C. Guy Suits, George R. Harrison, and Louis Jordan, eds. (Boston, 1948), pp. 3, 17, 123–125, 127. (Hereafter *Applied Physics*).

[48] Baxter, p. 428.

[49] *Applied Physics,* pp. 10–11.

[50] *Applied Physics,* p. 10–11; Baxter, p. 158–159; Compton to Terman, February 16, 1942, Terman Papers.

[51] Sam Morris to Terman, Jan. 12, 1942, Terman Papers; *Applied Physics,* p. 11.

[52] Terman, "Presidential Address," *Proc. IRE,* 29 (1941), p. 406.

[53] Notes on Personnel, January 16, 1946, Terman Papers.

[54] Terman Interview, April 22, 1975, Terman Papers; *Applied Physics,* pp. 12–14, 18; Baxter, p. 157; see also, *Proc. IRE* (May 1946).

[55] *Applied Physics,* pp. 15–17.

[56] Baxter, p. 164–167; Memorandum on "Countermeasures," Hogan to Bush, January 8, 1944, Terman Papers.

[57] Oswald G. Villard, Jr., "Administrative History of the Radio Research Laboratory" (March 21, 1942, and January 1, 1945), March 21, 1946, Terman

Papers; Terman to William H. Claflin, July 19, 1943, Terman Papers.
[58] The phrase is Thomas P. Hughes, "The Development Phase of Technological Change," *Technology and Culture,* 17 (July, 1976), pp. 423–431.
[59] Radio Research Laboratory and ABL-15, "NDRC Review Meeting—February 21, 1945," Terman Papers.
[60] Baxter, pp. 160–161; "Radio Research Laboratory Contributions to the War," September 24, 1945, Terman Papers.
[61] Terman to W. H. Claflin, Jr., "Demobilization of the Radio Research Laboratory," September 6, 1945; "Administrative History of the Radio Research Laboratory" (OSRD, March 21, 1946), pp. 8–1 to 2–3, Terman Papers.
[62] Terman to All Laboratory Personnel, "Demobilization of RRL," August 31, 1945.
[63] Terman Interview.
[64] Bush to Harvey H. Bundy, August 8, 1944, Terman Papers.
[65] For the JNW, see *The Papers of Dwight David Eisenhower, The Chief of Staff: VII,* Louis Galambos, ed. (Baltimore, 1978), note, p. 708; Herbert F. York and G. Allen Greb, "Military Research and Development: A Postwar History," *Bulletin of the Atomic Scientists* (January, 1977), pp. 13–26. Louis A. Turner to Terman, January 10, 1946, and attachment, "Functions and Organization of VTDC," Terman Papers.
[66] Bush, "Memorandum to Scientists and Engineers Now or Formerly Associated with the Office of Scientific Research and Development," with Eisenhower's "Memorandum for Directors and Chiefs of War Department General and Special Staff Divisions and Bureaus and The Commanding Generals of the Major Commands," April 30, 1946, Terman Papers; also see *The Papers of Dwight David Eisenhower: VII,* pp. 840, 1046–1050.
[67] Bush, "Research and Strategy" (c. 1948), Bush Papers, MIT Archives; York and Greb, p. 194.
[68] Edward Bowles, "National Security and a Mechanism for its Achievement," *Proc. IRE,* 34 (1946), pp. 154P–155P.
[69] Roosevelt to Bush, November 17, 1944, pp. vii–viii, in Bush, *Science—The Endless Frontier* (U.S. Government Printing Office, 1945).
[70] *Science—The Endless Frontier,* pp. 1, 59.
[71] York and Greb, p. 198; Truman to Oliver Buckley, April 19, 1951, and Buckley to Killian, April 25, 1951, Office of the President, Killian Papers, MIT Archives; York and Greb, p. 198.
[72] F. B. Llewellyn, "Research and Development Installations in Department of Defense" (c. October 1951), Killian Papers; National Science Foundation, "National Patterns of R&D Resources: Funds and Personnel in the U.S.," 1953–1978/9 (Washington, D.C.: GPO, 1978).
[73] Buckley to Killian, October 26, 1951, with graphs; Llewellyn, "Department of Defense Expenditures for Research and Development," October 1951, Office of the President, Killian Papers, MIT Archives.
[74] National Science Foundation, Scientific Personnel Resources (NSF 1955), pp. 11, 14.
[75] For this and the following paragraph: Reich, "Research, Patents, and the Struggle to Control Radio: A Study of Big Business and the Uses of Industrial Research," *Business History Review,* 51 (Summer 1977), pp. 210, 215, 235.
[76] See Nathan Reingold, "The Case of the Disappearing Laboratory," *American Quarterly,* 29 (Spring, 1977), pp. 70–102, especially p. 100.

[77] Federal Laboratory Review Panel, Report of the White House Science Council—May 1983 (Washington, D.C.: Office of Science and Technology Policy, 1983), p. 1.
[78] For this and the next two paragraphs: Dorothy Nelkin, *The University and Military Research: Moral Politics at MIT* (Ithaca: Cornell University Press, 1972), pp. 19, 27, 28–32 (table 4), 49. For a detailed study of the complex relationships in one research program that involved electrical engineers, see Joan Lisa Bromberg, *Fusion: Science, Politics, and the Invention of a New Energy Source* (Cambridge, MA: MIT Press, 1982).
[79] Terman, "Science Legislation and National Progress," *Proc. IRE,* 34 (1946), p. 860; Pratt, "Society's Hopes for the Engineer," ibid., p. 48P; see also, P. R. Mallory, "The Electronics Engineer," ibid. 33 (1945), p. 141.

Chapter 7

[1] Espenshied to Frederick B. Llewellyn, July 30, 1946, IEEE Archives.
[2] Interview with John D. Ryder, IEEE Archives.
[3] Layton, "Science, Business, and the American Engineer," *Engineers and the Social System,* Robert Perrucci and Joel E. Gerstl, Eds., (New York, 1969), p. 51; see also Layton, "Scientists and Engineers: The Evolution of the IRE," *Proc. IEEE,* 64 (1976), pp. 1390–1391.
[4] Terman to Raymond F. Guy, September 6, 1946, Terman Papers.
[5] K. B. McEachron to L. F. Hickernell, May 8, 1953, Quarles Papers; Van Dyck to Terman, August 8, 1942, Terman Papers.
[6] "Report of the Secretary, 1953," *Proc. IRE,* 42 (1954).
[7] H. H. Henline, "Decentralization of Institute Technical Activities," February 1951, IEEE Archives.
[8] W. R. G. Baker, "The IRE Professional Group System," *Proc. IRE,* 40 (1952), p. 1028; Everitt to John D. Ryder, April 14, 1982, in possession of Ryder.
[9] Ibid.
[10] "Institute on the March," *Proc. IRE,* 36 (1948), p. 570; Baker, "Professional Group System."
[11] Ibid.
[12] Baker, "The Professional Group Activity of the IRE," *Proc. IRE,* 39 (1951), p. 92; for this and the next three paragraphs, see also Baker, "Professional Group System," especially the table on p. 1029; Ryder Interview, IEEE Archives; Ryder to author, September 8, 1980.
[13] Weber, "Why Transactions?" *Proc. IRE,* 42 (1954), p. 1221.
[14] *Electronics,* 13 (July 1940), p. 12; *Electrical Engineer* (February, 1940); "Radio Research in Wartime," (typescript, n.d.), RCA Archives, New York.
[15] Quarles to Ralph L. Goetzenberger, March 25, 1952, Quarles Papers, Dwight D. Eisenhower Presidential Library.
[16] T. A. Marshall, Jr. to AIEE, February 2, 1953; Quarles to Henline, February 20, 1953; Cisler to Quarles, April 14, 1953; Quarles to Cisler, April 30, 1953; Quarles to Prentice, May 6, 1953, with Louisville statement of April 23, 1953, "Atomic Industrial Forum"; Quarles to J. L. Atwood, May 27, 1953; Quarles Papers.
[17] Prentice to Quarles, *et al.,* July 29, 1953, with "Statement of Methods of Encouraging Industrial Development of Atomic Energy by Engineers Joint

Council before the Joint Committee on Atomic Energy of the U.S. Congress," July 27, 1953, and Sporn to Prentice, May 9, 1953, Quarles Papers.

[18] For this and the following paragraph, Bruce A. Smith, *Technological Innovation in Electric Power Generation, 1950–1970* (Michigan State University, 1977), pp. 73, 74–77; Newsclipping, June 18, 1953 and Quarles, "Technology in Defense," October 11, 1954, Quarles Papers.

[19] Richard G. Hewlett, in "Beginnings of Development in Nuclear Technology," *Technology and Culture,* 17 (July, 1976), pp. 465–578, explains how difficult it was for the physicists to let go and for the engineers to gain a voice in the early AEC; J. A. Lane, "The Contributions of Engineering to Nuclear Energy Development," in *U.S. Atomic Energy Commission: The Role of Engineering in Nuclear Energy Development* (Oak Ridge, TN, December, 1951), pp. 73–74.

[20] Smith, Technological Innovation, p. 58 ff.

[21] R. S. Gardner to L. F. Hickernell, August 11, 1952, Quarles Papers; John D. Ryder Interview, 1978, IEEE Archives; Hooven, "Where We Stand," *Elec. Eng.*, 75 (August 1956), p. 679.

[22] Berkner, "Communications," *The Age of Electronics,* Carl F. J. Overhage, Ed. (New York, 1962), p. 47; Quarles, "Technology in Defense."

[23] Stanley V. Forque, "Electronics and the Research Physicist," *Proc. IRE,* 34 (1946), p. 595; Everitt, "The Institute Looks to the Future," *Proc. IRE,* 33 (1945), p. 73.

[24] The following discussion of postwar IRE conventions comes from reports and announcements in *Proc. IRE,* 34 (1946), p. 149W; ibid., 35 (1947) and 37 (1949).

[25] For this and the following paragraphs, Ernest Braun and Stuart MacDonald, *Revolution in Miniature: The History and Impact of Semiconductor Electronics* (London: Cambridge University Press, 1978), pp. 54 ff., 104–105; Shockley, "Transistors," *The Age of Electronics,* pp. 134–162; Robert W. Holcomb, "John Bardeen, Walter Brittain, William Shockley," pp. 208–216, in Robert L. Breeden, ed., *These Inventive Americans* (National Geographic Society: Washington, D.C., 1971).

[26] *Proc. IRE,* 40 (1952), p. 1287; for a discussion of the RDB's Committee on Electronics, which Quarles chaired and on which Everitt served, see Edwin A. Speakman, "Research and Development for National Defense," *Proc. IRE,* 40 (1952).

[27] Braun and McDonald, p. 78, 91. Two recent histories on the first computers are Paul E. Ceruzzi, *The Reckoners: The Prehistory of the Digital Computer, 1935–1945* and Nancy Stern, *From ENIAC to UNIVAC: An Appraisal of the Eckert-Mauchly Computers* (Bedford, MA, 1981).

[28] The following discussion is based on the Special Issues in *Proc. IRE:* "The Transistor Issue," 40 (November, 1952); "The Computer Issue," 41 (October, 1953); "The Transistor Issue," 46 (June, 1958); "The Computer Issue," 49 (January, 1961); "Solid-State Electronics," 43 (December, 1955); and "Space Electronics," 48 (April 1960).

[29] J. W. McRae, "Transistors in Our Civilian Economy," and I. R. Obenchain, Jr. and N. J. Galloway, "Transistors and the Military," *Proc. IRE,* 40 (1958), pp. 1285–1288; Simon Ramo, "A New Field," *Proc. IRE,* 40 (1952), p. 3.

[30] For the following discussion, see *Proc. IRE,* 40 (1952), p. 899; 41 (1953), pp. 667–668; also, see William R. Hewlett, "An Evaluation of the IRE Professional Group Plan," 41 (1953), pp. 964–965.

[31] Ryder Interview, IEEE Archives.

[32] F. Alton Everest and Arthur L. Albert to Conference Participants, July 31, 1939, Terman Papers.
[33] Jackson, "Engineering Education," *SPEE J. Engineering Education,* 29 (June, 1939), pp. 823–830; "Report," *SPEE J. Engineering Education,* 30 (March, 1940).
[34] Palo Alto Times, November 24, 1947, Terman Papers.
[35] Terman Interview, April 17, 1978.
[36] "Report of Committee on Engineering Education After the War," *SPEE J. Engineering Education,* 34 (May, 1944), pp. 589–614.
[37] Engineers Council for Professional Development, "Report of Committee on Adequacy and Standards of Engineering Education," reprint from 1951 Annual Report, Quarles Papers.
[38] "Report of the Committee on Evaluation of Engineering Education," *ASEE J. Engineering Education* (September, 1955), pp. 25–59.
[39] Terman to A. V. Loughren, May 4, 1956, Terman Papers; *Proc. IRE* (June, 1956), pp. 738–740.
[40] Fink to Terman, March 9, 1956, Terman Papers; IRE Board of Directors Minutes, May 10, 1956, IEEE Archives.
[41] Ryder, "What Kind of Engineers?" *Proc. IRE,* 37 (1949), pp. 1168–1170 and "Electrons, Engineers, and Education," *Proc. IRE,* 40 (1952), p. 259.
[42] Ryder to Terman, May 17, 1956; Ryder to Terman, February 22, 1957, with proposed resolution; Terman to Ryder, March 6, 1957, Terman Papers; IRE Board of Directors, "Response to ASEE Report," April 11 and 12, 1957, Wickenden Papers.
[43] Ryder, "Engineering Education: A View Ahead," *Proc. IRE,* 45 (1957), pp. 1459–1462; *Proc. IRE,* 47 (1959), p. 1; *Proc. IRE,* 46 (1958), p. 1947.
[44] IRE Board Minutes, January 6, 1960, IEEE Archives; Norman Balabanian, Ed., "Undergraduate Physics and Mathematics in Electrical Engineering" (Proceedings of the Sagamore Conference, Syracuse, September, 1960), and Balabanian and Wilbur R. Lepage, Ed., "Electrical Engineering Education" (Proceedings of the International Conference at Syracuse and Sagamore, September, 1961).
[45] Terman, "A Brief History of Electrical Engineering Education," *Proc. IEEE,* 65 (September, 1976), p. 1407.
[46] Van Dyck to Terman, August 8, 1942, and Terman to Raymond F. Guy, September 6, 1946, Terman Papers.
[47] K. B. McEachron to L. F. Hickernell, May 8, 1953, Quarles Papers.
[48] Fred O. McMillan to W. J. Barrett, April 1, 1952, Minutes, AIEE Membership Committee, January 21, 1953, Quarles Papers.
[49] Barrett to L. F. Hinckernell, September 29, 1952, Quarles Papers; for another account of the merger, see George Carlyle Sell, "The Historical Origins of the Institute of Electrical and Electronics Engineers" (unpublished master's thesis, 1980), esp. pp. 68–94.
[50] Barrett to McMillan and Quarles, March 14, Quarles to Barrett, March 19, Hooven to Quarles, September 19, 1953, Quarles Papers; L. F. Hickernell, "Farewell Remarks to AIEE Board of Directors," June 23, 1961, IEEE Archives.
[51] Robertson to Barrett *et al.*, December 11, 1953, Quarles Papers.
[52] Ryder to W. Reed Crone, June 12, 1978, in Ryder's possession.
[53] IRE Minutes: Board of Directors, September 10, 1958; Executive Committee, March 8 and July 13, 1960, IEEE Archives.
[54] Lloyd V. Berkner, "Open Letter From the President," *Proc. IRE,* 49 (1961),

pp. 1743–1746; Berkner, "Open Letter . . . II," ibid., 50 (1962), pp. 4–5; *Proc. IRE*, Part II, 50 (1962), pp. 1–40.
[55] Ryder to Crone, June 12, 1978, IEEE Archives.
[56] McEachron to Hickernell, September 2, 1953, Quarles Papers; "An Evaluation of the I.R.E. Professional Group Plan," *Proc. IRE*, 41 (1953), pp. 964–965.
[57] Terman to Arthur V. Loughren, February 6, 1959, Terman Papers.
[58] Terman, "IEEE—Impossible—Except for Electrical Engineers," March 27, 1963, Terman Papers.
[59] *Proc. IRE*, Part II, p. 39.

Chapter 8

[1] D. G. Fink, "For Electronics, the Past is Prologue," *IEEE Spectrum* (January 1972), p. 66.
[2] William C. White, "A Look at the Past Helps to Guess at the Future in Electronics," *Proc. IRE*, 40 (May 1952), pp. 523–524.
[3] Eckert, "The Integration of Man and Machine," *Proc. IRE*, 50 (1962), p. 612–613.
[4] R. M. Page, "Man–Machine Coupling—2012 A.D.," p. 613, and J. M. Bridges, "There Will Be No Electronics Industry in 2012 A.D.," p. 580, *Proc. IRE*, 50 (1962).
[5] Terman, "Education in 2012 for Communication and Electronics," *Proc. IRE*, 50 (1962), pp. 570–571; Everitt, "Engineering Education—Circa 2012 A.D.," ibid. pp. 572–572; also see John D. Ryder, "Engineering Education: A View Ahead," ibid. 45 (1957), pp. 1459–1462.
[6] Tank, "Some Thoughts on the State of the Technical Science in 2012 A.D.," *IEEE Spectrum*, p. 623.
[7] Robert H. Tanner, "Democracy and the IEEE," *IEEE Spectrum* (September, 1972), pp. 7–8.
[8] Biographical material in Election Files, IEEE Archives.
[9] See the Annual Reports for 1964, 1967, and 1968 in *IEEE Spectrum* (June, 1965, 1968, and 1969).
[10] Oliver, "The IEEE Looks Ahead," *IEEE Spectrum* (April, 1965), p. 34.
[11] Shepherd, "The Uses of a Professional Society," *IEEE Spectrum* (December, 1966), pp. 35–38; Herwald, "Message from the President," *IEEE Spectrum* (January, 1968), p. 46.
[12] "How IEEE Plans to Solve its Technical Communications Problems," *IEEE Spectrum* (December, 1969), pp. 122–124.
[13] Ibid.; "A Progress Report . . . . ," *IEEE Spectrum* (August, 1970), p. 113–114.
[14] Granger, "Birth of a Concept," *IEEE Spectrum* (January, 1971), p. 21.
[15] Fink, "Electronics and the IRE—1967," *Proc. IRE*, (1957), p. 1190.
[16] S. W. Herald, "The IEEE Membership Attitude Survey—What it Means to You," *IEEE Spectrum* (December, 1968), pp. 33–38.
[17] Young, "The Road to Professionalism," *IEEE Spectrum* (January, 1973), pp. 71–72.
[18] Seymour Tilson, "Report on the ABM," *IEEE Spectrum* (August, 1969), p. 31.
[19] "Concerned Scientists," *IEEE Spectrum* (April, 1969), p. 8.
[20] "The Forum," *IEEE Spectrum* (July, August, 1969).
[21] "Forum," *IEEE Spectrum* (September, 1970).
[22] "Spectral Lines," *IEEE Spectrum* (October, 1969), p. 27; see also Dorothy Nelkin, *The University and Military Research* (Ithaca, NY: Cornell University Press, 1972) for an account of the reaction at MIT.

[23] Ibid.
[24] See Wald, "A Generation in Search of a Future," *IEEE Spectrum*, (June, 1969), pp. 34–37, and Tilson, "Report on the ABM."
[25] *IEEE Spectrum* (August, 1969), p. 24.
[26] J. D. Ryder, "The Genesis of an Editorial Policy," *IEEE Spectrum* (June, 1964), pp. 147–150.
[27] "Spectral Lines," *IEEE Spectrum* (October, 1969), p. 27.
[28] *IEEE Spectrum* (September, 1970), p. 21.
[29] *IEEE Spectrum* (September, 1969), p. 6 and (March, 1970), p. 6.
[30] *IEEE Spectrum* (September, 1970), p. 11; ibid. (October, 1969), p. 27.
[31] William O. Fleckenstein, "Economic Conditions in the U.S. Electrical, Electronics, and Related Industries: An Assessment," *IEEE Spectrum* (December, 1971), pp. 63–71.
[32] Fink, "For Electronics, the Past is Prologue," *IEEE Spectrum* (January, 1972), pp. 63–66.
[33] "The Layoff Problem," *IEEE Spectrum* (May, 1970), p. 6.
[34] Granger, "President's Address at the 1970 Convention Banquet," *IEEE Spectrum* (May, 1970), p. 22.
[35] "Spectral Lines," *IEEE Spectrum* (July, 1969), p. 29.
[36] "The Forum," *IEEE Spectrum* (September, 1969).
[37] *IEEE Spectrum* (October, 1969), pp. 10–11; also see ibid. (December, 1969), pp. 11–15.
[38] William E. Cory, "EE's and Public Needs," *IEEE Spectrum* (February, 1972), pp. 33–35; ibid., (May, 1972), p. 11; ibid., (February, 1972), p. 7.
[39] "President's Report," *IEEE Spectrum* (February, 1971), p. 113 and Harry C. Simrall, "If You Are a Concerned Engineer," p. 69.
[40] *IEEE Spectrum* (August, 1971), pp. 7–8.
[41] "Inside IEEE," *IEEE Spectrum* (May, 1971 and February, 1972).
[42] *IEEE Spectrum* (June, 1971), p. 7.
[43] "Report of the Secretary for 1971," *IEEE Spectrum* (June, 1972), p. 60; see also "1972: A 'Threshold' Year," *IEEE Spectrum* (July, 1973), pp. 56–60.
[44] Fink, "Blueprint for Change," *IEEE Spectrum* (June, 1972), p. 38–39.
[45] "Report from IEEE's President," *IEEE Spectrum* (December, 1971), p. 7.
[46] "Inside IEEE," *IEEE Spectrum* (December, 1971), p. 7.
[47] *IEEE Spectrum* (April, 1972), p. 7; Fink, "Blueprint for Change," ibid., (June, 1972), p. 39.
[48] The proposed amendment is in "Blueprint for Change," p. 39.
[49] Harold S. Goldberg, "A New Focus for IEEE's Professional Goals," *IEEE Spectrum* (June, 1973), pp. 62–64.
[50] Ibid.
[51] "Road to Professionalism," p. 74.
[52] D. Christiansen, "The New Professionalism," *IEEE Spectrum* (June, 1972), p. 17.
[53] *IEEE Spectrum* (February, 1983), p. 25.

INDEX

CREDITS

Bell Laboratories: pp. 180, 225, 228 (top).
Consolidated Edison: p. 25.
Edison National Historic Site: pp. 38, 106, 107, 141, 165.
Franklin Institute: p. 2.
IEEE Center for the History of Electrical Engineering: pp. 5 (right), 41, 49, 84, 94, 113, 131, 144, 148, (top), 158, 187 (3).
MIT Museum: pp. 72, 197, 201, 251, (2), 256.
NASA: p. 230 (bottom).
Smithsonian Institution: pp. 5 (left), 14, 23, 148 (bottom), 227.
Texas Instruments: p. 228 (bottom).
U.S. Army: p. 135.
Westinghouse Electric Corp.: p. 221.

The quotations from the Frederick E. Terman interview are used by courtesy of The Bancroft Library, University of California, Berkeley.

AUTHOR'S BIOGRAPHY

A. Michal McMahon received the bachelor's degree from Southern Methodist University, Dallas, TX, and the doctorate degree in history from the University of Texas, Austin.

He has taught at Kansas State University, and at both Temple and Drexel Universities in Philadelphia, PA. During the 1970's, he worked at the Franklin Institute in Philadelphia as Historian and Archivist in the library and as an Exhibit Developer and Curator of Collections for the Science Museum. He has published a number of articles on the history of technology in nineteenth- and twentieth-century America.